Excel

SUCCESS ONE® HSC

BIOLOGY

Past HSC Questions & Answers
2001–2003 by Topic
2009–2017 by Paper

PASCAL
PRESS

ISBN 978 1 74125 666 6

Pascal Press
PO Box 250
Glebe NSW 2037
(02) 8585 4044
www.pascalpress.com.au

Publisher: Vivienne Joannou
Commissioning and project editor: Mark Dixon
Chapters 1–6 edited by Brendan Atkins, Big Box Publishing Pty Ltd and Anna Beth McCormack
Chapter 7 edited by Anna Beth McCormack
Chapters 8–10 edited by Christine Eslick
Chapters 1–10 answers checked by Chris Greef
Chapter 11 edited by Leanne Poll
Chapters 12–14 edited by Karen Pearce
Chapters 15–16 edited by Rosemary Peers
Chapters 11–14 answers checked by James Mansfield
Chapter 15 answers checked by James Mansfield and Diane Alford
Chapter 16 answers checked by Sarah Barrand

Cover design by Michael Sherman
Page layout and typesetting by Grizzly Graphics (Leanne Richters), DiZign Pty Ltd and Midland Typesetters
Cover photo by Janie Barrett, Fairfax Syndication
Printed by Vivar Printing/Green Giant Press

Foreword

Congratulations on choosing the Biology course for your HSC. In undertaking this course you have the opportunity to gain valuable knowledge, understanding and skills, and to extend yourself to meet new challenges.

Excel Success One HSC Biology is a valuable learning tool that has been developed to assist students with their HSC preparation, and this edition has been kept up to date with the inclusion of the 2017 HSC Examination paper with sample answers.

On behalf of the Science Teachers' Association of New South Wales, we hope that your year will be enjoyable and productive, and the results a just reward for your efforts.

STANSW thanks those members who continue to be involved in the production of this book.

Margaret Shepherd
STANSW President
December 2017

The Mark Maximizer Guide was prepared by Diane Alford.

HSC EXAMINATION PAPERS

SAMPLE ANSWERS

The sample answers to HSC questions contained in this publication are examples of answers which the authors believe would score full marks. However, they are not endorsed by NESA. Contributors to the sample answers include Monika Khun, Mary Cahill, David Randall, Dianne Shaw, Col Harrison and Diane Alford.

COMMENTS

The Science Teachers' Association of NSW (STANSW) and Pascal Press welcome constructive comments on the questions and answers for future editions of this book.

Contents

Mark Maximizer Guide

This Mark Maximizer Guide will arm you with strategies and tips so you can maximize your marks by making every minute of your HSC Examination count. And it has some great tips for your exam preparation too. Read it. Use it.

The structure of the HSC Examination paper

All Science exams consist of two sections. The following table summarises the structure, which changed in 2010.

Section	*Part*	*Compulsory/ Optional*	*Marks*	*Question type*	*Answers*	*Recommended time*
I	A	All questions are compulsory.	Each question is worth one mark – a total of 20.	Objective-response – select one alternative from A, B, C or D.	Fill in the response oval completely on the answer sheet.	About 35 min
I	B	All questions are compulsory.	55	Short-answer questions with different mark values, each indicated on the question: some questions may be in parts; some may require a more integrated approach.	Answer the questions in the space provided in the exam paper.	About 1 hour 40 min
II		Attempt only one question.	25	The question is divided into parts, with each part having the mark clearly indicated.	Answer the questions in a writing booklet.	About 45 min

Time allocation

The examination begins with 5 minutes reading time. You will not be able to write in this time. Rather than starting to read the objective-response questions at the beginning, you could:

- spend this time reading some of the short-answer questions in Part B and Section II,

  or

- select some of the easy questions to attempt earlier in the exam after completing the objective-response questions, to build your confidence for the more difficult questions later.

Time management

A planned approach to time management in the exam is important. There are several strategies you could use:

- Prepare an exam timetable
  The table above shows the suggested times for each section as recommended by NESA. If you choose to proceed through the exam in an orderly sequence of sections or parts, you can work out an exam timetable before the exam. You will then know when you should be changing sections and check that you are on schedule. Try to be a bit ahead of schedule so that you have time to check answers, make corrections or return to more difficult questions.

- Calculate time allocated per mark
  If you do not choose to work through the exam in a fixed sequence, you may wish to calculate time allocations for questions based on the 100 marks and 180 minutes in which to record answers. Each mark should take 1 minute 48 seconds to earn. The following table gives calculations of approximately how long you should spend on questions of a particular mark value.

Question value (marks)	*Suggested time*
1	1 min 45 s
2	3 min 30 s
3	5 min 20 s
4	7 min 00 s
5	9 min 00 s
6	10 min 45 s
7	12 min 45 s
8	14 min 30 s
9	16 min 00 s
10	18 min 00 s

- Don't panic!
  Worrying about time too much will only add to your stress and waste time. There is no need to time yourself for each question, and particularly not for the low-value questions. However, if you panic or lose track of time or waste too much time on a question of relatively small mark value, you may be severely disadvantaged.

Analysing the questions

Questions in the HSC are designed to enable students to show their achievement of a range of outcomes over a range of levels. They are not trying to trick you, and the question will state clearly all that is required of you. If marks are to be awarded for giving an answer in the correct units, then the question will clearly state that you need to give the answer in the correct units.

- Highlight key terms
  Highlight or underline the key terms or things you have to do (the verbs) and the key ideas or concepts that make up the question — a useful strategy.

- Learn the lingo
  The glossary of terms that are often used in questions will help you analyse what is required of the question (see the Glossary of key words on page ix).

- Remember that questions are planned to vary in their level of difficulty
  Don't panic or waste excessive time on a question that you find particularly difficult. The following table may give you some ideas on how to find the easier questions that you can attempt first. Then, when you have finished everything that you can do, you can move on to spend time with the very difficult questions.

Question type	*Key words that indicate level of difficulty*
Difficult or complex questions often demanding high levels of skills to use the facts	assess, construct, critically analyse, critically evaluate, evaluate, justify, propose, recommend
Questions of medium levels of difficulty requiring some thinking and using the facts	account for, analyse, apply, assess, calculate, classify, compare, contrast, deduce, demonstrate, discuss, examine, explain, estimate, extract, extrapolate, how…, interpret, investigate, predict or recommend, sketch, why...
Simple or direct questions of low levels of difficulty based on remembering and communicating the facts	account, clarify, define, describe, identify, outline, recall, recount or summarise, state, what..., which...

  Also, there may be variation in difficulty within the topics you have studied. A relatively easy remembering-type question may become a little more difficult if it is about a particularly difficult idea.

- Look at question structure and mark value
  The structure of the question may also contribute to differences in levels of difficulty. The mark value for the question may give you some clues. For example if the question is only two lines long but is worth six marks, that is a clue that it may require extra care to ensure all aspects of the question have been covered.

  Many students find the questions that are divided into subsections easier to attempt. They feel that each small portion of the question can be 'bitten off' at a time. Often the answer to one subsection helps to provide some ideas to help with the other subsections.

  The more difficult integrated questions may seem a bit more difficult to swallow, and when you do, they may give you indigestion. Practising these questions using examples from this book will really help you develop your skills and help avoid the need for antacid tablets!

- Answer all parts
  A common error is that students often only answer one part of a two-part question. This is very easy to do if the parts are not clearly numbered but are written as sentences following each other. It is worth making a note on the paper of any multiple-part questions and make sure you return to the question to finish all sections.

Glossary of key words

For each subject you should prepare summaries and a glossary of the key definitions. You cannot answer questions if you are unfamiliar with the content of the subject.

You also need to be familiar with the language of exam questions. This involves understanding the verbs that are the keywords that tell you what you have to do to earn the marks. Below is a table of the key verbs that may be contained in questions and the definitions of those words.

Key word	*Definition*
Account for	state reasons for; report on
(Give an) account of	narrate a series of events or transactions
Analyse	Identify components and the relationship among them; draw out and relate implications
Apply	Use, utilise, employ to a particular situation
Appreciate	Make a judgment about the value of
Assess	Make a judgment of value, quality, outcomes, results or size
Calculate	Ascertain/determine from given facts, figures or information
Clarify	Make clear or plain
Classify	Arrange or include in classes/categories
Compare	Show how things are similar or different
Construct	Make; build; put together items or arguments
Contrast	Show how things are different or opposite
Critically (analyse/ evaluate)	Add a degree or level of accuracy, depth, knowledge and understanding, logic, questioning, reflection and quality to
Deduce	Draw conclusions
Define	State meaning and identify essential qualities
Demonstrate	Show by example
Describe	Provide characteristics and features
Discuss	Identify issues and provide points for and/or against
Distinguish	Recognise or note/indicate as being distinct or different from; to note differences between
Evaluate	Make a judgment based on criteria; determine the value of
Examine	Inquire into
Explain	Relate cause and effect; make the relationships between things evident; provide why and/or how
Extract	Choose relevant and/or appropriate details
Extrapolate	Infer from what is known
Identify	Recognise and name
Interpret	Draw meaning from

Key word	*Definition*
Investigate	Plan, inquire into and draw conclusions about
Justify	Support an argument or conclusion
Outline	Sketch in general terms; indicate the main features of
Predict	Suggest what may happen based on available information
Propose	Put forward (for example, a point of view, idea, argument, suggestion) for consideration or action
Recall	Present remembered ideas, facts or experiences
Recommend	Provide reasons in favour
Recount	Retell a series of events
Summarise	Express concisely the relevant details
Synthesise	Putting together various elements to make a whole

From Senior Science Support Document © NSW Education Standards Authority 1999

Answering the question

Objective-response questions

Many people find multiple-choice questions relatively easy. Recording the answer is certainly easy, but they can be deceptive in their level of difficulty. Some make up for the time saved recording the answer by requiring time to think about the answer. Some of the distracters (i.e. the wrong answer options) may be based on common mistakes or misconceptions and may lead you away from the correct answer. In one 2000 HSC Science exam only 15% of students got one particular objective-response question correct, showing how effective the distracters can be.

Careful reading and interpretation of questions are critical. Sometimes you may be left with two similar answers and you have to dig into your deeper understanding to work out the implications of the subtle difference. The more practice you get on objective-response questions, the more skilful you will become. In this book we have provided explanations for making the correct choices; you do not have to do this in an exam.

Short-answer questions

- Core modules
  The short-answer questions from Part B relate to the three core HSC modules. After you have analysed the question, link it to the specific module and the concepts of that module. If the question is in parts, look at the parts and see if they give you clues about the other sections of the question. If the question gives you the freedom to select the format of your answer, choose a format that you feel most comfortable with. For example, you may prefer to express some ideas in a table, flowchart or diagram, or in writing. This may be especially important in those questions with a high mark value that require you to integrate ideas.

  Use the mark value and space provided to give you a rough guide to the number of points in the answer and the length of the answer. If it is worth six marks, you may need to include at least six key points or one point with a set of arguments, supportive statements, reasons or counter-arguments.

- Options
  Short-answer questions from Section II all relate to the specific option you have chosen. For the integrated questions, it may be worth taking a minute to do a mental revision about the major ideas of that option and checking which sections relate to the particular question. Again, if you are given the opportunity, choose an answer format that best expresses the key ideas. Answers should be as succinct as possible.

- Read through your answers
  It is important to read through your answers to short-answer questions to check that you have clearly communicated your understanding. Check that any assumptions you have made have been stated and that your thinking is clear. If you are running out of time it is important to attempt questions in a very concise manner; even if you remain uncertain as to what the question is about, think back to the option and try to make some links to key ideas.

- Attempt all questions
  It is essential you attempt all questions. You do not lose marks for guessing or recording wrong answers. Remember, however, if you ramble on with a lengthy, confused or contradictory answer you may make a satisfactory answer meaningless and therefore worth no marks. Clear and concise answers are always preferable. If you are running out of time, writing key points, tables or diagrams may help communicate an answer or earn partial marks.

  Avoid persevering with one question to make it perfect if it means you leave out several others. The mark value of questions can serve as a guide for prioritising your attempts at questions when running out of time. You will do yourself a great disservice if you do not attempt questions worth six or more marks.

- Show your working

Common mistakes

Examiners provide reports after each HSC Exam that indicate areas of strengths and weaknesses in student responses. From these common mistakes, it is clear that students generally can benefit by:

- more precise and scientific use of terminology, especially that specified in the syllabus;
- more precision with graphs and labelling of diagrams;
- avoiding answers that are too long or detailed or that use ambiguous terms;
- correct interpretation of questions, e.g. listing instead of describing, or giving a description not an explanation;
- answering in specific terms not generalities;
- ensuring all parts of multi-part questions are attempted;
- being able to clearly describe procedures, apparatus, etc. for mandatory practical activities as a first-hand experience.

Golden rules

Preparing for the exam

- Obviously the exam will reward those students who have a thorough knowledge of the whole course. Do all assignments and assessment tasks to your best ability and use them in your revision.
- Get as much practice at completing and getting feedback on exam-type questions as you possibly can. This book is perfect for this!
- Make certain you complete, know and revise the mandatory practicals from the syllabus so you can describe the procedure or evaluate its methods.
- Revise a question about an open-ended investigation you have done and be prepared to use it as an answer.
- Make sure you have learnt a few examples that show something about the history, nature and practice, applications and uses, and the implications of science as it relates to your specific subject.

In the exam

- Don't give up! If you find it difficult, so will most other students. Attempt all questions with concise, clear answers.
- Have a plan that ensures you manage time well. If the option is your particular strength, you may start with it, but do not spend more than 45 minutes on it. Alternatively, start with the objective-response because the correct answer must be there, and it can help you get over your nerves.
- Know yourself and work to your strengths. If you want to experiment with different strategies for tackling the exam, do that in the trial exam and reflect on how the plan worked.
- At the end re-read the questions and the answers you have written. Check that you have really communicated what you intended to write.
- Don't panic! It can lead you to make some silly decisions. There will be some difficult questions in the exam but there will also be some easy ones.
- Even if the option you have studied looks difficult, and another one looks like it has some 'easy' questions, you are far better off answering the question from the option you have studied in class than one you have not studied.

Finally, GOOD LUCK!

CHAPTER 1

Core Topic 1
Maintaining a Balance

Multiple-choice Questions

Past HSC Questions

1 The Australian hopping mouse, *Notomys alexis*, is a desert animal. It produces urine that is very concentrated.

Why is this an advantage for the animal?

(A) It needs to conserve water.
(B) It is nocturnal and only drinks at dusk.
(C) It has a high intake of salt in its specialised diet.
(D) It needs to excrete large amounts of water to survive.

2 What is the role of ADH (vasopressin)?

(A) It increases the amount of water reabsorbed in the kidney.
(B) It increases the amount of sugar reabsorbed in the kidney.
(C) It decreases the amount of water reabsorbed in the kidney.
(D) It decreases the amount of sugar reabsorbed in the kidney.

3 The flowchart represents one example of homeostasis in an endotherm.

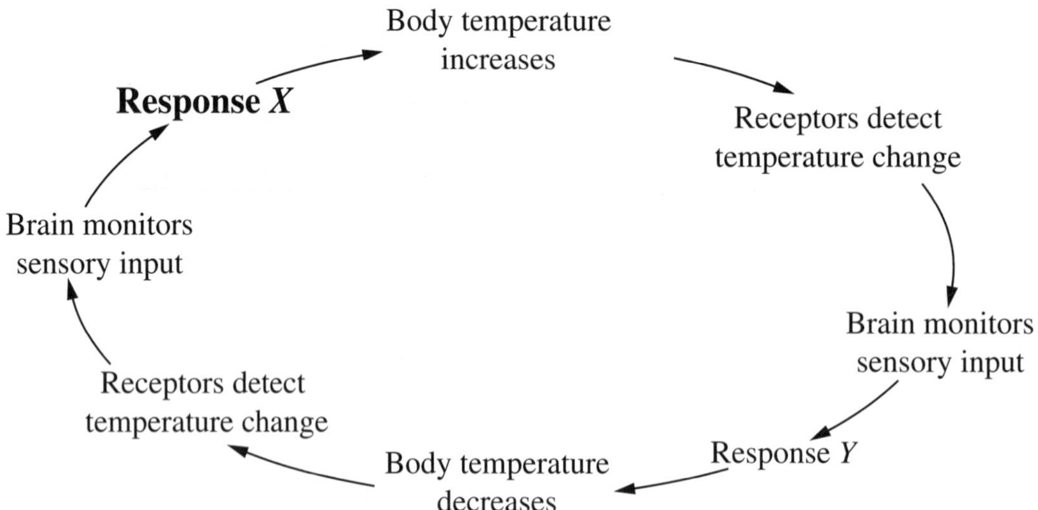

Which of the following does Response *X* represent in this cycle?

(A) Increased rate of sweat production
(B) Increased rate of urine production
(C) Decreased rate of sweat production
(D) Decreased rate of urine production

4 Four students were asked to design a first-hand investigation to determine the effect of pH on the activity of an enzyme.

Their designs of the investigation are shown in the tables.

Design *A*

Tube number	*Contents*	*pH*	*Temp.* (°C)
1	E	3	20
2	E	7	20
3	E	12	20
4	S	3	20
5	S	7	20
6	S	12	20

Design *B*

Tube number	*Contents*	*pH*	*Temp.* (°C)
1	E + S	3	20
2	E + S	7	20
3	E + S	12	20
4	S	3	20
5	S	7	20
6	S	12	20

Design *C*

Tube number	*Contents*	*pH*	*Temp.* (°C)
1	E + S	3	10
2	E + S	7	20
3	E + S	12	30
4	S	3	10
5	S	7	20
6	S	12	30

Design *D*

Tube number	*Contents*	*pH*	*Temp.* (°C)
1	E + S	7	10
2	E + S	7	20
3	E + S	7	30
4	S	7	10
5	S	7	20
6	S	7	30

Key: E = enzyme S = substrate

Which investigation is the most appropriate?

(A) Design *A*
(B) Design *B*
(C) Design *C*
(D) Design *D*

5 The sweet taste of freshly-picked corn is due to the high sugar content in the kernels. Enzyme action converts about 50% of the sugar to starch within one day of picking. To preserve its sweetness, the freshly-picked corn is immersed in boiling water for a few minutes, then cooled.

Which of the following explains why the boiled corn kernels remain sweet?

(A) Boiling destroys the sugar molecules so that they cannot be converted into starch.
(B) Boiling inactivates the enzyme responsible for converting sugar into starch.
(C) Boiling kills a fungus on the corn that is needed to convert sugar into starch.
(D) Boiling activates the enzyme that converts starch into sugar.

6 In an endotherm, which of the following homeostatic responses would be produced by a sudden and prolonged decrease in ambient temperature?

(A) Decreased uptake of oxygen
(B) Decreased muscular activity
(C) Decreased blood flow to the skin surface
(D) Decreased rate of internal metabolic processes

7 The graph shows information about four species of bacteria and their reproductive rates at different temperatures.

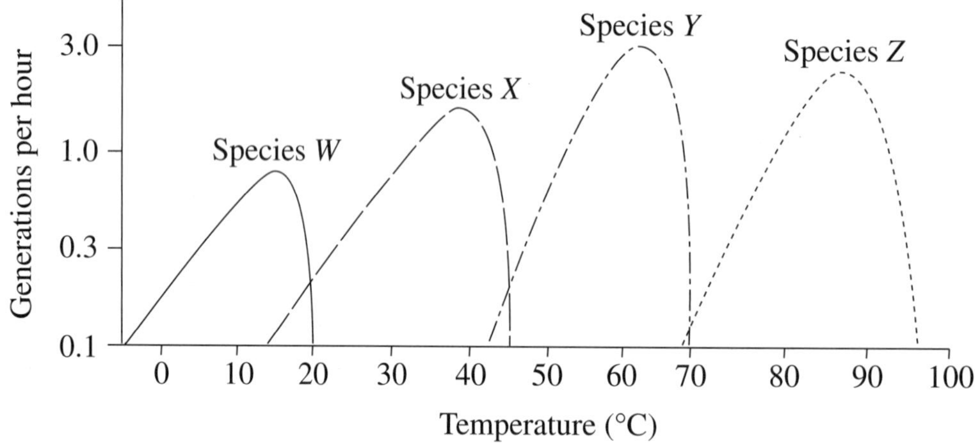

What conclusion can be drawn from this graph?

(A) All bacterial species can adapt to a broad range of temperatures.
(B) Individual species can reproduce in a broad range of temperatures.
(C) All bacterial species are limited to a range between 0°C and 100°C.
(D) Individual species reproduce in a relatively narrow range of temperatures.

8 Metabolic processes in cells produce waste substances. These wastes are constantly removed from the cells.

Why is waste removal essential for metabolic activity to continue?

(A) Waste products prevent the entry of essential nutrients into the cell.
(B) Metabolism of waste products produces chemicals that kill cells.
(C) Waste products alter the internal chemical environment of the cells and the metabolic processes would stop.
(D) Retention of waste products causes cells to lose water by osmosis and they become dehydrated.

9 A biologist studied the concentration of urine produced by a terrestrial mammal, a freshwater fish and a marine fish.

Which row of observations would be the most likely for these organisms in their natural environment?

	Terrestrial mammal	*Freshwater fish*	*Marine fish*
(A)	Produces dilute urine	Produces concentrated urine	Produces dilute urine
(B)	Produces concentrated urine	Produces dilute urine	Produces dilute urine
(C)	Produces dilute urine	Produces concentrated urine	Produces concentrated urine
(D)	Produces concentrated urine	Produces dilute urine	Produces concentrated urine

10 Spinifex is also called porcupine grass because its leaves can curl up into a needle shape. The stomates are located in sunken grooves on the underside of the leaf and are enclosed as the leaf curls up.

Which process do these adaptations best reduce?

(A) Conduction
(B) Pollination
(C) Translocation
(D) Transpiration

11 A small Australian mammal that lives in the alpine regions of New South Wales has specific features that enable it to retain body heat. Identify the features that are most likely to be present in the mammal described.

(A) Long ears, rounded body, long legs
(B) Short ears, rounded body, short legs
(C) Short ears, slender body, long legs
(D) Short ears, slender body, short legs

12 What are three products extracted from donated blood?

(A) Oxygen, water and urea
(B) Red blood cells, salts and oxygen
(C) Plasma, platelets, and red blood cells
(D) Platelets, hormones and amino acids

13 The graph represents the relationship between substrate concentration and enzyme activity for a metabolic reaction.

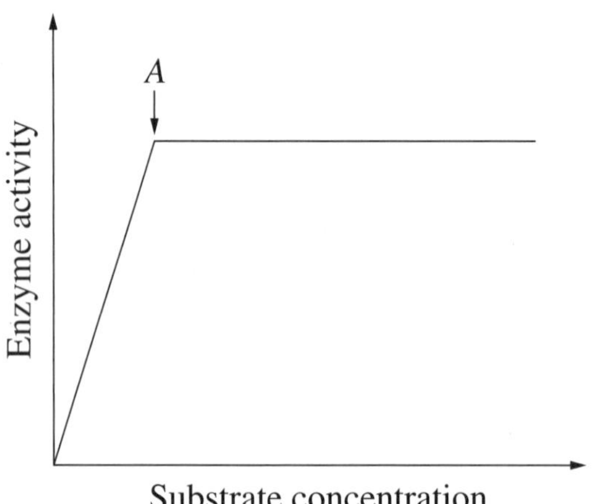

Which statement is an accurate interpretation of the graph?

(A) Point *A* is the maximum level of enzyme concentration.
(B) Increasing substrate concentration beyond Point *A* will not increase the rate of reaction.
(C) Increasing substrate concentration beyond Point *A* will increase the rate of reaction.
(D) Increasing the enzyme activity beyond Point *A* will not increase the rate of reaction.

14 Which alternative correctly describes the process involved in the movement of materials in the cells labelled *M* and *W*?

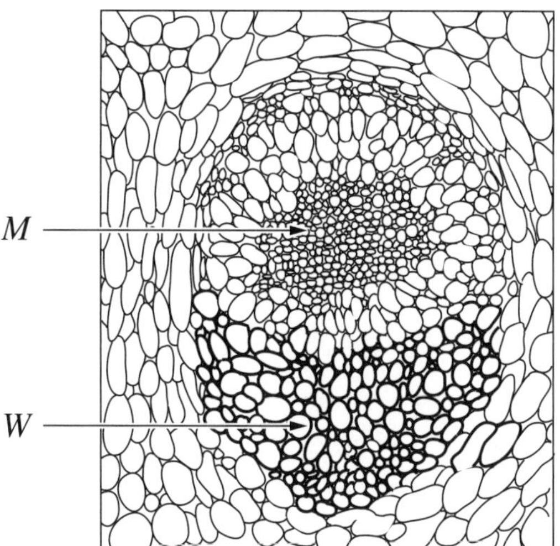

A typical vascular bundle from a stem

Mauseth, James D, 1988, *Plant Anatomy*, Benjamin/Cummings Pub. Co., California

	M	*W*
(A)	Active transport	Active transport
(B)	Active transport	Passive transport
(C)	Passive transport	Active transport
(D)	Passive transport	Passive transport

Free-response Questions

Question 1 (3 marks) **Marks**

(a) Label, on the diagram, ONE structural feature of the artery shown. **1**

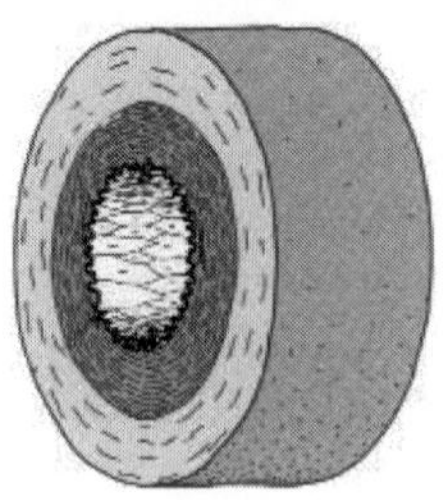

(b) How does the labelled feature relate to the artery's function? **2**

*(6 lines)**

Question 2 (5 marks)

(a) Name TWO products extracted from donated blood, and state their uses. **4**

(4 lines)

(b) Propose ONE reason why scientists have now begun to develop artificial blood. **1**

(2 lines)

Question 3 (6 marks)

In your Biology course, you performed a first-hand investigation to gather information about structures in plants that assist in the conservation of water.

(a) Describe the procedure you followed. **4**

(8 lines)

(b) Identify TWO safe work practices needed during this investigation. **2**

(4 lines)

* Shows the number of lines available in the HSC answer booklet for this question.

Question 4 (7 marks) **Marks**

Name ONE example of an Australian endothermic animal and ONE example of an Australian ectothermic animal, and summarise their responses to the following environmental changes. Give your answer in the form of a table. **7**

Change 1: The ambient temperature rises well above the average daily temperature range.

Change 2: The ambient temperature drops well below the average daily temperature range.

Endothermic animal: ....

Ectothermic animal: ....

Question 5 (6 marks)

The diagram shows the arrangement of tissues in a young plant root.

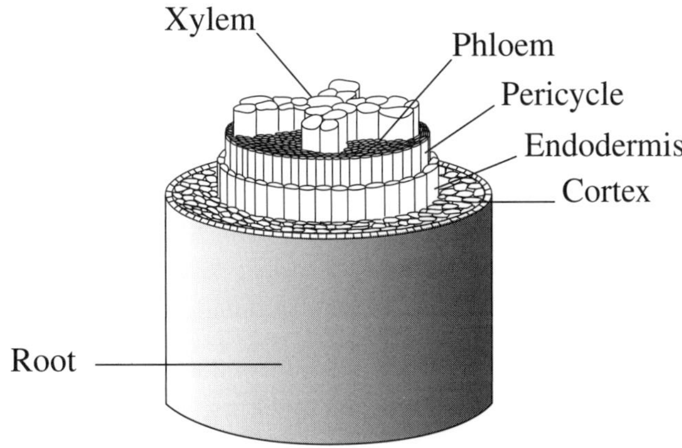

(a) From this diagram, draw a transverse section of the plant root. Clearly label the location of xylem and phloem tissue. **3**

(b) Describe ONE current theory about the processes responsible for the movement of materials through phloem tissue in plants. **3**

(6 lines)

Question 6 (8 marks) **8**

Describe a first-hand investigation used to estimate the size of red blood cells on a prepared microscope slide.

In your description include:

- a list of equipment used;
- a safety precaution needed;
- the step-by-step method used;
- a scaled diagram of a red blood cell.

(21 lines)

Question 7 (6 marks) **Marks**

The diagram represents a nephron which is the functional unit of the kidney.

Nephrons make urine by:

- filtering small molecules and ions from the blood;
- reabsorbing the needed amounts of useful materials.

Surplus or waste molecules and ions flow out as urine.

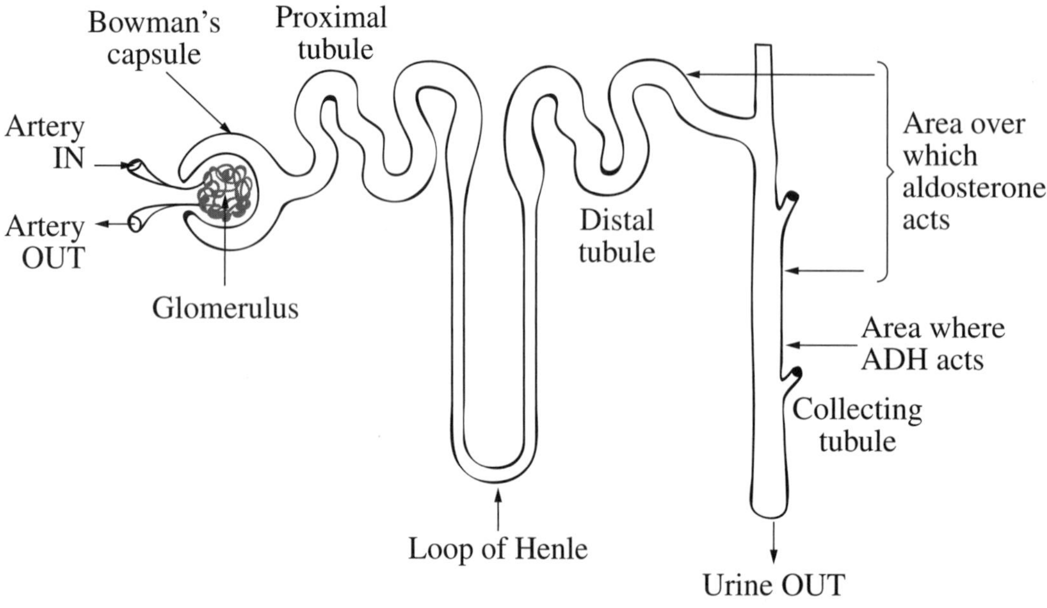

(a) Identify the area where filtration occurs, by marking it with an *X* on the diagram. **1**

(b) Identify the area where reabsorption occurs, by shading it on the diagram. **1**

(c) Discuss the importance of hormone replacement therapy for people who cannot secrete aldosterone. **4**

(8 lines)

Question 8 (4 marks) **Marks**

The graph shows the variation in body temperature for two different organisms over a range of ambient temperatures.

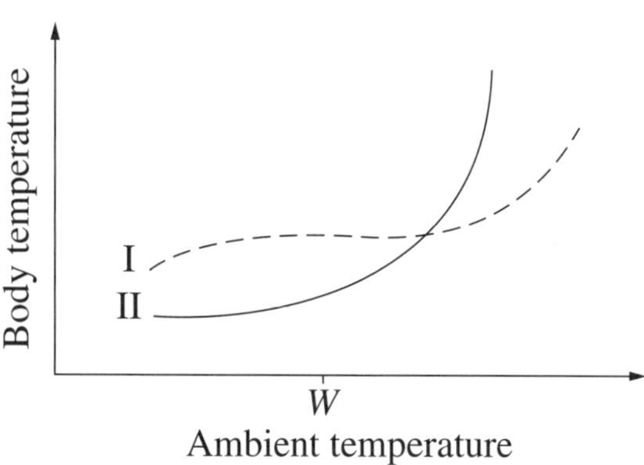

(a) State the term applied to animals that exhibit the type of body temperature response shown by organism I. **1**

(1 line)

(b) Name an Australian native animal that would exhibit a response similar to curve II? **1**

(1 line)

(c) Describe how a response by organism II would assist in temperature control when the ambient temperature increases beyond W. **2**

(4 lines)

Question 9 (5 marks)

(a) Describe an investigation that you have carried out to demonstrate the effect of dissolved carbon dioxide on the pH of water. **3**

(6 lines)

(b) Explain why the removal of carbon dioxide from living cells is important. **2**

(4 lines)

Question 10 (4 marks) **4**

Draw a labelled diagram to show how a specific feedback mechanism plays an essential role in homeostasis.

Question 11 (3 marks)

(a) Define the term *enantiostasis*. *(2 lines)* **1**

(b) Outline methods used by estuarine plants to maintain appropriate internal salt concentration. *(4 lines)* **2**

Core Topic
Maintaining a Balance
Worked Answers

Multiple-choice Questions

1 A Concentrated urine assists an animal to conserve water, especially in arid environments where water sources are scarce.

2 A ADH or vasopressin is a hormone that stimulates water absorption in the kidneys.

3 C Response Y causes the body to increase sweat production, enabling it to cool down. Once sufficiently cool, the body needs to stop excess perspiration and Response X decreases sweat production.

4 B This option utilises a control at each pH level, without the enzyme and while maintaining a constant temperature, making pH the only variable.

5 B Boiling will denature the enzyme, hence preventing the breakdown of the sugar and enabling the sweetness to be retained.

6 C Blood is directed away from the extremities to retain core body temperature.

7 D Each species is tolerant of a narrow temperature range, outside of which the species cannot exist.

8 C Wastes create an imbalance within the internal environment, and are often toxic, causing changes to body physiological processes.

9 D This is the only correct combination of urine concentration for each of the animals listed.

10 D Water loss occurs through leaves by the process of transpiration.

11 B The surface area to volume ratio needs to be small to reduce heat loss.

12 C Only possible combination of blood products.

13 B Optimum function of the enzyme has been reached.

14 B Correct method of transport for phloem and xylem tissue.

Free-response Questions

1 (a) Thick muscular wall

(b) The blood in an artery is under pressure as leaves the heart. The thick muscular wall of an artery enables it to withstand the stretching and contraction under pressure and it helps to transmit the pulse of blood around the body.

2 (a) Answers could include: red blood cells, platelets, granulocytes, Factors I–VIII and plasma.

For example: Red blood cells are used to increase the oxygen carrying capacity of blood. Factor VIII is used to promote clotting of blood.

(b) One reason for the development of artificial blood is the widespread shortage of donor blood.

3 (a) Collect a variety of leaves from different species, recording type and location. Use a hand lens to observe the leaves, recording water conservation features such as sunken stomates, surface hair, thickness and type of cuticle. Slides can be prepared from the leaves and observed under a microscope.

(b) Use of gloves to protect from possible toxic plants; careful use of scalpel when preparing slides.

4

	Ectothermic: Blue-tongue lizard	*Endothermic: Antechinus*
Change One	Seeks shade or hides under rocks	Seeks shelter out of direct sun
Change Two	Basks in sun to absorb heat	Increases activity to generate heat

5 (a)

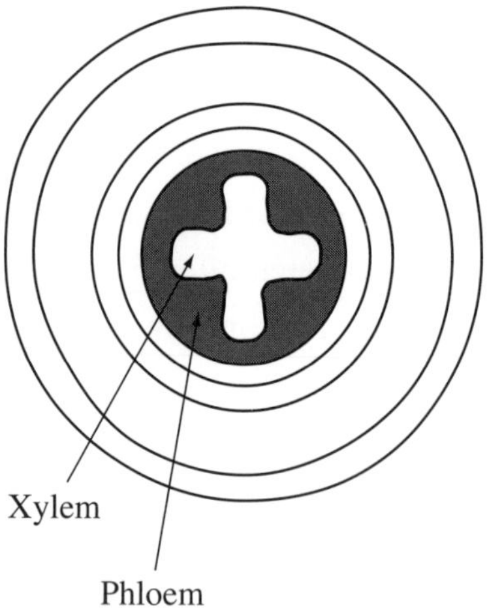

Xylem

Phloem

(b) Materials are transported in the phloem by translocation. Energy is used as sugars are actively loaded against the concentration gradient into the phloem at the site of photosynthesis (source). Sap is carried along the phloem sieve tubes by water that enters by osmosis. The sugar is then unloaded by diffusion, which is passive, at the storage (sink) site.

6 Equipment: Microscope, ocular lens with scale, microscope lamp, minigrid and prepared slide of red blood cells.

Safety precaution: Carry microscope using two hands to avoid dropping it; do not look directly down objective lens whilst adjusting light source into microscope; do not touch hot lamps to avoid burns; avoid smashing the slide by starting in the closest position and focusing away from slide.

Method:

1. Set up microscope and lamp.
2. Direct light into the microscope by adjusting mirror.
3. Place mini grid onto stage 4.
4. Wind the objective as close as possible to the stage whilst observing from the side.
5. Looking down the objective, bring the microscope into focus by winding away from the stage until the view is in focus.
6. Use the mini grid to calculate the size of the field of view using the scale.
7. Place the prepared red blood cell slide onto the stage and repeat steps 3–5.
8. Use the scale established in step 6 to calculate the size of a red blood cell.

Observations: The red blood cell appears as a biconcave disc with variable thickness across the cell. The cell is thickest around the perimeter as it lacks a nucleus and it appears less opaque in the centre.

Scale diagram of red blood cell

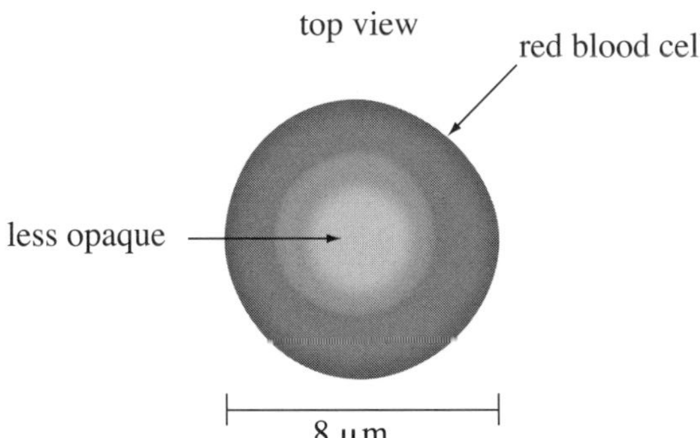

7 (a) and (b)

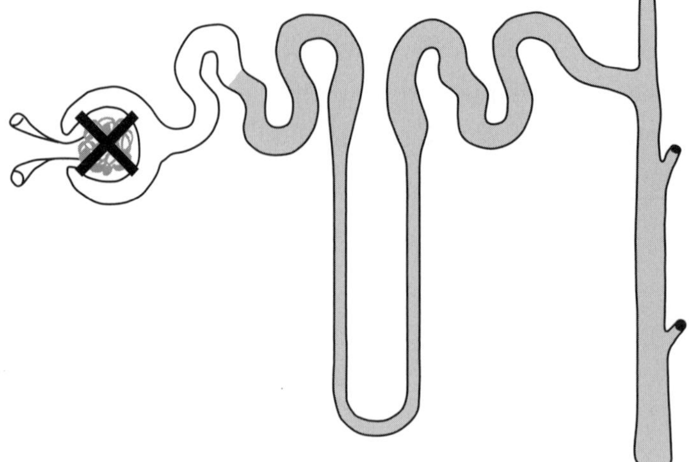

(c) Hormone replacement therapy (HRT) is used by patients who cannot secrete for themselves hormones required by the body. The role of aldosterone in the nephron is to control the permeability of the tubule to regulate the levels of sodium. This results in water reabsorption, ensuring that the water, and hence blood pressure, is maintained constantly. In the case of insufficient aldosterone secretion by the body, HRT can be used. This will enable the blood pressure to stabilise, thereby enabling appropriate functioning of the nephron by promoting reabsorption.

8 (a) Ectothermic

(b) Any named Australian ectotherm, e.g. Central Netted Dragon.

(c) Beyond W, the organism needs to maintain its body temperature in order to maintain correct biological function. The Central Netted Dragon would move out of the sun, reducing its exposure to heat or burrow in shaded sand to reduce its body temperature.

9 (a) Carbon dioxide can be produced by reacting marble chips with HCl in a test tube. The gas produced (CO_2) then needs to be transferred to another test tube filled with distilled water and universal indicator. The universal indicator is then observed and recorded for the duration of the reaction as the colour changes with decreasing pH. Alternatively a data logger can be used to measure the concentration of the CO_2 in the collecting test tube by using a pH probe in the solution to record the pH as it decreases for the duration of the reaction.

(b) The removal of carbon dioxide is important for cell function because it lowers the pH of the solution in which it is dissolved thereby affecting the enzyme function and possibly the shape of the haemoglobin molecules. This causes greater oxygen release and thereby decreases the oxygen saturation. Both are undesirable conditions necessitating the removal of carbon dioxide from living cells to ensure correct cellular processes.

10

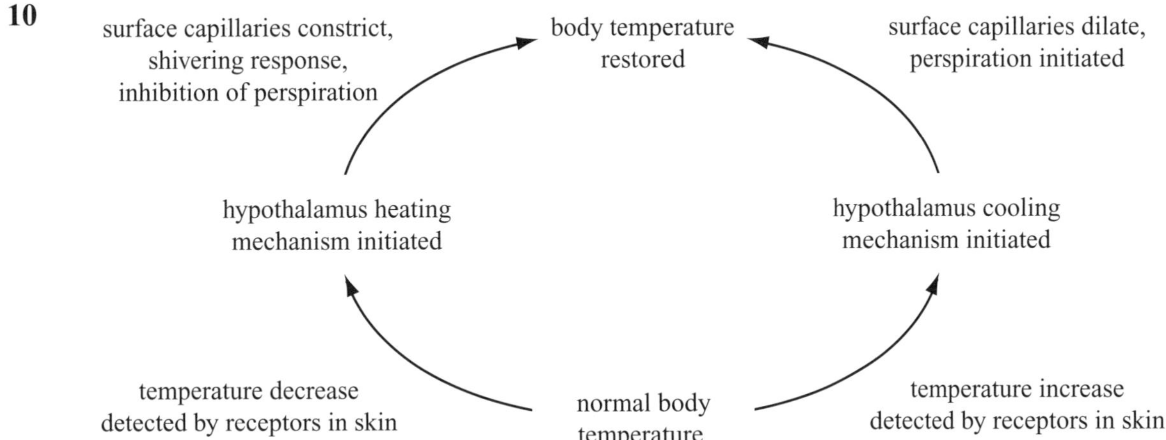

Homeostasis – role of feedback mechanism in maintaining body temperature

11 (a) Enantiostasis is the term used to describe the maintenance of physiological and metabolic functions of organisms in response to changing conditions in their environment.

(b) Organisms in estuarine environments such as mangroves have developed adaptations that enable them to maintain internal salt concentration by excreting excessive salt from their leaves and then shedding the leaves. Other organisms maintain similar salt concentrations in cellular fluid to that of their surroundings thereby preventing osmosis and regulating the amount of salt uptake.

CHAPTER 2

Core Topic
Blueprint of Life

Multiple-choice Questions

Past HSC Questions

1 Fossil evidence indicates that the Australian environment in the past supported a large and diverse range of megafauna. The megafauna has now been displaced by a variety of smaller marsupials.

What is the best explanation for this?

(A) Smaller marsupials coped with climatic changes, and survived.
(B) Larger marsupials reduced in size so as to cope better with climatic changes.
(C) A meteorite collision caused a mass extinction of the megafauna.
(D) Introduced plant species were not a suitable food for the megafauna.

2 Which of the following is true of a mutation that produces an allele that is dominant?

(A) It would be expected to cause death.
(B) It would be expected to spread more quickly through a population than a recessive mutation.
(C) It could give an observable phenotype in a heterozygous genotype.
(D) It could give an observable phenotype only in a homozygous genotype.

3 In 1940, Beadle and Tatum developed the *one gene–one protein* hypothesis. This has now been modified to the *one gene–one polypeptide* hypothesis.

Why was this modification needed?

(A) All proteins are comprised of more than one type of polypeptide.
(B) Most proteins are comprised of more than one copy of the same polypeptide.
(C) Many proteins are comprised of more than one polypeptide that may be the same or different.
(D) The number of polypeptides in proteins is always the same as the number of genes specifying those polypeptides.

4 Haemophilia is a human disease in which the blood of an affected individual does not clot. The disease is known to be caused by a sex-linked recessive allele.

The family pedigree shows the pattern of inheritance of this disease in a family.

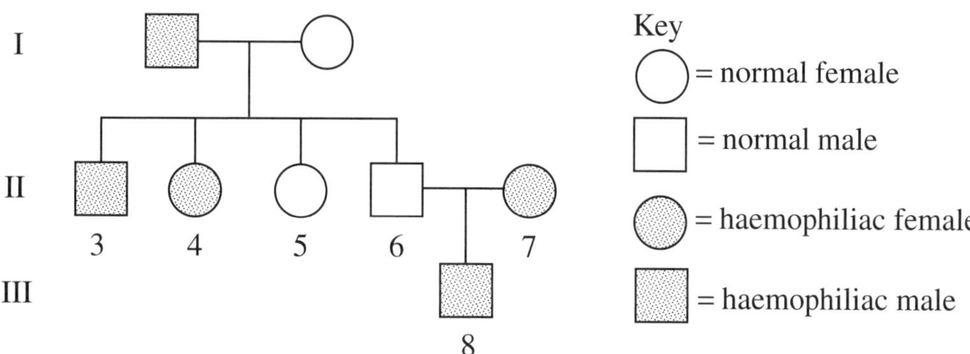

If X^h is the allele for haemophilia and X^n is the allele for normal clotting, what is the genotype of individual 5?

(A) XhXh
(B) XhXn
(C) XnXn
(D) XnY

5 The diagram represents one pair of homologous chromosomes during meiosis. Crossing-over occurs and random segregation takes place.

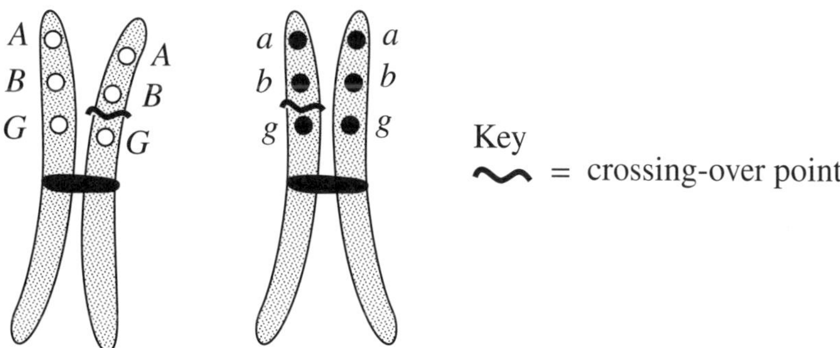

What genotypes are produced?

(A) *ABG*, *abG*, *ABg*, *abg*
(B) *ABG*, *aBG*, *Abg*, *abg*
(C) *ABG*, *ABG*, *abg*, *abg*
(D) *ABG*, *aBg*, *Abg*, *abg*

6 In a variety of garden peas, the allele for tall plants (T) is dominant over the allele for short plants (t). A cross between a tall plant and a short plant resulted in 50% of the offspring being short.

What were the genotypes of the parents?

(A) *Tt* and *tt*
(B) *Tt* and *Tt*
(C) *TT* and *Tt*
(D) *TT* and *tt*

7 An experiment was conducted to examine the effect of ultraviolet radiation on the development of antibiotic resistance in a strain of bacteria. The table summarises the outcomes of this experiment.

	Antibiotic resistance				
Treatment	*Antibiotic P*	*Antibiotic Q*	*Antibiotic R*	*Antibiotic S*	*Antibiotic T*
No exposure to ultraviolet radiation	✓	✓	✗	✗	✗
Exposure to ultraviolet radiation	✓	✓	✗	✓	✗

✓ = resistant ✗ = not resistant

Which of the following statements best summarises the stages in the development of the new strain of bacteria that was resistant to antibiotic *S*?

(A) Hybridisation → Mutation → Natural Selection
(B) Replication → Mutation → Natural Selection
(C) Mutation → Natural Selection → Replication
(D) Mutation → Hybridisation → Natural Selection

8 Polycystic kidney disease is a rare disorder affecting 1 : 1250 live-born infants. The pedigree shows the incidence of this disease in a family.

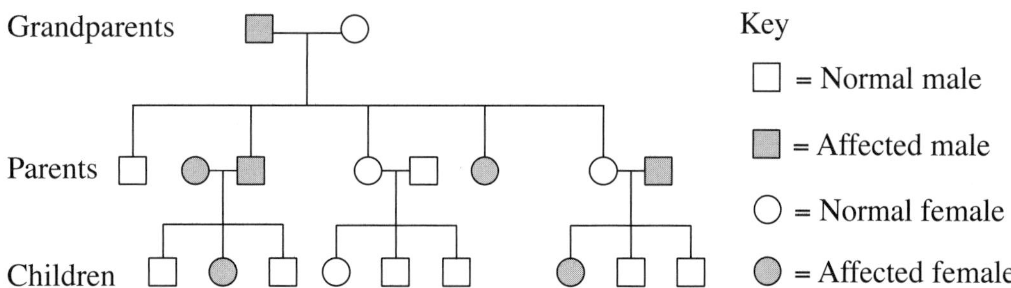

Which of the following statements best describes the mechanism of inheritance of this disorder?

(A) Sex-linked and dominant
(B) Sex-linked and recessive
(C) Non sex-linked and dominant
(D) Non sex-linked and recessive

9 Which of the following statements best describes the process of hybridisation frequently used in agriculture?

(A) The transfer of a gene from one species to another
(B) The crossing of two genetically different strains of a species
(C) The production of genetically identical offspring by cloning
(D) The artificial selection and breeding of suitable offspring within a species

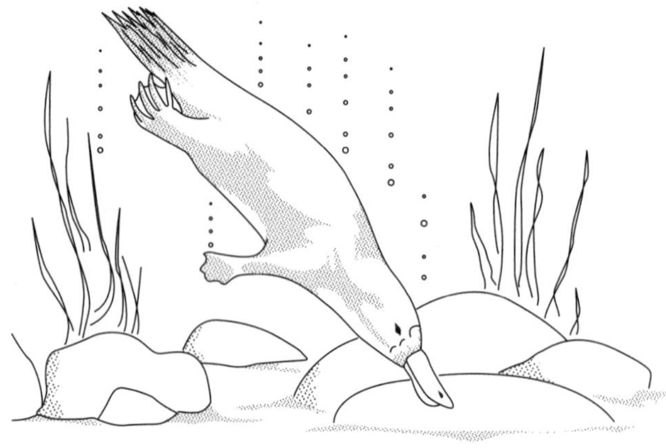

O! Thou prehistoric link,
kin to beaver, rooster, skink,
duck, mole, adder, monkey, fox,
palaeozoic paradox!

HARRY BURRELL,
The Mud-Sucking Platypus
with permission of
Harper Collins Publishers

10 Which technique would be used to measure the extent of the evolutionary relationship between the platypus and the eight other animals mentioned in the poem above?

(A) Identifying fossils that are transitional between the platypus and the animals listed
(B) Sampling DNA and identifying similarities between the animals listed
(C) Comparing the anatomical features of each animal
(D) Comparing the embryos of each animal

11 The steps involved in DNA replication and protein production are summarised below.

Step *A*:	DNA is copied and each new cell gets a full copy.	DNA replication
Step *B*:	Information is copied from DNA and taken to cytoplasm.	Protein synthesis
Step *C*:	Ribosome reads information and assembles protein.	
Step *D*:	Protein formation is completed.	

In which step would a mutation lead to the formation of a new allele?

(A) Step *A*
(B) Step *B*
(C) Step *C*
(D) Step *D*

12 How many sex chromosomes does a normal human female inherit from her mother?

(A) 1
(B) 2
(C) 23
(D) 46

13 Deoxyribonucleic acid (DNA) is a double-stranded nucleic acid molecule. For all double strands of DNA, which one of the following statements is true?

(A) The numbers of adenine (A) and guanine (G) bases are equal.
(B) The numbers of guanine (G) and cytosine (C) bases are equal.
(C) The numbers of thymine (T) and cytosine (C) bases are equal.
(D) The numbers of adenine (A), guanine (G), thymine (T) and cytosine (C) bases are equal.

14 Following birth, each baby in Australia has a sample of blood taken that is tested for the genetic disease phenylketonuria (PKU). This disease affects both genders equally and can be found in babies of parents who do not show the disease.

Which of the following best describes the mechanism of inheritance for phenylketonuria?

(A) Co-dominant
(B) Dominant
(C) Recessive
(D) Sex-linked

15 The pedigree chart below shows a possible pattern of inheritance for human albinism. Albinism is a condition in which people do not produce pigment in their skin, hair and eyes.

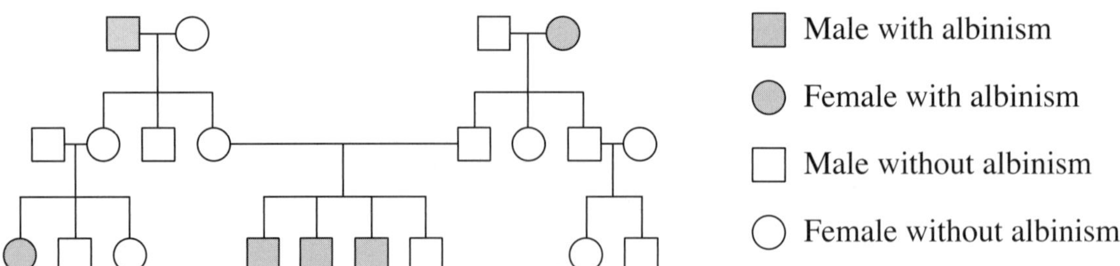

Which of the following statements is correct?

(A) People with albinism are homozygous for albinism.
(B) The gene for normal skin pigmentation is recessive.
(C) There are two genes that code for albinism.
(D) Albinism is a sex-linked characteristic.

16 The tortoiseshell cat has a combination of black and orange fur. The gene for black fur is represented by X^b and the gene for orange fur is represented by X^o. A tortoiseshell female cat ($X^b X^o$) mates with an orange male cat ($X^o Y$).

Which alternative shows the probable percentages of coat colours in the litter of kittens?

(A) 50% tortoiseshell females and 50% orange males
(B) 50% orange females and 50% tortoiseshell males
(C) 25% black females, 25% orange females, 25% black males, 25% orange males
(D) 25% tortoiseshell females, 25% orange females, 25% black males, 25% orange males

Free-response Questions

Past HSC Questions

Question 1† (4 marks) **Marks**

> **Genetically modified food on menu**
>
> Australians are growing more accepting of genetically modified (GM) food, a federal agency survey has found.
>
> This survey found that 44 per cent of Australians now believe GM food would become more widely accepted and less risky in the next few years.
>
> An earlier survey revealed that people thought the risks outweighed the benefits, but that the situation would change.

(a) State ONE opinion held by Australians about genetically modified food, according to this article. **1**

(2 lines) *

(b) Justify ONE piece of information you would need in order to determine the validity of the survey results. **3**

(6 lines)

Question 2 (4 marks)

Sutton, Boveri and Morgan worked in the field of genetics. **4**

Describe the contribution made by TWO of these scientists to the understanding of the chromosomal nature of inheritance.

(8 lines)

Question 3 (6 marks)

(a) Cloning is a technique that could be used to increase numbers in an endangered species. What effect would cloning have on the genetic diversity of the species? **2**

(4 lines)

(b) Explain TWO possible evolutionary effects of a disease entering an endangered population containing some cloned individuals. **4**

(8 lines)

---

† This is also a Biology Skills (9.1) question.
* Shows the number of lines available in the HSC answer booklet for this question.

Question 4 (8 marks) **Marks**

Evaluate the impact of major advances in scientific understanding and technology, in the field of genetics, on developments in reproductive technologies. **8**

(26 lines)

Question 5 (4 marks)

The diagram shows a cell containing three pairs of chromosomes just prior to a *meiotic* division. **4**

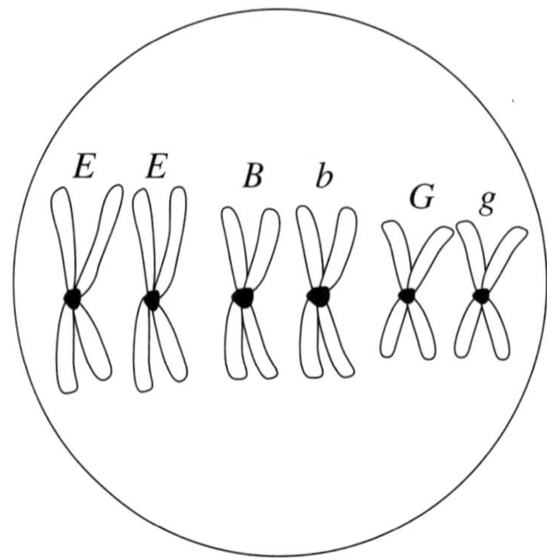

Assuming that random segregation occurs, construct a table that lists the possible genotypes that could be produced and states the expected frequency for each genotype.

Question 6 (4 marks) **Marks**

Traditionally, banana plants in Australia have been propagated asexually by cutting out and planting suckers from the adult plant. **4**

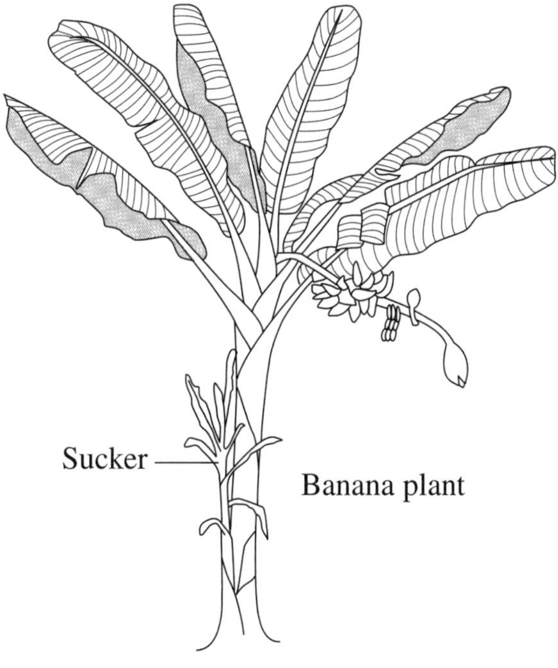

There is a growing trend to produce disease-free plants in laboratories through a process of cloning from disease-free tissues from existing plants.

Assess the potential impact of this cloning process on the genetic diversity of banana plants in Australia.

(11 lines)

Question 7 (4 marks)

(a) Define the concept of *punctuated equilibrium* in evolution. **1**

(2 lines)

(b) How does punctuated equilibrium differ from the process proposed by Darwin? **3**

(10 lines)

Question 8* (5 marks) **Marks**

The following is an extract from a gardening website. **5**

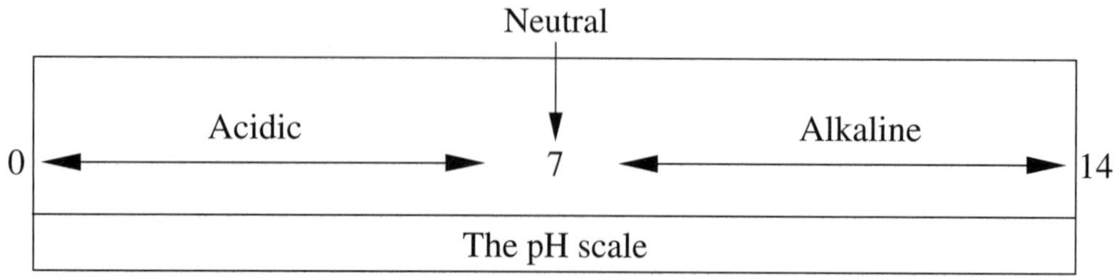

Hydrangeas are amazingly versatile in that you can alter the flower colour by changing the pH of the soil. In acid soils, hydrangeas produce blue flowers. In alkaline soils, hydrangeas produce mauve, pink and red flowers.

Describe a first-hand investigation that could be used to verify the effects of pH on the colour of hydrangea flowers.
(12 lines)

Question 9 (3 marks)

The widespread use of antibiotics for the treatment of bacterial infections has led to the development of antibiotic resistance in some species of bacteria. From your studies of evolution and the mechanisms of inheritance, explain how resistance has developed in bacteria. **3**
(5 lines)

Question 10 (3 marks)

The diagram shows various forms of radiation that are part of the electromagnetic spectrum.

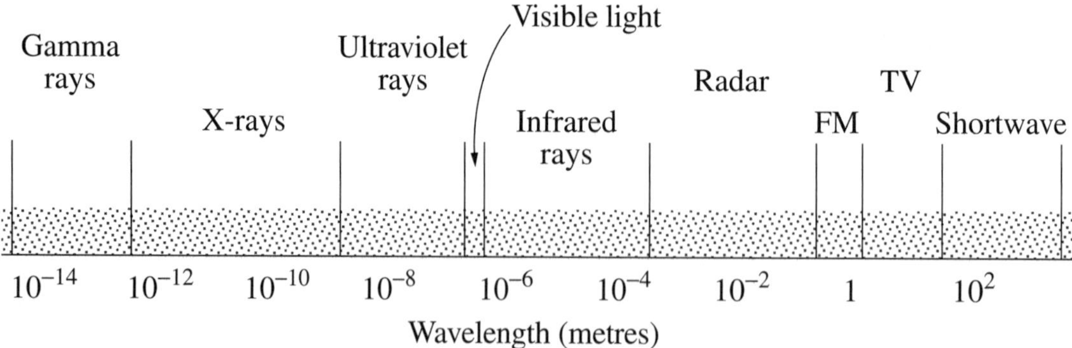

(a) Select ONE form of radiation that is considered to be a mutagen. **1**
(1 line)

(b) Describe evidence that supports the mutagenic nature of the selected form of radiation. **2**
(4 lines)

* This is also a Biology Skills (9.1) question.

Question 11 (3 marks) **Marks**

(a) Name ONE technology that could be used to establish evolutionary relationships. **1**

(1 line)

(b) Explain how the data revealed by this technology can be used to establish evolutionary relationships. **2**

(4 lines)

Question 12 (4 marks)

Discuss some ethical considerations arising from the development of genetically modified animals. **4**

(8 lines)

Question 13 (8 marks)

(a) Describe how the important work carried out by Rosalind Franklin enabled James Watson and Francis Crick to identify the structure of DNA. **2**

(4 lines)

(b) In 1962, James Watson, Francis Crick and Maurice Wilkins were awarded the Nobel Prize in Medicine for identifying the structure of DNA. Using examples, evaluate the impact of this discovery on current genetic technology. **6**

(12 lines)

Question 14 (4 marks)

You have carried out first-hand investigations that have attempted to model biological concepts. **4**

Discuss the use and limitations of models when illustrating biological concepts.

(8 lines)

Question 15 (3 marks)

Clarify, using examples, the difference between the terms *allele* and *gene*. **3**

(8 lines)

Core Topic
Blueprint of Life

Worked Answers

Multiple-choice Questions

1 A There is not enough evidence to support any of the other options.

2 C There is no reason to suggest that all mutations are lethal. All alleles have a 50% chance of being passed on and can produce an observable phenotype in both a heterozygous and homozygous genotype.

3 C Proteins can be made up of multiple polypeptides that may or may not be identical, and so this is the only possible option.

4 B Individual 5 (a female) does not suffer from haemophilia, so must be heterozygous for the characteristic, as the father would have passed on an affected allele X(h) to the daughter and the mother a normal unaffected allele X(n).

5 A These are the only possible combinations from this crossing-over point.

6 A One of the parents had to be heterozygous for height if 50% of the offspring were short and one of the parents was homozygous recessive.

7 C This is the only possible sequence as the bacteria had to become different first before resistance could develop and then be selected for before passing on the new characteristics.

8 C It is non sex-linked because the females in line two do not produce affected sons.

9 B All other options involve artificial manipulation techniques.

10 B Sampling DNA will enable the comparison of different proteins in the organisms and hence identify similarities at a very basic bio-chemical level.

11 A This represents a change in the sequence of bases and hence a mutation.

12 A Only one sex chromosome is passed on to offspring from each parent.

13 B Only correct combination of complementary bases.

14 C PKU must be recessive if parents do not show the disease.

15 A Pedigree shows a recessive condition, so albino phenotype must be a homozygous genotype.

16 D The fur colour is co-dominant and as such 25% for each type of offspring (including tortoiseshell) is the only possible alternative.

Free-response Questions

1 (a) Australians now believe that genetically modified food will become more widely acceptable to people.

(b) Answers could include information about: questions asked, survey sample size, structure and composition of the sample.

For example: The sample size used for the survey is important. Validity of the survey increases with greater sample size. This enables a better representative opinion for the population being surveyed. The most valid surveys would include the whole population.

2 Any two of the following suggestions:

Sutton proposed the chromosomal theory of inheritance, and suggested that Mendel's inheritance factors are located on chromosomes. He demonstrated the principles of segregation and random assortment using grasshoppers.

Boveri showed that chromosomes are separate, continuous entities within the nucleus of a cell and that cells with the correct number of chromosomes were able to develop normally.

Using Drosophila, Morgan demonstrated that some genes are carried on the X chromosome and others on the Y chromosome, leading to the awareness that some genes could be sex-linked.

3 (a) Cloning results in genetically identical offspring. This can reduce the frequency of diversity within a species' gene pool.

(b) Example: A disease entering a population of endangered organisms in which there are some cloned individuals and some normal individuals may have no effect at all if all of the individuals (the species and its clones) are resistant to it. However if only the cloned individuals have resistance to this disease, then they would increase in the population through their greater survival and reproduction while the non-resistant individuals would die off. While disease resistance would be selected for, there would be a decrease in genetic diversity because only the cloned individuals having identical genotypes to the parent would survive.

4 The answer should include more than one advance in scientific understanding and technology. These should be clearly identified and evaluated in terms of their impact on developments in reproductive technology. Some advances include: identification of chromosomes and the mechanism of inheritance; DNA structure; mapping of the genome; and improved microscopes. Some technological improvements in reproductive technology include: artificial insemination, cloning, sex selection, and transgenic species.

Example: Advances in tissue culture and in understanding DNA structure have enabled the development of transgenic species. Transgenic species have had a huge impact on ethical considerations in the area of genetics and reproductive technologies. For example the production of BT cotton raised many questions and opened up many possibilities in the area of agriculture, where the use of disease-resistant crops is seen as a huge advantage.

5

Genotypes possible	*Frequency (%)*
EBG	25
EBg	25
EbG	25
Ebg	25

6 Banana plants reproduce asexually, which means they have little genetic diversity because asexual reproduction does not result in variation of offspring from the parents. Hence the genetic diversity of banana plants currently in existence will not be different to that of the originally introduced plants. Cloning is a process that produces genetically identical offspring to the parents. So if the banana plants were to be cloned there would be no impact on the diversity of the banana plants; they would remain genetically identical to the parent plants.

7 (a) Punctuated equilibrium involves evolution in a series of sequential bursts or episodes of rapid change between long periods of stability or little change.

(b) Darwin proposed evolution as occurring over an extended period of time where changes occurred gradually from existing species. After a very long period of time the species will be very different and more complex than those from which they originated. Punctuated equilibrium on the other hand may have a similar result but occurs in a series of bursts in which rapid change occurs between episodes of stability as opposed to the long period of gradual change described by Darwin.

8 The main elements of an investigation that must be included are the identification of the variables; control; repetition and sample size; risk assessment; equipment list; and a logical sequence of steps in the method, including the data collection method.

Example: Investigating the effects of pH on the colour of hydrangea flowers.

Equipment:
- 10 identical pots
- Bag of potting mix
- Five blue and five pink hydrangeas of the same type and size
- Graduated watering can
- Acid, to increase soil acidity
- Base, to reduce soil acidity
- Tape measure
- Graph paper
- Camera with film

Take 10 equally sized pots and mark two of each with letters from A to E. Place identical amounts of the potting mix in them. The pH of the soil (independent variable) should be altered for one of each of the plant groups. 'A' plants should have the pH increased to 8, and 'B' plants should have the pH increased to 10. 'C' plants should have the pH decreased to 5, and 'D' plants should have the pH decreased to 3. The 'E' plants should be left untreated as a control group.

Plant one pink plant each in one group of the pots labelled A to E. Plant one blue plant in each of the other pots labelled A to E. Each group should be identical except for the flower colour (dependent variable). These plants should be placed in the same position and watered regularly with the same volume of water (that is, all other variables should be kept the same).

The growth and colour of the plants should be recorded daily for a month on a graph and photographed weekly. Any colour change can be compared to samples in books or charts depicting colour variance.

Care must be taken to ensure that chemicals used are clearly labelled and kept out of reach of children. Gloves should be worn and hands must be washed to avoid ingesting chemicals or pathogens in the potting mix.

9 Bacteria may be killed by antibiotics, each antibiotic targeting a specific bacterium. Variation exists in any bacteria gene pool so some bacteria may not be killed by a particular antibiotic. These survivors pass on their genotype that manifests in a population creating a resistant strain by natural selection. Spontaneous mutations may also occur that may result in resistance to a particular antibiotic and these mutations may subsequently be passed on by replication.

10 (a) Example: UV radiation

(b) UV radiation is greater in areas where the ozone is thinner resulting in greater exposure and a higher incidence of skin cancer in those areas.

11 (a) Example: DNA sequencing through chemical hybridisation

(b) Heat is used to separate the strands of DNA from two different species. The separated strands are then mixed, forming hybrid molecules. The evolutionary relationships can be determined by the degree of base pairing in the hybrid, whereby the most recently diverged species have a high degree of correlation and those that diverged from their common ancestor in the distant past have a low correlation.

12 Genetically modified animals result from the artificial alteration of an organism's genetic blueprint whereby desired characteristics are inserted to enhance specific features. These features may then be passed on to offspring in the form of transgenic species.

Some ethical considerations arising from the development of genetically modified animals could include:

Positives: Transgenic organisms used for agricultural purposes can increase productivity and yields. They could possibly produce replacement human body parts and hormones for greater survival of humans.

Negatives: Who decides whether it is appropriate to modify animals? Modification reduces genetic diversity in a changing environment thereby affecting biodiversity. It can cause unnecessary killing of animals. The future consequences are unknown.

13 (a) Rosalind Franklin carried out X-ray crystallography to obtain diffraction patterns of crystallised DNA that indicated that DNA could be helical in structure. Watson and Crick were able to use this information to develop a model of DNA.

(b) The impact of the discovery made by Rosalind Franklin and later work carried out by Wilkins, Watson and Crick had a significant impact on genetic technology. Understanding the structure of DNA has resulted in the development of technologies currently used that assist in the agricultural industry, medical and forensic fields. Advances in these areas include the creation of transgenic species that are disease resistant (BT cotton), increasing yields (wool in sheep, milk in cows), cloning (production of identical plant species), overcoming genetic diseases (gene therapy), and DNA fingerprinting. The knowledge applied has resulted in very significant advances: the quality of human life can be improved by the possibility of treatment for some diseases such as cystic fibrosis, and agricultural products have the potential to be cheaper and of higher quality or quantity.

There are possible negative impacts as the technology described can result in reduced genetic diversity; monocultures can be exposed to detrimental environmental changes; the costs may be a burden to the community; and there are issues regarding control and regulation due to unforseen consequences.

14 Biological concepts are often depicted using models because they enable processes to be observed and investigated more easily. For example, biological processes that take extended lengths of time such as evolution and natural selection can be represented in more realistic and observable periods through the use of models using bacteria and antibiotics. Another model type uses known quantities of jellybeans with different colours being scattered in grass and re-collected to simulate natural selection. Complex processes can be simplified and studied more easily in models one stage at a time. Some problems associated with modelling biological concepts include the fact that a model may oversimplify the process resulting in misconceptions, or the relative time or size may be misrepresented.

15 The term *gene* refers to the position on a chromosome where a particular characteristic such as eye colour is coded. A gene can have many different possible alleles that represent the variation of that gene. An allele is the term used to describe a specific sequence of DNA that codes for a specific trait, such as the blue/brown allele for eye colour.

CHAPTER 3

Core Topic

The Search for Better Health

Multiple-choice Questions

Past HSC Questions

1 If campers have to drink water from a creek, which is the best way of making the water safe to drink?

(A) Boil the water for five minutes.

(B) Filter the water through a clean shirt.

(C) Collect the water and let it stand in a clean container.

(D) Expose the water to the sun's ultraviolet rays for two hours.

2 Overseas equestrian competitors brought their horses to Australia for the Sydney 2000 Olympic Games.

Why were the horses quarantined for a period of time before the Olympic Games began?

(A) To acclimatise them to Australian conditions

(B) To make sure that no horse diseases spread to the spectators

(C) To make sure that the horses did not contract Australian diseases

(D) To make sure that the horses did not have an infectious disease

3 What is a possible immune response to a pathogen?

(A) T lymphocytes produce antibodies.

(B) T helper lymphocytes are activated.

(C) B lymphocytes produce antigens.

(D) B lymphocytes phagocytose the pathogen.

4 How does immunisation against diseases such as diphtheria and polio limit the spread of these infectious diseases?

(A) Immunisation kills the relevant pathogens.
(B) Immunisation suppresses or reduces the immune response and associated inflammation.
(C) Immunisation strengthens first-line defence barriers and prevents the entry of the relevant pathogens into the body.
(D) Immunisation reduces the multiplication of the relevant pathogens in immunised hosts and this reduces the chance of other people becoming infected.

5 Eight sick animals were found to be suffering from the same symptoms. Blood tests showed that they were infected with the same type of bacterium.

Which of the following strategies would be the best to determine if this particular type of bacterium is the cause of the disease?

(A) Find other animals with the same symptoms. Attempt to isolate the same type of bacterium from their blood.
(B) Inject blood from animals with the symptoms into suitable host individuals. If they develop the same symptoms, this proves that this type of bacterium caused the disease.
(C) Use bacteria cultured from the blood of the animals with these symptoms to infect suitable host individuals. If they develop the disease, attempt to isolate the same type of bacterium from their blood.
(D) Treat all eight animals with an antibiotic known to kill this type of bacterium. They will recover if this type of bacterium is the cause of the disease.

6 The following paragraph describes a body response.

The response is protective, and it makes nearby blood vessels leak. Plasma and white cells move into the affected area, diluting and destroying the infectious agent. This is why the infection site swells, reddens and feels hot. Although we tend to think of this response in terms of annoyance, soreness and pain, it is actually a beneficial response.

What response does this paragraph describe?

(A) Inflammation

(B) Cell differentiation

(C) The action of antibodies

(D) The activation of helper T-cells

7 The table lists the causative agents for four different diseases.

Disease name	*Causative agent*
Influenza	Virus
Creutzfeldt-Jacob	Prion
Ringworm	Fungus
Food poisoning	Bacterium

For which of these diseases would treatment with antibiotics be most appropriate?

(A) Influenza

(B) Creutzfeldt-Jacob disease

(C) Ringworm

(D) Food poisoning

8 Which of the following is an example of quarantine used to control the spread of disease across regions of Australia?

(A) Killing weeds in infested forests using herbicides and direct removal

(B) Sterilisation of all food products that come from overseas

(C) Sterilisation of all food products before packaging

(D) Removal of fruit from cars travelling interstate

9 During the last 50 years, over-use of prescription drugs has led to the emergence of resistant strains of pathogens. Why is this a problem?

(A) Resistant pathogens will cause new diseases.

(B) Many diseases may become untreatable.

(C) Prescription drugs will cause the release of toxins by pathogens.

(D) A single prescription drug can no longer kill all strains of a pathogen.

10

Accident victim's fingers saved, stored and transplanted onto other hand

Doctors have successfully transplanted the fingers of a man's severed hand in the first operation of its kind in Australia.

The man was critically injured in a train accident. His left arm was severed and right arm crushed.

A team of medical staff operated to replace the crushed fingers of his right hand, using those that were saved from his severed left arm.

The man is expected to have almost normal use of his hand within nine months.

Transplanted organs and tissues are often rejected. Why was there no tissue rejection in the man described in the above paragraph?

(A) The man's skin was damaged so his first line defences were not functional.

(B) Antigens on the man's left hand fingers were the same as those on his right hand.

(C) The man lost so much blood that lymphocytes were not present in sufficient numbers to cause an immune response.

(D) There was no blood supply to the transplanted fingers so mixing of donor and recipient antigens did not occur.

11 What was Macfarlane Burnet's major contribution to science?

(A) Better understanding of the immune response

(B) Identification of complementary bases in DNA

(C) Proposal of the one gene–one protein hypothesis

(D) Identification of the importance of chromosomes

12 The following diagram summarises the steps of an experiment similar to that carried out by Louis Pasteur, which identified microbes as agents of decay.

Step 1 Two swan-neck flasks are filled partially with equal volumes of beef broth.

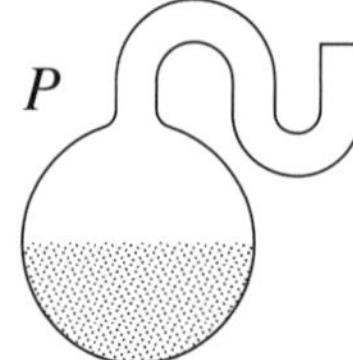

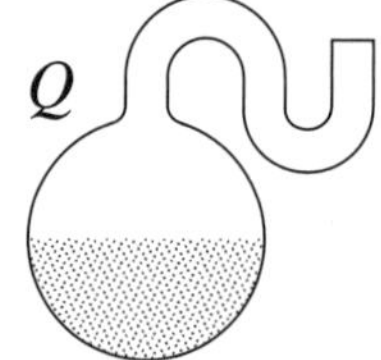

Step 2 The broth is boiled for at least 20 minutes.

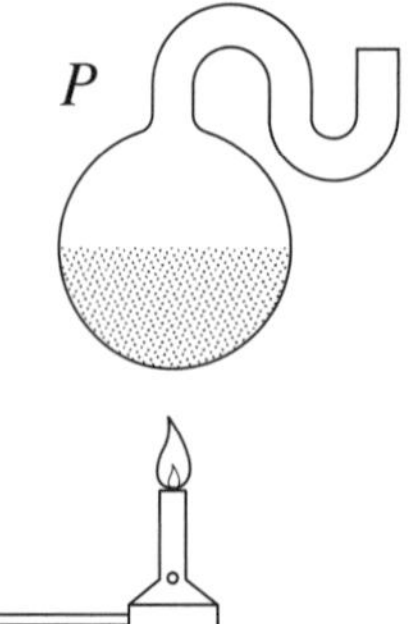

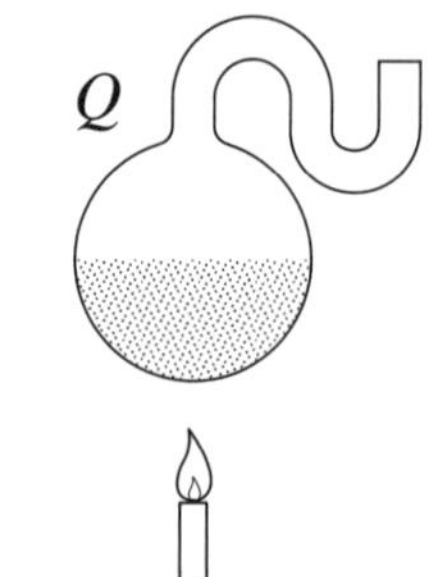

Step 3 The neck of Flask *P* is left intact whereas the other is broken.

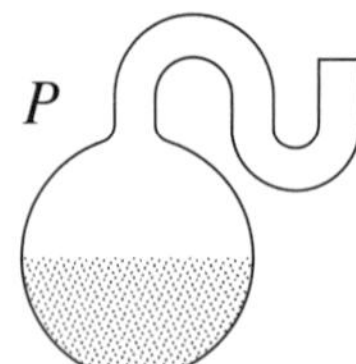

Step 4 The flasks are observed two weeks later for evidence of decay

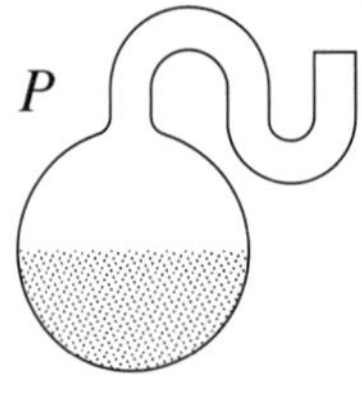

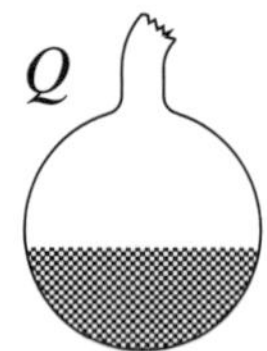

No decay Decay present

Which of the following statements best explains the results obtained?

(A) There were no microbes in the air around Flask *P* at Steps 3 or 4.

(B) There were no microbes in Flask *P* at the beginning of the experiment.

(C) Microbes in Flask *Q* were not all killed by boiling, and multiplied following the cooling down of the flask.

(D) Any microbes present in both Flasks *P* and *Q* were killed by the boiling process, and only Flask *Q* allowed microbes to re-enter.

13 The incidence of malaria is currently increasing world wide. Which of the following strategies is currently the most effective means of reducing the spread of malaria?

(A) Quarantine all infected people

(B) Reduce mosquito breeding grounds

(C) Treat all infected people with high doses of antibiotics

(D) Genetically modify human red blood cells to make them malaria-resistant

14 What is the function of T-helper cells?

(A) Initiation of inflammation

(B) Phagocytosis of bacteria and viruses

(C) Promotion of B-cell and T-cell activity

(D) Production of specific antibodies against pathogens

15 The table lists the types of microbes identified in a cheeseburger prepared at an outdoor market.

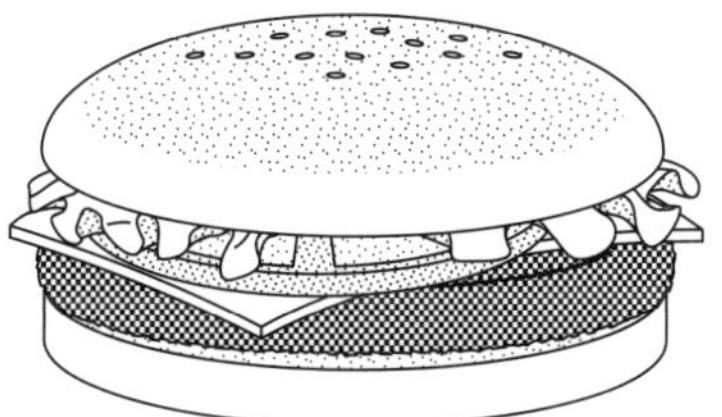

Type of microbe	**Description of microbe**
Staphylococcus epidermidis	Common skin organism
Lactobacillus bulgari	Organism present in dairy products
Saccharomyces cerevisiae	Baker's yeast
Bacillus subtilis	Non-pathogenic microbe with widespread environmental occurrence

Would it be safe to eat this cheeseburger?

(A) No, food should be completely free of microbes.

(B) No, lactobacillus and saccharomyces are highly pathogenic.

(C) Yes, organisms that grow in or on the human body do not cause disease.

(D) Yes, most of the food we eat is contaminated by different types of microbes.

Free-response Questions

Past HSC Questions

Marks

Question 1 (3 marks)

In twelfth-century China, people seeking protection from smallpox removed scabs from people mildly scarred from the disease. These scabs were then ground and inhaled as powder. Similarly, in the seventeenth century, an Englishwoman, Mary Montagu, injected bits of smallpox scabs into healthy children to protect them from the disease. **3**

In the light of our current knowledge about the immune response, explain why these practices were successful.

(6 lines) *

Question 2 (4 marks)

Explain the relationship between the cause and ONE symptom of ONE named non-infectious disease. **4**

(8 lines)

Question 3 (3 marks)

Antibiotics are drugs widely used in most industrialised societies. They are used to treat bacterial infections, are added to animal feed, and have been included in plastic products such as sandwich bags. **3**

Explain TWO possible effects of this widespread use of antibiotics on the likely spread of disease in the future.

(6 lines)

Question 4 (3 marks)

When a body organ is transplanted from one person to another, the immune system of the recipient is triggered.

(a) Patients who have an organ transplant are given drugs to suppress their immune response. State the reason for this. **1**

(2 lines)

(b) Explain a possible consequence for the general health of organ transplant patients as a result of suppressing the immune system. **2**

(4 lines)

---

* Shows the number of lines available in the HSC answer booklet for this question.

Question 5 (4 marks) **Marks**

Epidemiological studies have demonstrated a relationship between ultraviolet radiation exposure and the development of melanoma, a type of skin cancer. 4

The graph shows the rate of occurrence of melanoma in males and females between 1972 and 1997.

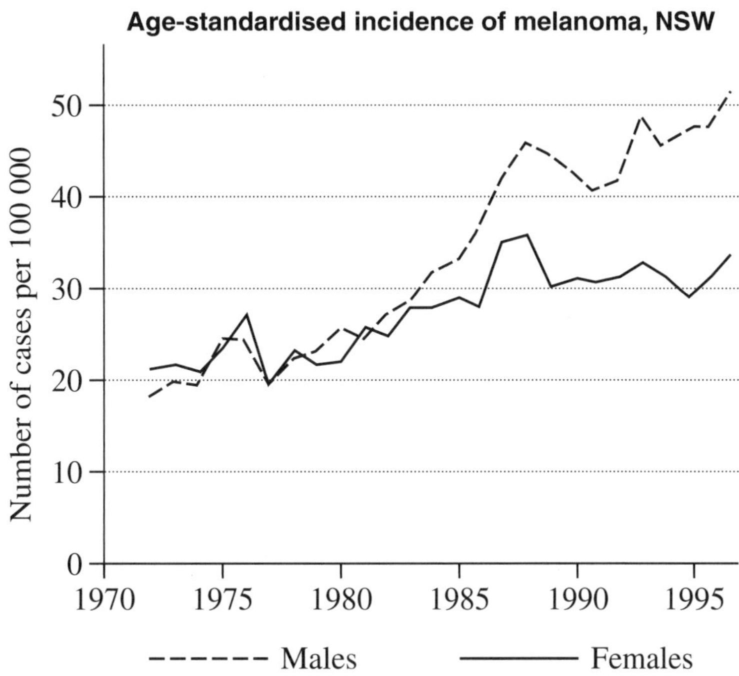

A student studying the graph made the following statement.

'The incidence of melanoma will continue to increase beyond 1997 at a greater rate in males than in females.'

Analyse the data in the graph to assess the validity of this statement.

(8 lines)

Question 6 (4 marks)

'Modern methods of disease control place more emphasis on prevention than on treatment.' 4

Discuss this statement using at least ONE example in your answer.

(8 lines)

Question 7 (4 marks)

Investigators gathered data on a group of 100 smokers for a period of 10 years. During this time 12 people in the group developed lung cancer, two died in traffic accidents, and three died of heart attacks. The investigators used this data to state that smoking caused lung cancer. 4

Describe how this investigation could be improved.

(8 lines)

Question 8 (3 marks) **Marks**

The diagram summarises one method used to treat water to make it suitable for drinking. **3**

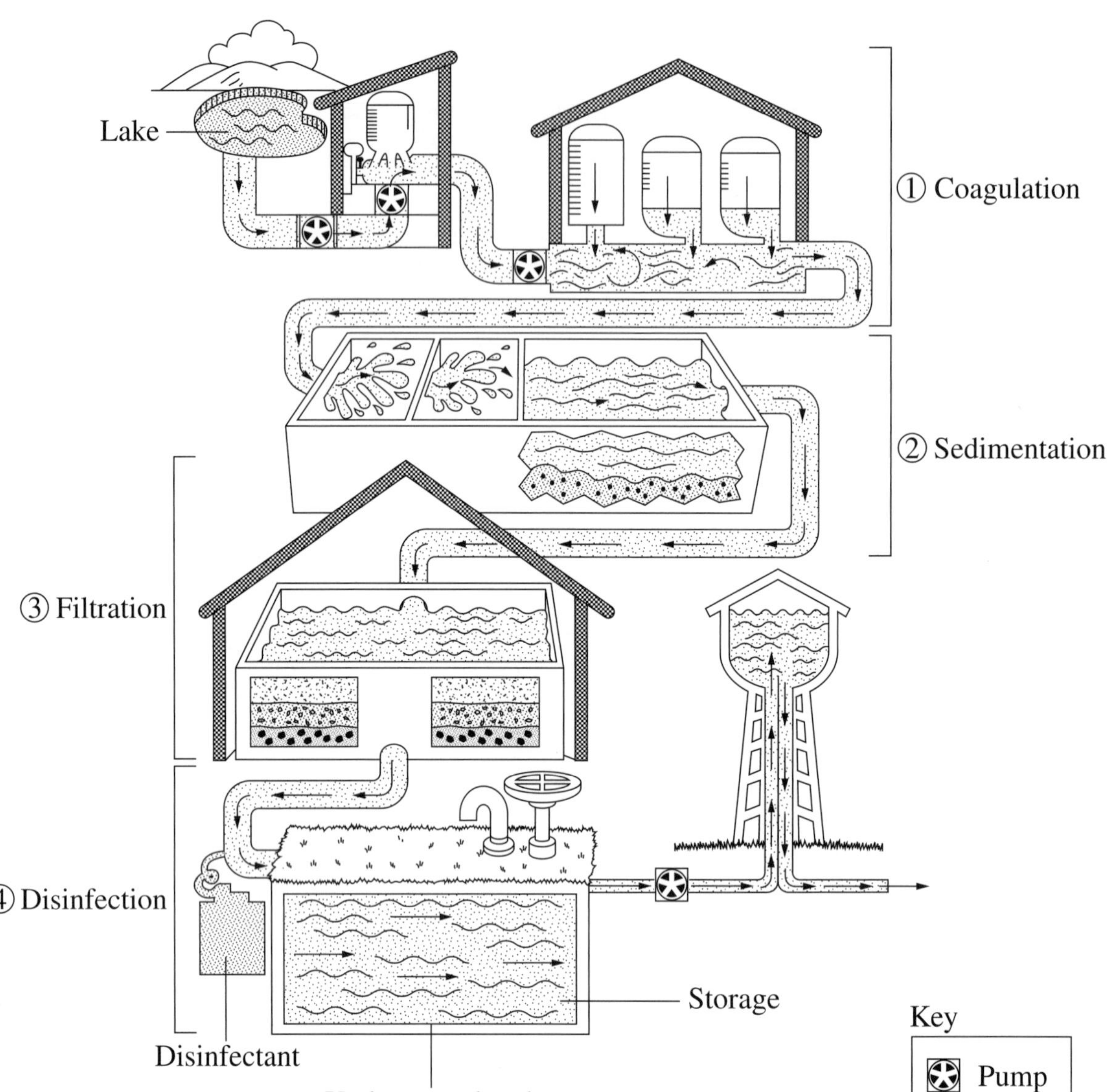

The treatment method illustrated has four processes which can remove the contaminants that may be found in water extracted from the lake.

Select ONE of the four processes and explain how this process reduces the risk of infection from pathogens.

(9 lines)

Marks

Question 9 (4 marks)

Outline how mitosis and cell differentiation assist in the maintenance of health. **4**

(10 lines)

Question 10 (8 marks)

Evaluate the contributions made by both Louis Pasteur and Robert Koch to our present understanding of the causes and possible prevention of infectious diseases. **8**

(25 lines)

Question 11 (4 marks)

Old age is not a disease. Discuss the difficulties in defining the terms *health* and *disease*. **4**

(8 lines)

Question 12 (4 marks)

A standard NSW vaccination schedule for diphtheria/pertussis/tetanus (DPT) is shown. **4**

Age
2 months
4 months
6 months
18 months
4 years

Propose reasons for the frequent vaccination between 2 months and 4 years.

(8 lines)

Question 13 (8 marks) **Marks**

The following diagram shows a rural coastal area and the associated towns, rivers and industry for each of the townships. **8**

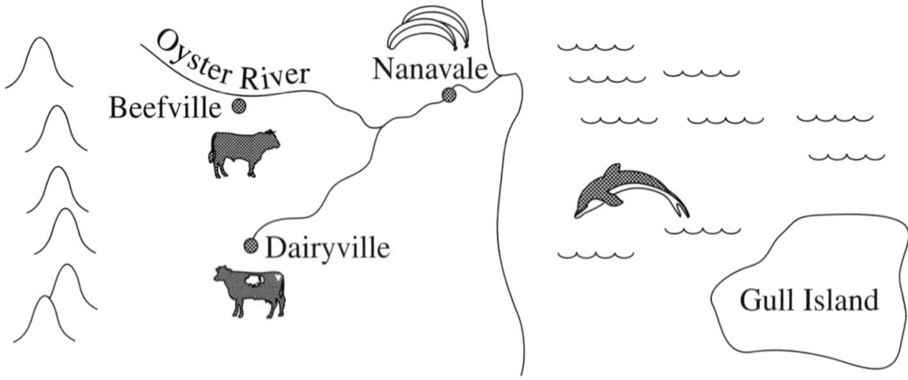

An epidemic of a disease has broken out in Nanavale. The symptoms are stomach ache, vomiting and tiredness. Many families in Nanavale have only one member with the disease, therefore it is apparently non-infectious. The symptoms appear worse in infants than adults.

Isolated cases of this disease have occurred in the nearby towns of Dairyville and Beefville. No cases have been reported on Gull Island.

Design an epidemiological study to investigate the origin of the disease.

(16 lines)

Core Topic

The Search for Better Health

Worked Answers

Multiple-choice Questions

1 A Only extended boiling will destroy (kill) any disease-causing micro-organisms in water.

2 D Quarantine is essential for imported livestock as some diseases have an extended incubation period and may not have shown any symptoms of disease prior to departure.

3 B This is the only correct possibility, because the question is asking for a secondary line of defence mechanism.

4 D Immunisation enables resistance to build in an organism, thereby reducing the frequency of infection and the spread of the disease.

5 C This is the only option that correctly follows Koch's postulates in identifying the disease.

6 A The paragraph describes the second-line defence mechanisms.

7 D Antibiotics are effective only against bacterial diseases.

8 D The question refers to quarantine measures for preventing the spread of disease between states within Australia rather than from overseas.

9 B Many diseases will become untreatable as the production of new drugs will not be able to keep pace with the changing nature of the disease-causing pathogens.

10 B The use of the man's own fingers means that there is no foreign tissue being used and so tissue rejection does not occur.

11 A MacFarlane Burnet's area of study was immunology.

12 D Describes Pasteur's experiment correctly.

13 B A realistic method of treatment.

14 C Correct definition.

15 D Most correct answer, as all food eaten will have some microbes.

Free-response Questions

1 Such practices were successful because the scabs contained the smallpox virus and led to an immune response. The introduction into the body through injection resulted in a primary immune response mechanism being initiated, which led to the production of antibodies. On subsequent exposure to the virus, the immune system's memory response produces the correct antibody and protects the body against the disease.

2 Example: Scurvy is a non-infectious disease caused by a deficiency of vitamin C. It results in sore or bleeding gums. A lack of dietary vitamin C results in these symptoms.

3 The use of antibiotics controls the spread of disease and bacterial growth. The widespread use of antibiotics in the community can reduce the effectiveness of antibiotic control of the spread of bacteria, because the disease-causing bacteria can become resistant to the antibiotics. As bacteria become more resistant to antibiotics it becomes more difficult to kill them and prevent or control the spread of disease.

4 (a) The drugs prevent the rejection of the donor organ by the recipient's own immune system.

(b) The general health of an organ transplant patient is compromised as their own immune system is suppressed, resulting in the decreased activity of the lymphocytes. This makes the patient very susceptible to infection generally.

5 The graph depicts an increase in the incidence of melanoma, although the rate of increase is generally greater for males than for females between 1983 and 1995. The statement predicts that this trend will continue; however, it is difficult to state this conclusively as other variables that may affect this pattern have not been identified and only one State has been considered. Two factors not considered in this graphical representation include the location and age of the participants, factors that would have considerable impact on the validity of such a statement.

6 Prevention of a disease means that measures can be taken to stop a person contracting the disease. Methods for preventing disease include vaccination, quarantine or education, as opposed to treating the symptoms of disease, where a patient may only receive relief from the symptoms. For example, some diseases are preventable with the use of vaccinations such as smallpox, measles or tetanus. Other diseases, such as various forms of cancer cannot be prevented as such and patients can only be treated surgically by removing the cancer, or affected cells can be exposed to radiation or chemotherapy to try and destroy them.

7 The investigation could be improved by increasing the sample size, and then repeating the study using other groups of people with similar characteristics but in different locations or with differing levels of age, health or fitness. The use of a control group of non-smokers in the area would also serve to verify results. The data would need to be analysed statistically to verify any trends or patterns in the study groups.

8 Water can be treated by using filtration, depicted in step 3. In filtration liquid or water is passed through a filter, grill/gauze, charcoal, gravel or fine-grade membrane to eliminate any solid matter suspended in the water. This is a primary source of treatment only as microscopic solids or pathogens may still get through this stage.

9 Both mitosis and cell differentiation assist in the maintenance of health because mitosis produces new cells required for the growth of tissues as well as replacing damaged or old cells that can no longer function. Cell differentiation is important because without this there would be no specialised cells produced to maintain health such as occurs during the increased production of lymphocytes in an infection.

10 Both Louis Pasteur and Robert Koch contributed significantly to the understanding of the cause and prevention of infectious disease in modern medicine. Each contributed independently by way of research in different areas.

Pasteur identified that micro-organisms develop from pre-existing micro-organisms rather than from spontaneous generation. He developed sterilisation as a method of destroying pathogenic micro-organisms. This enabled further links between micro-organisms and disease to be made, including the destruction of the micro-organisms and the understanding for the need for basic hygiene, sanitation and water treatment.

Koch is attributed with the identification of a series of postulates that could be used to identify a micro-organism and determine whether it was responsible for a particular disease. This contribution led to the identification of a range of specific disease-causing micro-organisms and subsequently to their treatment.

Together the work of these men has been directly responsible for the procedures used in identifying and treating diseases today, whether by eliminating the micro-organisms from food and water or isolating the causative agent of a disease making it possible to develop a treatment or vaccinations to prevent its spread.

11 Health is defined as the complete physical, physiological, mental and social wellbeing of an organism. It is not merely the absence of disease. Disease can be defined as a state which prevents correct physical functioning or that impairs the function of any part of the body. It is difficult to define these terms because aspects of each are very subjective and vary from person to person: what is healthy to one person may not be to the next. Old age is an example of how the definitions can be misleading because the elderly may have reduced body function and yet not be sick or diseased. Where the body is just slowing down or wearing down older people may not consider themselves sick; similarly children are not considered diseased just because they do not have complete body function.

12 Frequent vaccinations are required between the age of 2 months and 4 years because very young children have immature immune systems and require vaccination to assist in the development of their immune response. The first vaccination initiates the primary immune response by the stimulation of the production of antibodies and memory cells. The frequency of the subsequent vaccinations enables the secondary immune response to activate the production of increased numbers of B- and T-memory cells and antibodies thereby protecting the child through longer-lasting memory. Should they be exposed to the active antigen they will produce the required antibodies faster.

13 Epidemiological studies are used to study diseases that affect large numbers of people in a population. The key features of an epidemiological study include the collection of large amounts of relevant data, statistical analysis of the information generated and continued monitoring of the population over an extended length of time. The information is then used to isolate the cause of a disease and develop a treatment/control measure.

To determine the cause of the disease in the location depicted the following steps should be carried out:

1. Interview all persons affected by the disease. Information about their age, sex, diet, occupation, lifestyle and general movements in the area will need to be gathered.
2. Interview a similar sample of people in the area who have not been affected by the disease. These will need to be compared to communities that are not directly associated with the area such as Gull Island. Both surveys should include a large number of people and include the collection of the same data gathered from the disease-affected group.
3. The data need to be analysed statistically to identify any patterns or trends and identify similarities and differences between the non-affected groups and compare these with the affected groups.
4. A possible cause for the disease will be identified and a management plan developed to prevent the disease. This may include various tests of possible cures or prevention strategies and education of the populations and the development of management strategies for the environment.
5. The population will need to be monitored periodically.

CHAPTER 4

Option Topic

Communication

Past HSC Questions

Question 1 (25 marks) **Marks**

(a) (i) Where are photoreceptor cells located in the eye? **1**

(ii) State ONE function for each of the structures labelled in the diagram below. **2**

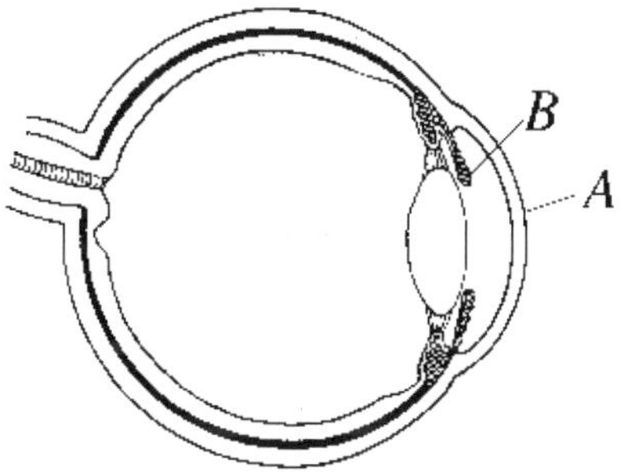

(b) (i) How would you gather information on the structures used by animals to produce sound? **2**

(ii) How would you assess that the information you collected was relevant and reliable? **2**

(c) Describe ways in which technology can be used to overcome the effects of cataracts. **5**

(d) In your study of Communication, you performed a first-hand investigation to model the process of accommodation. **6**

Justify the procedure used and the conclusions drawn.

(e) Evaluate the appropriateness of TWO devices designed to assist people with different types of hearing impairment. **7**

Question 2 (25 marks) **Marks**

(a) (i) Outline the function of the organ of Corti in hearing. **1**

(ii) State the relationship between wavelength, frequency and pitch of a sound. **2**

(iii) Compare TWO structures used by animals to produce sound. **2**

(b) (i) During a first-hand investigation, a student acquired three photographs of cross-sections of the human brain. The student then lost the labels and mixed up the photographs. **3**

What structural features of the cerebrum, cerebellum and medulla oblongata could the student use to identify each photograph correctly?

(ii) Draw a sketch of the brain and clearly label those regions involved in speech. **3**

(c) The data in the table gives the focal length of six lenses which have the same diameter but different thickness.

Thickness of lens (mm)	*Focal length* (cm)
10	12.5
9	14.0
8	15.5
7	18.0
6	21.0
5	25.0

(i) Graph the data on the graph paper provided. **4**

(ii) State the relationship between lens thickness and focal length. **1**

(iii) Refer to the information above in explaining how human eyes can focus on objects at different distances. **2**

(d) The light signal reaching the retina is transformed into electrochemical signals. **7**

Describe the different structures and processes in the retina that can achieve the energy transformation described above.

Question 3 (25 marks) **Marks**

(a) For each stimulus in the table, name an appropriate receptor. Clearly label your answers to (1), (2) and (3) in your writing booklet.

Receptor	*Stimulus*	*Sense*	
(1)	Black-and-white photograph	Vision	1
(2)	Pin prick	Touch	1
(3)	Concert band	Hearing	1

(b) Following an industrial accident, a person was found unconscious with no apparent injury to the face and eyes. Upon recovery the person experienced blindness in the right eye.

(i) Based on the evidence provided, propose ONE possible cause for the loss of sight in the right eye. **2**

(ii) Discuss how loss of sight in one eye would affect other aspects of vision. **3**

(c) The chart shows the range of frequencies that selected animals can hear.

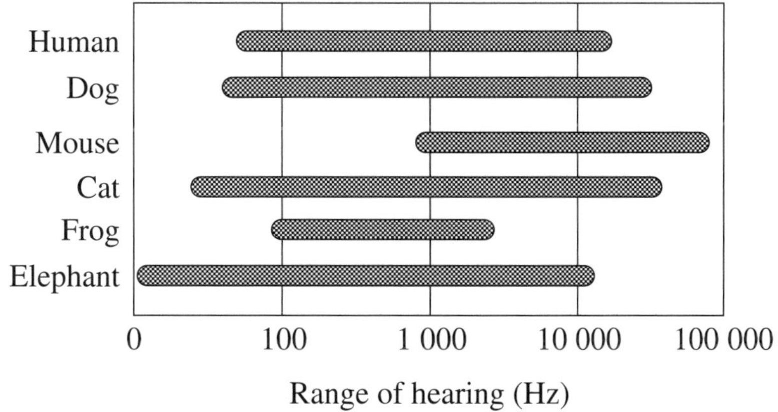

(i) Identify the animal that has the narrowest range of hearing. **1**

(ii) Explain the possible reasons for the differences in sound frequencies detected by humans and other animals. **4**

Question 3 continues

Question 3 (continued) **Marks**

(d) During your study of Communication you gathered secondary information relating the structure and function of a mammalian ear.

(i) Describe how you processed and analysed the gathered information. **4**

(ii) State how you assessed the reliability of the data obtained. **1**

(e) Recent advances in health care have significantly extended life expectancy. As people age, both vision and hearing become impaired. Evaluate identified technologies that have been developed to deal with these problems. **7**

Option Topic
Communication

Worked Answers

Question 1

(a) (i) Retina

(ii) *A*: the cornea enables light to enter the eye; it is the first point at which refraction occurs.

B: the iris controls the amount of light entering the eye.

(b) (i) Information about the structures used by animals to produce sound could be gathered from secondary sources such as books and the Internet. The information would need to be found in these sources by using the index or a search engine to find words associated with the subject matter, such as 'sound production' and 'animals'.

(ii) The reliability of the information would need to be considered by assessing the validity and reliability of the source; for example, is it from a journal/textbook publisher and are references cited in the articles? For the Internet, what is the source of the Internet page? Is it from a university lecture or formal journal article, and who produced it? The relevance of the information would be assessed by considering whether it addresses the content area and whether the information is supported by that found in other sources.

(c) Cataracts can be overcome using a number of methods. For example surgical techniques have enabled the physical removal of cataracts from patients by cutting the cataract or using technology such as a cryoprobe or lens phacoemulsification. Other technological advances that have assisted cataract patients include the use of spectacles for improving vision both before and after the operation. Pre-operation spectacles have a very high magnification; sometimes a hand lens may be used. Other technological advances have made the insertion of intraocular lenses or artificial lenses possible and these enable cataract sufferers to see well.

(d) Accommodation refers to the ability of the eye to focus light from objects at different distances, ensuring that the focus point occurs on the retina. Accommodation can be modelled using a ray box and biconvex lenses of different sizes. Firstly set up the ray box so that a light beam is aimed at a flat surface (representing the retina), then place different biconvex lenses between the ray box and the 'retina', noting the distance (focal length) required to produce a sharp image. A relationship was established between the thickness of the lens and its distance from the eye because the image was produced at different distances when different thicknesses of lens were used. It was concluded that thicker lenses were required for producing sharp images the closer the image was to the eye.

(e) The answer should include two devices, clearly identified. As an evaluation is sought it is important to include both advantages and limitations of the device and its application. Examples could include cochlear implants, various types of hearing aids, ossicle replacement and telephone amplification devices.

Example: Cochlear implants are used to replace the damaged cochlea. The cochlea normally is responsible for the transmission of different sound frequencies to the auditory nerve. If the cochlea is damaged amplified hearing aids do not work and so implants can be used to overcome inner ear problems. Some limitations of using cochlear implants can include the inability to detect all frequencies resulting in sound distortion. The recipient is required to wear a permanent device attached to the outside of the skull, which is a hindrance.

Question 2

(a) (i) The organ of Corti contains the receptor cells that enable sounds to be detected in the inner ear.

(ii) The wavelength of a wave decreases as the frequency increases. The pitch of a sound increases as the frequency increases.

(iii) Humans and cicadas are able to produce sounds by vibrating air columns. In cicadas, timbals are used to produce vibrations and clicking sounds. In humans, the larynx is used to produce sound. As exhaled air passes over the vocal cords within the larynx it causes them to vibrate producing distinct speech patterns as the shape of the lips changes.

(b) (i)

Cerebrum	Folded surface with white matter in the inner layer and grey in the outer layer
Cerebellum	Smaller folds on surface
Medulla	Grey matter in the inner layer and white matter in the outer layer, extension of spinal cord

(ii)

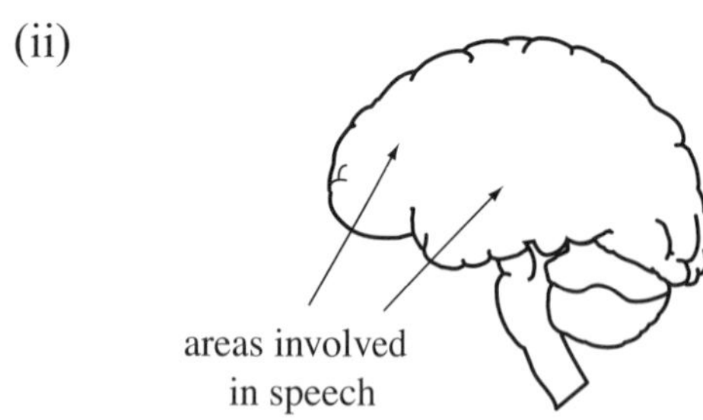

(c) (i)

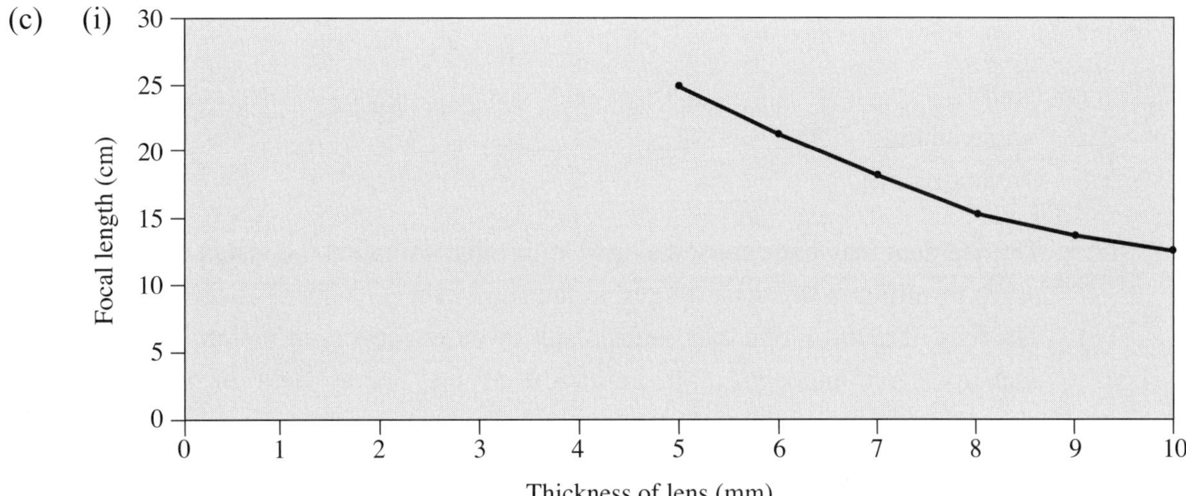

(ii) The focal length decreases as the thickness of the lens increases.

(iii) Humans are able to accommodate objects at different distances due to the ability of the ciliary body to change its tension thereby altering the shape of the lens. If an object is close to the eye, the lens thickness increases to enable the light rays to come to a focus on the retina. When a distant object is being observed, the lens shape becomes much flatter to enable the rays to extend to a focal point on the retina.

(d) The retina contains two different types of photoreceptor cells: the rods and cones. They each contain different pigments that enable them to absorb different wavelengths of light. There are three different types of cone cells each responding to three colour wavelengths: red, blue and green. They are unable to detect these wavelengths under low light intensities. The rods are scattered around the perimeter of the retina and are sensitive to light intensity, but are not sensitive to colours.

The visual pigments absorb light and change their structure resulting in an action potential variance. In this way, the light detected is changed into an electrochemical message that is transferred to the brain via the optic nerve.

Question 3

(a) 1. Rods
2. Nerve ending
3. Organ of Corti

(b) (i) The accident may have caused a head injury that damaged the visual cortex or optic nerve resulting in the loss of sight in the right eye.

(ii) The loss of sight in one eye could result in other aspects of vision being affected such as depth perception or limitation of the lateral field of vision. Three-dimensional vision relies upon the brain receiving and processing messages from both eyes, the two images produced from the two eyes enabling depth to be perceived. The field of vision is a result of peripheral vision from both eyes combined – with the loss of one eye this is significantly reduced.

(c) (i) Frog

(ii) Within the cochlea are located numerous nerve endings that each detect different sounds. The sounds detected by humans are related to the vast amounts of different sound frequencies that are produced by the larynx in order to facilitate the complex communication that is conducted between humans. Other animals do not necessarily rely on complex communication and so do not produce a wide variety of sound frequencies. Other animals may have a smaller range of sound detection related to their protection, such as being able to detect low-frequency movements of other animals or higher frequencies in the case of echolocation or sonar transmission.

(d) (i) Information was gathered and processed by reading through the material, making notes about keywords or points in order to summarise the required information. The information source was identified in each case, as was the date of publication to ensure that the most recent material was being utilised. Once a summary was completed, the relevant information was analysed by ensuring that any inconsistencies were identified and corrections were sought from different sources. Comparisons could be made through the use of tables and keyword matches.

(ii) The reliability of the gathered information can be assessed by ensuring that the references cited in the article are reputable and that they provide similar conclusions. The author should have appropriate qualifications in the area for which they are writing. If the source is an Internet site, then there may be links to corresponding university websites or research institutions.

(e) The recent advances in health care have significantly increased life expectancy as is evidenced by the growing number of older people in the community. As people age the body simply begins to 'wear out' necessitating the intervention of technology to restore the quality of life and enable continued community interaction. Some of the technologies that have been used for this purpose include hearing aids that can amplify the sound waves from a person's environment. The use of such devices is limited to people who have damage to the middle ear, the hearing aid simply replacing the ear drum or ossicles in the transmission of sound waves. If the damage is within the inner ear, a hearing aid is of no use as the sound waves cannot be detected within the cochlea and transmitted to the brain. In such cases the use of cochlear implants may resolve this problem; however, such technology is very expensive and tends to be used for patients who are profoundly deaf or have congenital defects in the ear from birth.

Vision is another area that tends to deteriorate with age. Once again technology does provide some assistance through the use of spectacles or more invasive surgery including cataract removal. Cataracts are a common cause of sight impairment in the elderly resulting in opaque or cloudy growths over the lens. These can be removed surgically and plastic replacements can be inserted. Lenses made of glass or plastic can also be inserted to correct vision impediments such as far-sightedness and myopia.

The use of such advances enables the elderly in particular to lead independent and productive lives without reliance on others. Their quality of life is restored as they are able to read and write, participate in conversation, watch TV and interact with their families. The technologies are a great investment in the community as the elderly give so much back through experience as they represent a growing and productive sector of society.

CHAPTER 5

Option Topic
Biotechnology

Past HSC Questions

Question 1 (25 marks) **Marks**

(a) The flowchart shows the major steps in the production of recombinant DNA.

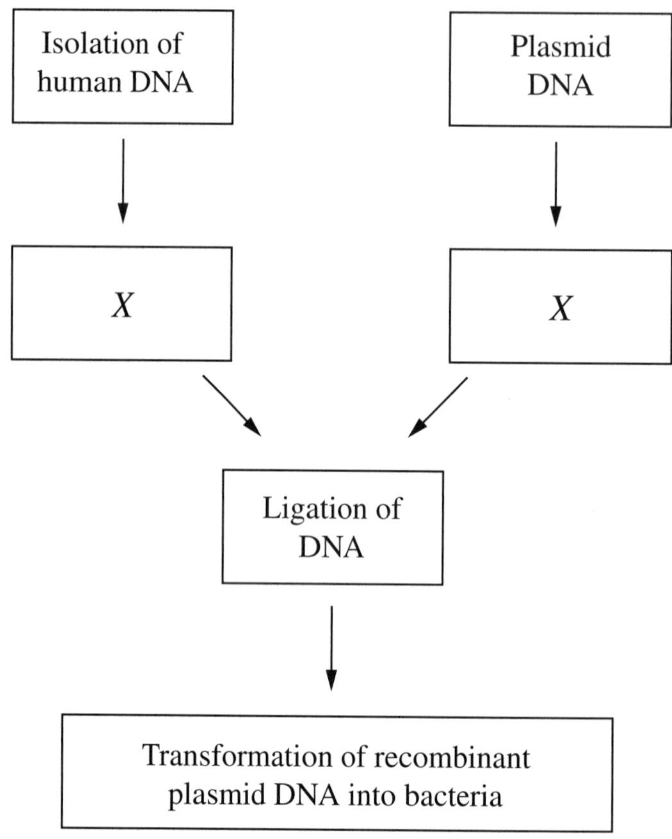

(i) Name the process labelled X. **1**

(ii) Outline the process of *ligation of DNA*. **2**

(b (i) How would you gather information on an ancient use of aquaculture? **2**

(ii) How would you assess that the information you collected was relevant and reliable? **2**

(c) Explain how changes in technology and scientific knowledge have modified traditional uses of biotechnology. **5**

Question 1 continues

Question 1 (continued) **Marks**

(d) In your study of Biotechnology, you performed a first-hand investigation of the use of the fermentation process in bread or alcohol production. **6**

Justify the procedure used and the conclusions drawn.

(e) Assess the efficiency of ONE modern application of biotechnology that you have studied. **7**

Question 2 (25 marks)

(a) (i) State ONE difference between RNA and DNA. **1**

(ii) The diagram illustrates three steps involved in the Polymerase Chain Reaction (PCR).

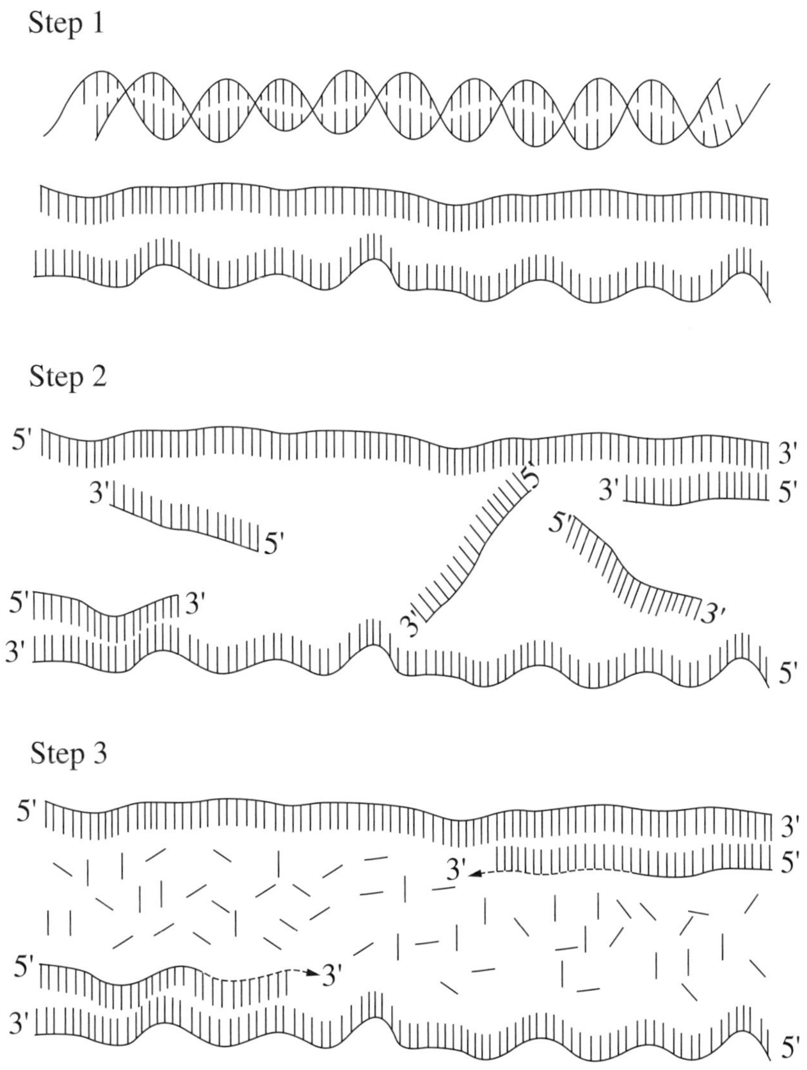

Step 1

Step 2

Step 3

(1) Give ONE use of PCR. **1**

(2) State what is happening in each of the steps shown in the diagram above. **3**

Question 2 continues

Question 2 (continued) **Marks**

(b) (i) Describe a first-hand investigation that you have carried out to test ONE condition that influences the rate of activity of enzymes. **4**

(ii) What variables need to be controlled in this investigation? **2**

(c) The data in the table compares the rate of fermentation for two different strains of yeast. This rate is measured as a decrease in specific gravity of the mixture, where a lower specific gravity indicates a higher alcohol content.

Time (hours)	*Specific gravity*	
	Yeast *A*	Yeast *B*
0	11	11
50	10	8
100	8	4
150	6	3
200	5	2
250	5	2

(i) Graph the data on the graph paper provided. **4**

(ii) Compare the alcohol production of the two different yeasts. **1**

(iii) Describe a process that could have been used to produce yeast strain *B*. **2**

(d) There are many applications and areas of research in biotechnology. **7**

Select ONE application of biotechnology. Describe both the process used and the outcome(s) of this process.

Question 3 (25 marks) **Marks**

(a) (i) Define the term *artificial selection*. **1**

(ii) Describe a change that has been produced in ONE named animal species or ONE named plant species as a result of artificial selection of characteristics suitable for agricultural stock. **2**

(b) (i) Outline the fermentation processes involved in either bread or alcohol production. **2**

(ii) Explain how changes in scientific knowledge have modified the traditional methods of bread or alcohol production. **3**

(c) (i) The following table summarises the activity of three different enzymes *A*, *B* and *C* that can join amino acids to form polypeptide chains. **2**

pH	*Temperature* (°C)	*Enzyme A activity*	*Enzyme B activity*	*Enzyme C activity*
2.0	4	low	low	none
	37	low	high	none
	72	low	none	none
5.5	4	low	low	none
	37	medium	medium	none
	72	low	none	low
7.5	4	medium	low	none
	37	high	low	low
	72	low	low	high
11.0	4	none	none	none
	37	none	none	none
	72	none	none	none

Explain which of the enzymes, *A*, *B* or *C* would be most likely to function in a human cell (human cell pH range = 7.36–7.44).

(ii) Describe the function of ONE enzyme involved in recombinant DNA technology. **3**

(d) During your study of Biotechnology you gathered secondary information to identify that complementary DNA is produced by either reverse transcribing RNA or the Polymerase Chain Reaction.

(i) Describe how you processed and analysed the gathered information. **4**

(ii) State how you assessed the reliability of the data obtained. **1**

(e) Evaluate current uses of industrial fermentation biotechnology. **7**

Option Topic
Biotechnology

Worked Answers

Question 1

(a) (i) Cutting of DNA using restriction enzymes

(ii) Ligation of enzymes involves the matching up of segments of DNA using the enzyme DNA ligase, which acts as an enzyme glue. In this process the enzyme DNA ligase is used to join the 'sticky' ends of the matching DNA segments.

(b) (i) Information about an ancient use of aquaculture could be gathered from secondary sources such as books and the Internet. The information would need to be found in these sources by using the index or a search engine to find words associated with the subject matter, such as aquaculture, water, fish/crustacean farming, ancient farming, and biotechnology.

(ii) The reliability of the information would need to be considered by assessing the validity and reliability of the source; for example, is it from a journal/textbook publisher and are references cited in the articles? For the Internet, what is the source of the Internet page? Is it from a university lecture or formal journal article, and who produced it? The relevance of the information would be assessed by considering whether it addresses the content area and whether the information is supported by that found in other sources.

(c) Example: *Penicillium* is fermented to produce penicillin. This represents a modified use of the traditional fermentation process. Scientific knowledge has advanced over the years to make it possible to understand the nature of microbes, fungus culture and their life cycles. Modern technology such as the use of specially designed fermenters that are able to stir, control temperature, and monitor nutrients and growth conditions, have enabled control over growth rates of the culture and the ability to produce large quantities of penicillin. It is possible to produce large quantities of pure penicillin today for use in the fight against disease.

(d) Example: Alcohol production was investigated and the correlation between the quantity of sugar to the amount of alcohol produced was determined. Six identical vessels were used, each containing an identical amount of solution and the same amount of yeast. The vessels ranged in amount of sugar from low at 10 g, then 20 g, 30 g, 40 g, 50 g and a high of 60 g. A control was established in a seventh vessel similar to the others but with no sugar. All the vessels remained under identical conditions of temperature with no stirring for the duration of 24 hours. After this time the alcohol content of the solutions was measured. As all conditions in the experiment were identical (except for the control), the amount of alcohol would be the only variable that could have changed and this could be directly attributable to the amount of sugar in the initial solution. The vessel with 30 g of sugar produced the largest amount of alcohol and represents the ideal combination of sugar and volume under the experimental conditions. This shows that there is a relationship between alcohol production and quantity of sugar if all other factors remain the same.

(e) The answer should include a modern application of biotechnology, including a description of its process and outcome. To assess its efficiency the answer must include supporting evidence of its effectiveness or success. Examples could include: applications related to the field of aquaculture, medicine, animal husbandry, or agriculture.

Example: Medicine and the development of vaccine.

A recent advance in medical biotechnology is recombinant vaccine production. This enables the recombinant DNA technology to introduce genetic information into bacterial cells for the production of viral antigens. The synthesised gene product is released from the bacteria that have expressed the introduced genetic information. The synthesised gene product is then used for the production of a vaccine. The quality and quantity of the proteins produced is greater, hence this technology has greatly improved the way in which vaccines are produced.

Question 2

(a) (i) One difference between RNA and DNA is the fact that RNA is single stranded and DNA is double stranded.

(ii) (1) Polymerase chain reaction (PCR) can be used to make copies of genes for analysis.

(2)

Denaturation	DNA is uncoiled
Annealing	Primers attach to DNA template
Extension	Gene is copied and the new copy extended

(b) (i) Answers will vary depending on the investigation carried out. Answers should include: controls used; all variables identified; enzymes and substrate identified; and a complete method with a logical sequence.

Example: The aim of the investigation was to determine if temperature influences the rate of activity of enzymes. The substrate used was milk, containing the protein casein, and the enzyme used was rennin, found in junket tablets. The independent variable was the temperature of the milk, which was changed to a range of temperatures from 10°C to 80°C in 10-degree steps. The dependent variable was the time taken for the milk to clot. A control was established for each temperature. This was a test tube containing milk only and no enzyme.

Method:

1. Prepare water baths at temperatures of 10, 20, 30, 40, 50, 60, 70, 80°C using ice and water, tap water and hot tap water, and a hot water bath for the higher temperatures.
2. Label 16 test tubes A or B and then pipette 3 mL milk into each. Place one A and one B tube into each water bath.
3. Leave for 10 minutes to equilibrate to the temperature of the water bath.
4. Crush a junket tablet in 10 mL water in a small beaker.
5. When test tubes have all reached appropriate temperature add 3 drops junket solution to each test tube A. Shake to mix and record the time immediately rennin is added.
6. Examine each test tube every minute for 20 minutes by tilting test tube while maintaining water bath temperatures constantly.
7. Record the time when the milk clotted and graph the results, plotting activity (1/clotting time) against temperature.

(ii) Variables will include pH, temperature, and substrate concentration.

Example: The controlled variables in the experiment included the amount of milk used in each test tube; the test tube size; ensuring temperature was stable in each test tube before beginning; using the same amount of junket solution; ensuring stable temperature throughout; and not shaking the test tubes during the investigation.

(c) (i)

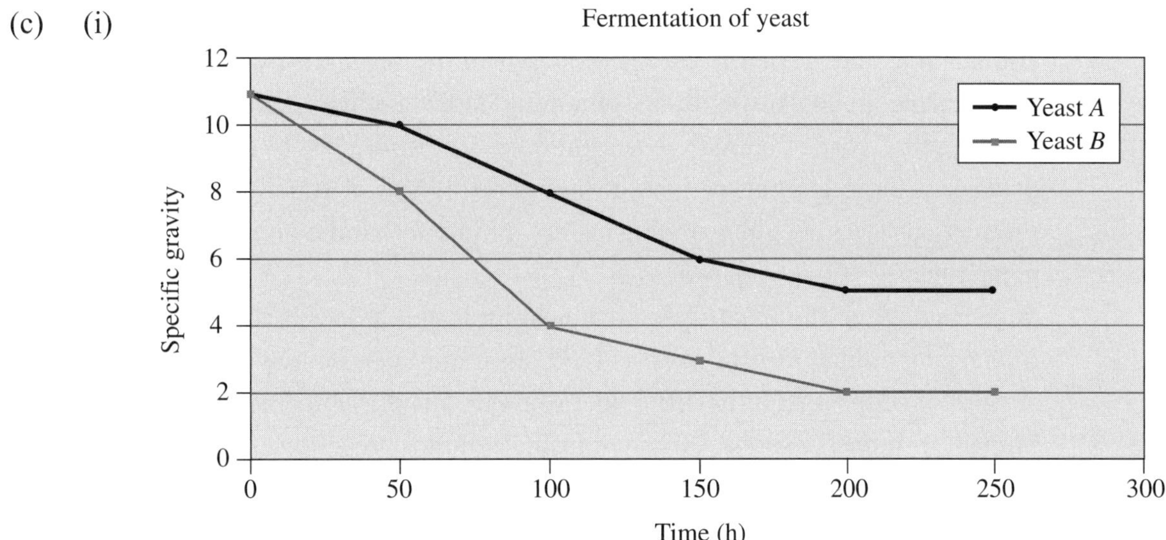

(ii) The rate of alcohol production by Yeast *A* is less than the rate of production of alcohol of Yeast *B*.

(iii) The yeast could be produced in a process such as strain isolation, involving screening various cultures or strains, and it would include a fermentation process that moves to completion more rapidly. The yeast strain would require isolation and repetition over several generations and would produce faster-working yeast strains.

(d) Answers will vary depending on the application selected. Better answers will include: the name and description of the chosen application; description of the organisms or tissue involved; the relationship between the process and the outcomes of the process; the process efficiency; and the advantages and disadvantages of the process.

Example: One application of biotechnology includes the manufacture of human insulin by the bacteria *Escherichia coli* using recombinant DNA technology. The steps in the process are:

1. Separate the plasmid (circular piece of DNA) from the *E. coli*
2. Cut the plasmid with a restriction enzyme ECOR1 leaving 'sticky ends'
3. Cut a human gene sequence for insulin production from a human pancreas cell (with the same 'sticky ends') using ECOR1 and isolating it from all the other genes
4. Join the open plasmid and the inserted gene with a ligase enzyme, forming a recombinant DNA plasmid
5. Place the recombinant plasmid in a cold calcium chloride solution with *E. coli* so the plasmid will be transformed into the bacteria
6. Bacteria that have been transformed successfully can be tested to monitor and isolate the recombinant strain and cultured to grow and reproduce in large numbers
7. The *E. coli* produces insulin, which can be harvested.

The outcome of this process is a large quantity of human insulin. Previously animals have been used to produce insulin for humans but this was an expensive method due to the difficulty is obtaining large amounts. Bacteria can produce insulin more cost effectively since they have a rapid growth rate and are able to produce a large amount of insulin in a short period of time. This leads to lower product costs. Unicellular organisms are simple to handle and are therefore more economically efficient to process.

An advantage of the *E. coli*-produced insulin is that it does not have the same risks of carrying animal viruses such as CJD, as animal insulin may do.

Animal insulin is slightly different to human insulin and this is detected by the human immune response often causing allergic reactions. Bacteria-produced insulin is a human product with no immune response triggered.

The disadvantage of the process is that it requires careful monitoring during the process to ensure the correct genes are spliced and to prevent the production of an accidental product. This monitoring is expensive.

The disadvantage of the product is that due to the ease and cheapness of treatment, patients may become complacent and continue with treatment rather than trying to address dietary habits to prevent diabetes. There is little warning of an approaching hypoglycaemic attack. There have not been sufficient human trials as yet using human insulin.

Overall the efficiency of the process is much greater than the previous process of obtaining insulin from a large number of animals. It also produces a greater amount with a much cheaper method of extraction.

Question 3

(a) (i) Artificial selection is the term used to describe the breeding of plants and animals selected for particular characteristics that are of benefit to humans, such as disease resistance, higher yields etc. to produce offspring that possess the desired characteristic.

(ii) Dairy cattle have been bred to produce milk with higher butterfat content. (Other examples could include, sheep and the texture of the wool, yield in wheat.)

(b) (i) Alcohol is produced through the fermentation of yeast and sugar in fruit or grain to produce carbon dioxide and alcohol.

OR

Sugars are fermented by yeast to produce carbon dioxide, causing bread to rise. Alcohol is produced but in this case would vaporise during baking.

(ii) Traditional methods of bread making have changed over time by the introduction of artificially selected yeasts that increase the amount of carbon dioxide resulting in more efficient rising of the bread.

OR

In alcohol production the naturally occurring yeasts are destroyed through the use of preservatives so yeasts with greater tolerance have to be used.

(c) (i) Enzyme *A* is the only possible match that could function in the human cell as the pH at which it functions is the closest and it has high activity at 37°C.

(ii) One enzyme involved in recombinant DNA technology is *ligase*, which is used to join fragments of DNA together. The strands of DNA then join by complementary base pairing.

(Other examples could include restriction enzymes used to cut sections of DNA.)

(d) (i) Information was gathered and processed by reading through the material, making notes about keywords or points in order to summarise the required information. The information source was identified in each case, as was the date of publication to ensure that the most recent material was being utilised. Once a summary was completed, the relevant information was analysed by ensuring that any inconsistencies were identified and corrections were sought from different sources. Comparisons could be made through the use of tables and keyword matches.

(ii) The reliability of the gathered information can be assessed by ensuring that the references cited in the article are reputable and that they provide similar conclusions. The author should have appropriate qualifications in the area for which they are writing. If the source is an Internet site, then there may be links to corresponding university websites or research institutions.

(e) Industrial fermentation processes utilise either aerobic or anaerobic fermentation on a large scale. These processes utilise microbes, usually yeasts, to produce various products such as bread, cheese, alcohol, yoghurt, penicillin antibiotics, organic acids and glycerol. These products have significantly assisted society in the areas of manufacturing, medicine, health and nutrition. For example organic acids have been used to improve processes such as food preservation, printing, and plastic production. The use of antibiotics has greatly increased the survival of patients with bacterial infections, and the more advanced production of bread has led to the creation of different varieties of bread with greater health benefits thereby improving nutrition. Increased yields will have financial advantages for the economy. Advances in the production of alcohol have not only produced greater yields of alcohol, but also made the production process more efficient. A downside of this might be that the alcohol is then more readily available and the alcoholic content of beverages can be increased. This could result in an increase in alcohol-related issues in the wider community.

Generally the use of these industrial fermentation processes whilst beneficial to society does carry some associated issues that require some thought in the long term.

CHAPTER 6

Option Topic

Genetics: The Code Broken?

Past HSC Questions

Question 1 (25 marks) **Marks**

(a) The diagram shows a simplified model of a DNA molecule.

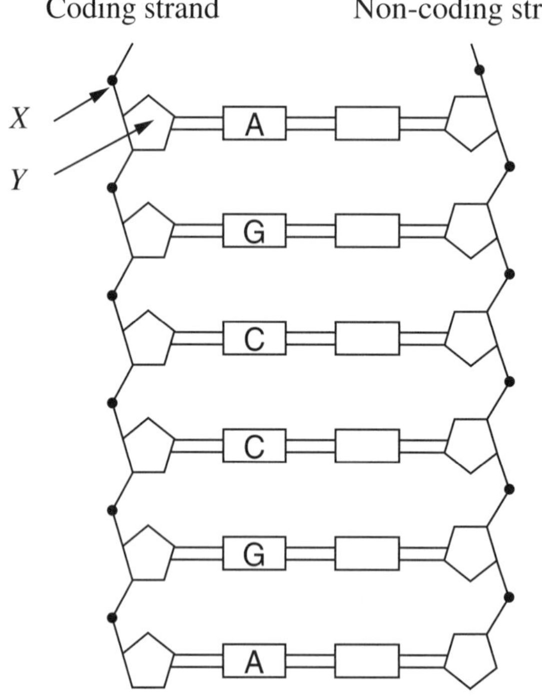

(i) Identify structures *X* and *Y*. **2**

(ii) What is the base sequence for the non-coding strand? **1**

(b) (i) How would you gather information on the processes used in tissue culture or animal cloning? **2**

(ii) How would you assess that the information you collected was relevant and reliable? **2**

(c) Discuss the role of public education in preventing cancers. **5**

(d) In your study of Genetics – The Code Broken?, you performed a first-hand investigation to model the processes involved in meiosis that relate to linkage. **6**
Justify the procedure used and the conclusions drawn.

Question 1 continues

Question 1 (continued) **Marks**

(e) Explain why the aims of the Human Genome Project could NOT be achieved by studying linkage maps. **7**

Question 2 (25 marks)

(a) (i) State ONE difference between a haploid cell and a diploid cell from the same species. **1**

(ii) Define the term *linkage* with reference to genes. **1**

(iii) Explain how linkage can be used to map chromosomes. **3**

(b) During your study of this option, you were required to construct a model of DNA.

(i) Describe the model you constructed. **3**

(ii) Justify the design of your model. **3**

(c) The data in the table shows the results of measuring the heights of 1000 adult humans.

Height group (m)	*Number of individuals in group*
1.60–1.64	58
1.65–1.69	92
1.70–1.74	178
1.75–1.79	322
1.80–1.84	190
1.85–1.89	110
1.90–1.94	50
1.95–1.99	0

(i) Graph the data on the graph paper provided. **4**

(ii) Define what is meant by *polygenic inheritance.* **1**

(iii) How does the pattern of polygenic inheritance of height in humans compare with the pattern of inheritance discovered by Gregor Mendel in his research on pea plants? **2**

(d) Selective breeding is different to gene cloning but both processes may change the genetic nature of species. **7**

Using appropriate examples, explain the above statement.

Question 3 (25 marks) **Marks**

(a) (i) Define the term *recombinant DNA*. **1**

(ii) Outline TWO steps used in the production of recombinant DNA. **2**

(b) (i) Outline the benefits of the Human Genome Project. **2**

(ii) Discuss the use of gene therapy to manage EITHER a named genetic disease, OR a named form of cancer, OR AIDS. **3**

(c) Paternity testing involves a number of procedures used to determine the biological father of a child.

The following data show results of two different procedures attempting to identify the biological father of a child. Maternity of the child has been verified.

Procedure 1: **ABO blood groups**

Mother	*Child*	*Male 1*	*Male 2*	*Male 3*
A	O	O	A	B

Procedure 2: **DNA fingerprint data**

Molecular weight	*Weight marker*	*Mother*	*Child*	*Male 1*	*Male 2*	*Male 3*
6000	■				■	
5500	■	■	■			
5000	■					
3000	■	■			■	■
2000	■			■		
1500	■		■			■
1000	■			■		

(i) From the ABO blood group data, identify which male(s), if any, can be excluded as the biological father of the child. **1**

(ii) From the DNA fingerprint data, identify which male(s), if any, can be excluded as the biological father of the child. **1**

(iii) Explain which of these two methods is more accurate in paternity testing. **3**

Question 3 continues

Question 3 (continued) **Marks**

(d) During your study of Genetics: The Code Broken, you gathered secondary information to assess the evidence that analysis of genes provides for evolutionary relationships.

(i) Describe how you processed and analysed the gathered information. **4**

(ii) State how you assessed the reliability of the data obtained. **1**

(e) Evaluate the current use of gene cloning in animals and plants. **7**

Option Topic
Genetics: The Code Broken?

Worked Answers

Question 1

(a) (i) X: phosphate, Y: deoxyribose sugar

(ii) TCGGCT

(b) (i) Information about the processes used in tissue culture or animal cloning could be gathered from secondary sources such as books and the Internet. The information would need to be found in these sources by using the index or a search engine to find words associated with the subject matter, such as tissue culture, cell division, cloning, DNA manipulation, reproductive technology.

(ii) The reliability of the information would need to be considered by assessing the validity and reliability of the source; for example, is it from a journal/textbook publisher and are references cited in the articles? For the Internet, what is the source of the Internet page? Is it from a university lecture or formal journal article, and who produced it? The relevance of the information would be assessed by considering whether it addresses the content area and whether the information is supported by that found in other sources.

(c) The role of public education is very important in the prevention of cancers. Public education includes the raising of people's awareness regarding the risks involved in sun exposure, preventative measures that could be taken to minimise risk, early detection and intervention options available and treatment for symptoms. Education programs may reduce the future incidence of cancer and enable those for whom it is too late to avoid the exposure to adopt appropriate management techniques such as awareness of early signs and access to detection and treatment in order to prevent large-scale surgery. If signs and symptoms go undetected the cancers can spread, leading to prolonged treatment, surgery and unnecessary suffering, all of which can add a huge burden to the cost of public health.

Cancer is a long-term disease so it is difficult to establish the effectiveness of campaigns because there is a time lag between the campaign and the development of the symptoms. For example, the sun bathers of the 1970s are only just now starting to develop symptoms of their past time, so the effectiveness of the 'slip, slop, slap' campaign will not be seen for several years yet.

(d) A model to show the process of meiosis related to linkage can be undertaken using coloured beads and cards. It is essential to start with a few homologous pairs of chromosomes, each placed on top of a card and the gene loci labelled.

Next label them so that each represents a heterozygous combination of each locus. Then model the process of meiosis, producing the result with identical beads, labelling each step on a new card. Repeat the process to demonstrate crossing over. The resulting genotypes of all crosses should be recorded.

The linked genes are represented by the labelled gene loci on the same chromosome, resulting in segregation only when crossing over occurs. This demonstrates that gene loci on different chromosomes are not linked and will segregate randomly. Those on the same chromosome will not necessarily segregate, particularly the closer together they are, but will move together as a linked pair.

(e) Example: The Human Genome Project (HGP) was undertaken to create a map of human DNA base sequences that can then be compared with the genetic maps of other species. The resulting map uses recombination rates with linked genes determining relative positions of gene loci. This is achieved with the use of advanced technology involving restriction enzymes; radioactive probes or autoradiography; use of fluorescent dye; computers to compile data; and reconstruction. The aims of the HGP could not be achieved by linkage mapping as this involves doing test crosses and studying pedigrees. In humans this is impractical due to the long generations in humans, the unavailability of phenotype records for different generations, the fact that relative gene locations will not always generate base sequences, and that some genes do not have observable phenotypes. Any recombinant data obtained may not be adequate for producing reliable linkage maps.

The ethical issues involved in obtaining such data from human generations has to be considered too and the impact it would have on the subjects' families would be significant. Hence the HGP expedient procedure to obtain the map is a welcome breakthrough as the information it provides for medical and reproductive technology is vast and would otherwise not be available for many more years.

Question 2

(a) (i) A haploid cell has only one set of each chromosome and a diploid cell has a pair of each chromosome, one from each parent.

(ii) The pattern of a sequence of two or more genes on the same chromosome to be inherited together is called linkage.

(iii) Genes can be mapped on chromosomes in a linear order because the probability of crossing over occurring between any two genes is directly related to the distance between the genes, so the frequency of a recombinant can be used as a measure of the genetic distance.

(b) (i) The models described will vary, but should identify all the molecular components used in the model.

Example: A double helix model can be a ladder made of pipe cleaners. The two main strands on either side are made of lengths of pipe cleaner (representing the sugar backbone) joined with knots (representing the phosphate bonds). Attached to the pipe cleaners are pegs representing the four bases (yellow and orange to represent adenine and guanine; blue and green to represent thymine and cytosine, respectively). Each of these is always joined to the complementary base: adenine (yellow) with thymine (blue) and guanine (blue) with cytosine (green). Each peg is attached to the pipe cleaner at the knots by a brass fastener creating a model of a complete nucleotide.

If a diagram is drawn, a key must be provided.

(ii) The justification will be directly related to the model related in part (i).

For example: The model described is composed of a series of nucleotides, each made up of a sugar, phosphate and base. These are arranged along the helix creating a ladder, with each base matching its complementary base on the opposite side. The whole double strand can then be twisted to represent a double helix DNA model.

(c) (i)

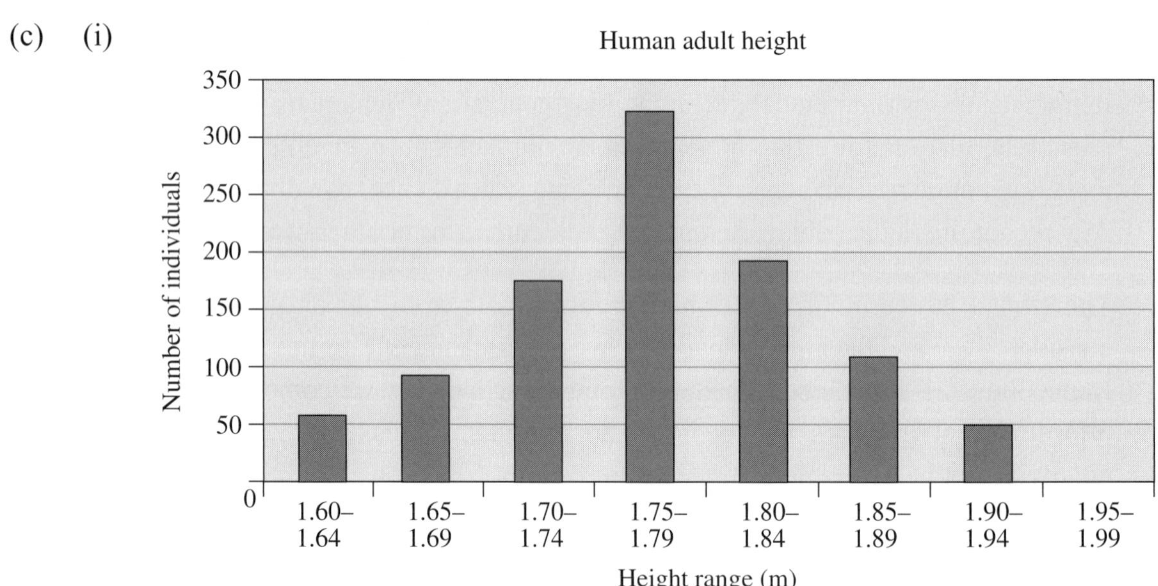

(ii) Polygenic inheritance refers to a form of inheritance in which a characteristic is controlled by more than two independent genes.

(iii) A wide range of variation is demonstrated by polygenic inheritance, whereas Mendel's work depicted a simple pattern of inheritance based on two alternatives in pea plants (such as tall or short). The inheritance of polygenic characteristics such as height in humans means that there can be a wide variety of different heights.

(d) Selective breeding involves selecting favoured organisms with desirable characteristics and breeding them. This can be done on a large scale with plants where pollen from one plant is used to pollinate seedlings obtained from previous germination, thereby reducing the genetic variety in the long run, as for example in the production of orchids.

Cloning involves selecting specific DNA and inserting the material into surrogate eggs to produce identical offspring, thereby reducing the genetic diversity. This method has been used to create maize plants that are disease resistant and have high yields.

Thus the two processes, even though entirely different, can reduce the genetic diversity in the long term, resulting in a change in the genetic nature of the species.

Question 3

(a) (i) Recombinant DNA is a term used to describe the manipulation of genetic material to produce specific characteristics.

(ii) Any two correct steps, for example:
- Extraction of plasmid from bacteria.
- Use of restriction enzyme to cut DNA.
- Use of identical restriction enzyme to cut the new DNA that will be inserted, ensuring compatibility between the ends of the respective DNA sections.
- Insertion of DNA into plasmid producing a new genetic combination.

(b) (i) The benefits of the Human Genome Project could include:
- Gene mapping has led to the better understanding of various diseases.
- The ability to locate genes in a particular order provides the ability to recognise changes to a genome.
- Ability to identify and isolate mutations.

(ii) Cystic fibrosis is a disease that causes increased mucous production in lungs. Gene therapy is one way that management of the disease has been attempted. In order to treat the disease, correct cystic fibrosis genes need to be removed from a human cell using restriction enzymes. These are then inserted into a plasmid within a bacteria and are replicated. The correct gene is then enclosed in liposomes and inhaled as an aerosol into the lungs. The liposome fuses with the lung tissue, thereby incorporating the correct gene into the genome of the diseased patient restoring normal function.

(c) (i) All of the males are possible fathers, as the blood group data does not eliminate any of them.

(ii) Using the DNA fingerprint data, males 1 and 2 can be excluded. Male 3 is the father.

(iii) From the data presented, it can be said that DNA fingerprinting is the more accurate method for testing paternity. Using blood groups alone is not conclusive – it may assist in eliminating some males, but cannot be used to positively identify the father. This is as a result of the patterns of inheritance of blood, a dominant/ recessive pattern with ABO alleles resulting in four different blood groups. DNA fingerprinting is more accurate in identifying the father as it is a map of the child's and possible father's genetic profile. There is a greater chance of matching the profiles accurately and provides greater comparison.

(d) (i) Information was gathered and processed by reading through the material, making notes about keywords or points in order to summarise the required information. The information source was identified in each case, as was the date of publication to ensure that the most recent material was being utilised. Once a summary was completed, the relevant information was analysed by ensuring that any inconsistencies were identified and corrections were sought from different sources. Comparisons could be made through the use of tables and keyword matches.

(ii) The reliability of the gathered information can be assessed by ensuring that the references cited in the article are reputable and that they provide similar conclusions. The author should have appropriate qualifications in the area for which they are writing. If the source is an Internet site, then there may be links to corresponding university websites or research institutions.

(e) Gene cloning refers to the replication of genetic information of plants and animals for insertion into other organisms, producing recombinant DNA. This is currently done for agricultural and medical purposes. This process is advantageous because it enables gene manipulation to take place, resulting in the production of proteins such as insulin and factor 8, used in the medical treatment of diabetes and haemophilia respectively. In agriculture this process has been used to develop frost-resistant strawberries and rust-resistant cereal crops. The advantages of this technology are substantial as there are many positive uses for humans and economic gain that can be had by way of increased yield, more economic use of resources, medical benefits for sick individuals and increased quality of life. Related advantages include decreased demand for medical treatment as people are treated for diseases that can eventually be eliminated from the gene pool.

There is another side to this argument, in that there are possible disadvantages such as unknown long-term results of using modified organisms, finding suitable vectors to insert the cloned genes into and disposing of experimental organisms. Risks associated with the research include the possible creation of unwanted effects and their subsequent release into the environment.

CHAPTER 7

Option Topic
The Human Story

Past HSC Questions

Question 1 (25 marks) **Marks**

(a) (i) Identify ONE feature that can be used to classify humans as mammals. **1**

(ii) The diagram shows a *Homo sapiens* skeleton. **2**

Homo sapiens skeleton

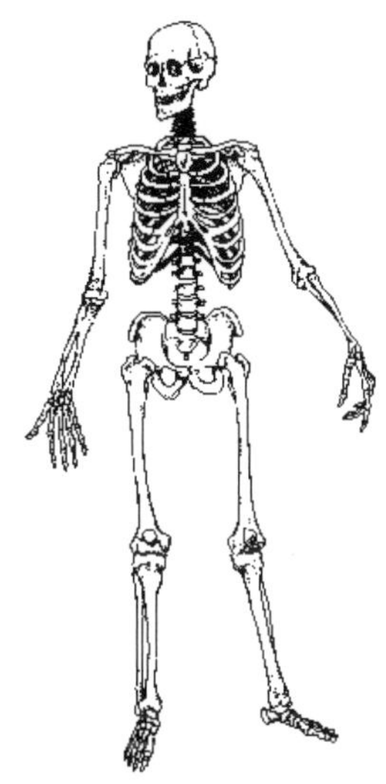

State TWO structural differences between this skeleton and the skeleton of *Australopithecus afarensis*.

(b) (i) How would you gather information about the use of radiometric data to date material collected from a fossil site?* **2**

(ii) How would you assess that the information you collected was relevant and reliable? **2**

(c) Analyse the evolutionary significance of the phenotypes displayed in ONE example of polymorphism in humans. **5**

Question 1 continues

* Due to modifications to the HSC Biology Syllabus, this question may no longer be within the current syllabus.

Question 1 (continued) **Marks**

(d) In your study of The Human Story, you performed an investigation to analyse the similarities and differences between prosimians, monkeys, apes and humans. 6

By referring to TWO features of these primates, outline the information you gathered to perform this analysis, and justify the conclusions drawn from this information.

(e) What would you predict will be the main factors affecting human biological evolution in the next one hundred years? Justify your predictions. 7

Question 2 (25 marks)

(a) (i) Name ONE significant hominid fossil. 1

(ii) Describe TWO structural features of this fossil that would identify it as a primate. 2

(iii) Use ONE of the features you described in part (ii), to compare humans with apes. 2

(b) Theories of hominid evolution have been developed by anthropologists such as the Leakey family, Johanson, Broom, Tobias, Dart and Goodall.

(i) Describe ONE major discovery made by ONE of the above scientists that has added to our knowledge of hominid evolution. 3

(ii) Explain how this discovery increased understanding of hominid evolution. 3

(c) The data shows the radioactive decay of carbon 14 (C^{14}) over time.

Time (years)	*Fraction of original radioactivity*
0	100
5000	54
10000	29
15000	16
20000	8
25000	5
30000	2
35000	1

Question 2 continues

Question 2 (continued) **Marks**

(i) Graph the data on the graph paper provided. 4

(ii) A fossil sample was found to contain 40% of its original radioactivity. Estimate the age of the fossil.* 1

(iii) Explain why measurement of carbon 14 (C^{14}) content in a hominid fossil may not be a useful radiometric dating technique. 2

(d) Cultural development has been a significant feature of human evolution. 7

Discuss the complex cultural development of humans in comparison with other primates.

Question 3 (25 marks)

(a) Complete the information missing in cells of the table marked (1), (2) and (3). Clearly label your answers to (1), (2) and (3) in your writing booklet. 3

Level of classification	*Classification*	*Distinguishing characteristics*
Kingdom	Animal	Heterotrophic; eucaryotic
Phylum	Chordate	Hollow dorsal nerve cord
Class	Mammalia	Mammary glands
Order	(1)	Opposable thumb
(2)	Hominidae	Upright stance
Genus	*Homo*	(3)
Species	*sapiens*	Large brain to body mass ratio; flat face; abstract thought; complex social structures

Question 3 continues

* Due to modifications to the HSC Biology Syllabus, this question may no longer be within the current syllabus.

Question 3 (continued) **Marks**

(b) The diagram shows possible evolutionary relationships between a number of fossils.

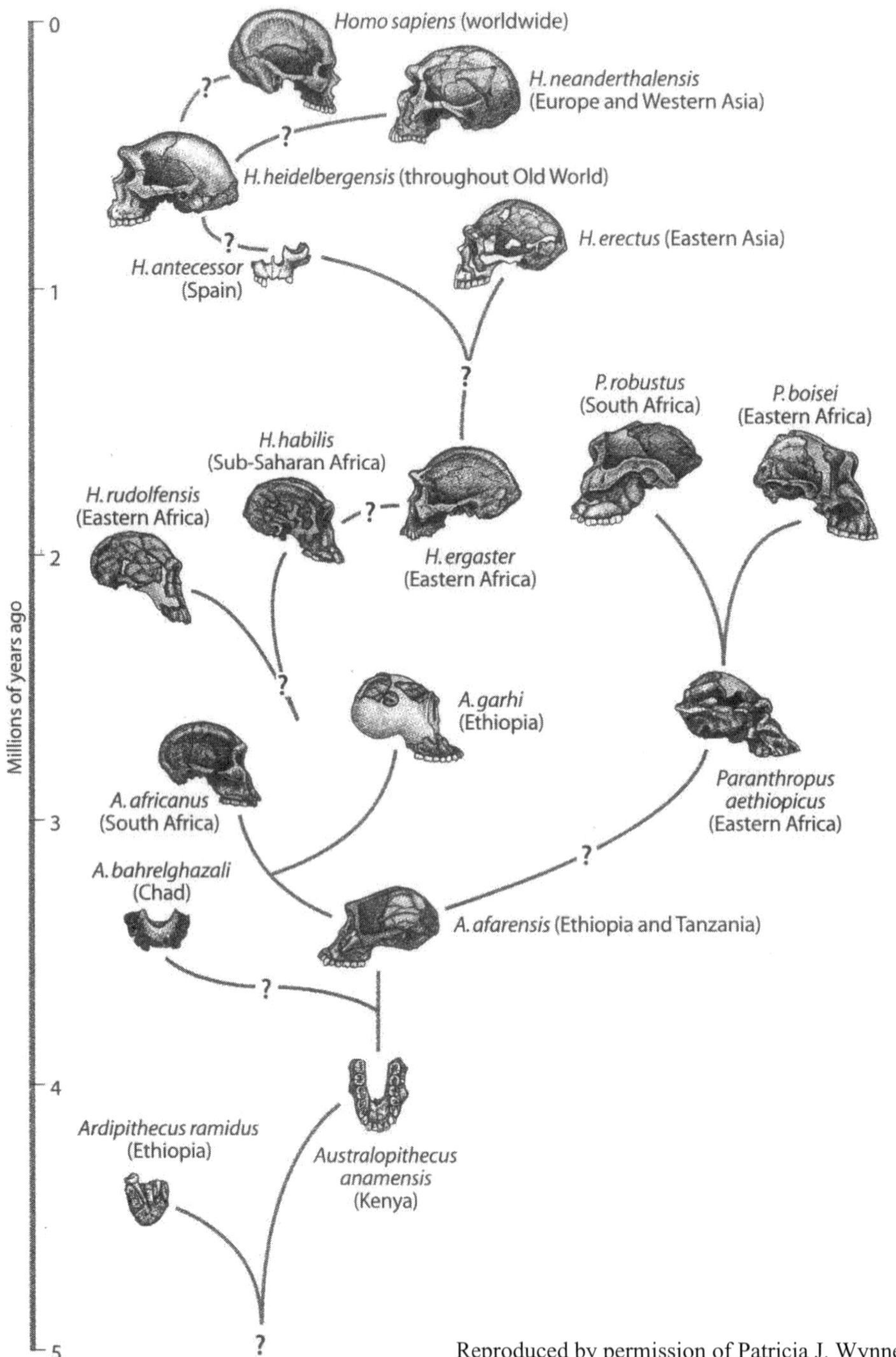

Reproduced by permission of Patricia J. Wynne

(i) Outline the differences between absolute dates and relative dates for the age of fossils. **2**

(ii) The diagram contains many question marks. Discuss TWO reasons for this uncertainty. **3**

Question 3 continues

Question 3 (continued) **Marks**

(c) The table and map show the frequency and distribution of human blood types.

Mourant A E, et al., 1976, *The Distribution of the Human Blood Groups and Other Polymorphisms*, 2nd ed.

Population group		*Blood group O* (%)	*Blood group A* (%)	*Blood group B* (%)	*Blood group AB* (%)
United Kingdom	①	47	42	8	3
France	②	43	47	7	3
Greece	③	40	42	14	4
Russia	④	33	36	23	8
Australian Aborigines	⑤	61	39	0	0
Peruvian Indians	⑥	100	0	0	0

Reprinted by permission of Oxford University Press.

(i) Propose an explanation for the variation in the frequency of blood group B shown in the table. **2**

(ii) Account for the limited variation in blood groups in the Australian Aborigines and Peruvian Indians. **3**

(d) During your study of The Human Story you gathered secondary information to account for changes in human population numbers in the last 10 000 years.

(i) Describe how you processed and analysed the gathered information. **4**

(ii) State how you assessed the reliability of the data obtained. **1**

(e) The 'Out of Africa model' and the theory of regional continuity (multi-regional hypothesis) are two different models used to explain human evolution that have been developed in recent years. **7**

Select ONE model, and use evidence to evaluate the model.

Option Topic
The Human Story

Worked Answers

Question 1

(a) (i) Example: Hair on the body

(ii) Example: Humans have reduced brow ridges and a chin, whereas *Australopithecus afarensis* has very prominent brow ridges and the jaw does not have a chin.

(b) (i) Information about the use of radiometric data to date material collected from a fossil site could be gathered from secondary sources such as books and the Internet. The information would need to be found in these sources by using an index or a search engine to find words associated with the subject matter, such as archaeology, fossils, dating, radioisotopes, radiometric, half-life.

(ii) The reliability of the information would need to be considered by assessing the validity and reliability of the source; for example, is it from a journal/textbook publisher and are references cited in the articles? For the Internet, what is the source of the Internet page? Is it from a university lecture or formal journal article, and who produced it? The relevance of the information would be assessed by considering whether it addresses the content area and whether the information is supported by that found in other sources.

(c) Answers could include: blood groups, hair type, skin colour, lactose tolerance, and eye colour.

Example: Skin colour varies significantly in humans, depending on the melanin concentration. Melanin absorbs ultraviolet (UV) radiation, which has been linked as a causative factor to skin cancer. The variation in skin colour is linked to geographical location around the globe and correlates to the exposure to UV radiation. Generally, darker-skinned people originate from tropical areas which are exposed to higher levels of UV radiation. Darker-skinned people would therefore have an evolutionary advantage over paler-skinned people in these areas because the higher levels of melanin pigmentation in their skin absorb more UV radiation than does paler skin. People with pale skin are found further away from equatorial regions where there is less evolutionary advantage from having darker skin. However, the body can manufacture vitamin D in sunlight in the skin, and paler skin has more of an advantage in temperate regions where light intensity is reduced and the body does not require as much protection from UV.

(d) Possible features could include: cranial size, thumb opposability, use of tools, care of young, body size, communication systems.

Example: Prosimians, monkeys, apes and humans were compared by collecting information from secondary sources and observations of animals at the zoo. The features that were compared included the cranial size and thumb opposability. It was found that the cranial volume increased disproportionally to body size; in other words, the prosimians had the smallest cranial size compared to body size and the humans the largest, with the monkeys and apes showing the same trend. The nature of the digits was also investigated, showing that all primates had an opposable toe, but only the humans possessed an opposable thumb, but not an opposable toe. From the evidence gathered it is possible to state that the groups investigated can be classified as primates; however, they possess substantial differences that can be used to place them into discrete classification groups.

(e) The answer should include at least two different factors that may affect human biological evolution in the next 100 years, with a justification for each of the factors selected. Factors could include: cloning, genetic engineering, mobility of population, reproductive technology, and medical technology.

Example: One factor that could affect human biological evolution in the next 100 years could be the use of genetic engineering. Genetic engineering is currently providing the technology to manipulate genes and the ability to test for the presence of deleterious genes in gametes and embryos. If undesired genes are identified, the choice has to be made as to whether the embryo should be removed or allowed to grow. This may be seen as advantageous where genetic diseases such as Tay Sachs and trisomy 21 can be identified, providing the option to terminate a pregnancy and thereby removing these genes from the population. It can also be used to determine the sex of offspring and the possibility of tailoring desired characteristics. This raises ethical and moral questions as to who should decide whether someone lives or dies, thereby changing the future evolutionary gene pool and removing the natural selection mechanism.

Question 2

(a) (i) *Australopithecus afarensis*

(ii) Opposable thumb on hand and foot; large cranial capacity compared to body size.

(iii) Humans have an opposable thumb only while apes have both an opposable toe and thumb.

(b) (i) *Australopithecus afarensis* was discovered by Donald Johanson. This fossil represents one of the earliest bipedal hominids. The features of the hominid include long arms compared to legs; bipedal gait; ape-like teeth; large cranial capacity compared to body size.

(ii) The features of *Australopithecus afarensis* indicate that the species was arboreal in habit, which was not previously thought of the hominids. The development of a bipedal gait prior to the increased brain size and complexity has changed the thinking about human evolution.

(c) (i)

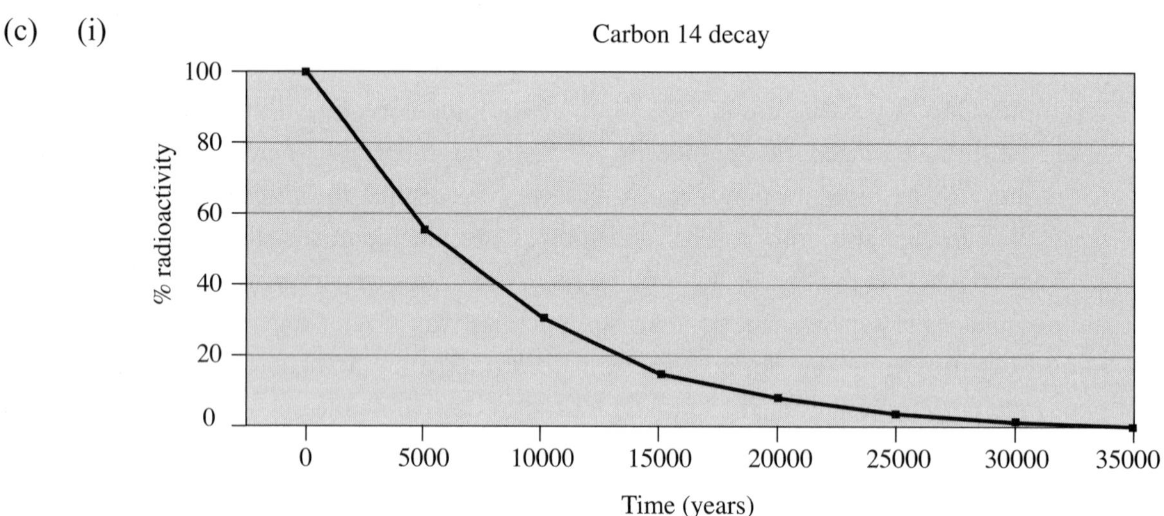

(ii) Approximately 7500 years

(iii) The majority of significant hominid fossils are aged over 100 000 years old. Carbon 14 dating as a radiometric dating technique is usually used for dating fossils up to an age of 50 000 years only.

Marks

(d) Answers will vary but should include a comparison of humans and apes in reference to the following features: 3

- Communication methods – humans have a very advanced method of communication with the use of body signals and complex language development; the apes, while they have the ability to communicate very well, do not possess the same level of complexity or diversity of expression.
- Social behaviours – humans possess a distinct set of social behaviours, as do the apes. However, in humans the social structure is more flexible and not as rigidly kept as in the ape hierarchy. The apes live in groups where a dominant male maintains the group order and protects them.
- Tool production and use – humans have the ability to make very advanced tools and can use them to modify their surroundings significantly and to find food. The apes are able to use objects such as sticks to obtain food, but do not make tools.
- Shelter construction – humans build very intricate shelters to protect themselves from the elements, while the apes do not actively build shelter.
- Length of juvenile stages and care of young – while primates have a long gestation period and a prolonged juvenile stage requiring adult care, it is longest in the humans. In apes, the care will extend for up to 3–6 years, whereas in humans it can be well into the teenage years.

Question 3

(a) (1) Primate

(2) Family

(3) Bipedal gait or small teeth/jaw or large brain

(b) (i) Absolute dates indicate the exact age of fossils in years and relative dates provide an indication of age based on a comparison of fossils relative to one another.

(ii) Reasons for uncertainty depicted within the speculative family tree include the fact that the existing fossil record is very incomplete giving rise to different interpretations and the fact the fossils that have been found are scarce and originate from an extended length of time, again causing differing opinions among anthropologists.

(c) (i) The frequency of the blood group B depicted in the table indicates that there has been little movement or migration between the populations depicted. For example, the low incidence of group B in Russia shows little movement west toward Greece and UK.

(ii) The limited variation of blood groups in the Australian Aboriginals and the Peruvian Indians appears to indicate that these populations have been relatively isolated from other groups, limiting their gene pool. The Peruvians show a uniform blood group, the implication being that they did not mix with any other groups and retained the homozygous recessive O type. This could be due to their remote location preventing movement into or out of the area. The Australian Aborigines by comparison did show a slight variation between types O and A, indicating that they may have mixed a little thereby increasing gene pool diversity. The country being closer to Indonesia may have resulted in a mixing of the population from the north.

(d) (i) Information was gathered and processed by reading through the material, making notes about keywords or points in order to summarise the required information. The information source was identified in each case, as was the date of publication to ensure that the most recent material was being utilised. Once a summary was completed, the relevant information was analysed by ensuring that any inconsistencies were identified and corrections were sought from different sources. Comparisons could be made through the use of tables and keyword matches.

(ii) The reliability of the gathered information can be assessed by ensuring that the references cited in the article are reputable and that they provide similar conclusions. The author should have appropriate qualifications in the area for which they are writing. If the source is an Internet site, then there may be links to corresponding university websites or research institutions.

(e) The 'Out of Africa' model suggests that the modern humans originated in Africa and moved or migrated into Europe and Asia, replacing the populations (non-modern) existing in those areas. The support for this model has been considerable and is based on stronger evidence than the evidence available for the 'regional continuity hypothesis'.

Evidence to support the 'Out of Africa' model includes the greater variation of mitochondrial DNA found in African populations, the discovery of transitional forms and the earliest dated modern human being found in Africa, all of which strengthen this argument.

The fact that there is such a variation in the mitochondrial DNA indicates that there had been more time to develop this diversity in the gene pool locally in Africa, pointing to the evolution being restricted to the region. If this was not the case, and for example transitional forms were found elsewhere, then the 'regional continuity hypothesis' would have better justification.

CHAPTER 8

2009

HIGHER SCHOOL CERTIFICATE EXAMINATION

Biology

General Instructions

- Reading time – 5 minutes
- Working time – 3 hours
- Write using black or blue pen
- Draw diagrams using pencil
- Board-approved calculators may be used
- Write your Centre Number and Student Number where required

Total marks – 100

Section I

75 marks

This section has two parts, Part A and Part B

Part A – 15 marks

- Attempt Questions 1–15
- Allow about 30 minutes for this part

Part B – 60 marks

- Attempt Questions 16–27
- Allow about 1 hour and 45 minutes for this part

Section II

25 marks

- Attempt ONE question from Questions 28–32
- Allow about 45 minutes for this section

Section I
75 marks

Part A – 15 marks
Attempt Questions 1–15
Allow about 30 minutes for this part

Use the multiple-choice answer sheet for Questions 1–15.

---

1 How are new alleles formed?

(A) By crossing over during meiosis

(B) By cloning a new variety in a population

(C) By mutation in the DNA of a gene

(D) By production of a new phenotype from the same DNA

2 Which scientist's work contributed to our understanding of the causes of infectious diseases?

(A) Frank Macfarlane Burnet

(B) Louis Pasteur

(C) James Watson

(D) Maurice Wilkins

3 Which of the following prevent entry of pathogens into the human body?

(A) The skin and phagocytosis

(B) The skin and chemical barriers

(C) Inflammation response and phagocytosis

(D) Inflammation response and chemical barriers

4 The potential for disease to spread through animal populations in intensive farming is heightened because the animals are kept close together.

A disease has been identified in animals in one enclosure on a farm.

Which procedure would best prevent the spread of the disease to animals in other enclosures on the farm?

(A) Isolate diseased animals from healthy animals then vaccinate all healthy animals.

(B) Vaccinate all animals so that healthy animals do not develop the disease and spread it further.

(C) Move the diseased animals into another enclosure to quarantine them from the healthy animals.

(D) Wash all animals with antiseptic solution so that the pathogen causing the disease cannot be spread from diseased animals to healthy animals.

5 Why was the importance of Mendel's work not widely accepted until some time after he completed his experiments?

(A) Mendel did not realise the importance of his own work.

(B) Mendel did not repeat his experiments so they were not seen as important by other biologists.

(C) Mendel published his results in a local scientific publication not accessed by many biologists.

(D) Mendel's description of sex linkage to explain his results was not understood by other biologists.

6 What is a role of the kidney in the excretory system of mammals?

(A) To remove salt from the body and to keep water in the body

(B) To remove water from the body and to keep salt in the body

(C) To remove nitrogenous waste from the body and to maintain water levels in the body

(D) To remove water from the body and to maintain levels of nitrogenous substances in the body

7 Thirty percent (30%) of the nucleotide bases in human DNA are adenine (A).

What is the percentage of guanine (G) bases in human DNA?

(A) 20%

(B) 30%

(C) 40%

(D) 70%

8 In humans, brown eye colour is dominant and blue eye colour is recessive. A brown-eyed boy and a blue-eyed girl have a blue-eyed mother.

What eye colour does the father have and why?

(A) Brown, because the gene for brown eye colour is sex linked.

(B) Brown, because at least one of the parents must have brown eyes.

(C) Blue, because at least two other members of the family have blue eyes.

(D) Blue, because at least one of the parents must be heterozygous for eye colour.

9 Which alternative best describes what happens to oxygen and carbon dioxide as blood travels through the lungs and muscles?

	Lung	*Muscle*
(A)	Oxygen dissolves in blood.	Carbon dioxide dissolves in blood.
(B)	Oxygen binds strongly to haemoglobin.	Carbon dioxide binds strongly to haemoglobin.
(C)	Oxygen binds strongly to haemoglobin.	Carbon dioxide dissolves in blood.
(D)	Carbon dioxide binds weakly to haemoglobin.	Oxygen binds weakly to haemoglobin.

10 Amylase is an enzyme that catalyses the breakdown of starch.

As the temperature rises, the activity of this enzyme will

(A) increase until the enzyme begins to denature.

(B) decrease until the substrate begins to denature.

(C) increase until the optimum pH level is reached.

(D) decrease until the substrate reaches a maximum concentration.

11 Experiments were carried out on plants living in different environments to measure the size of the leaf stomata at different times of the day. Previous investigations had shown that plants transpire more water when the size of the stomata is larger.

Which graph best represents a plant living in a dry environment?

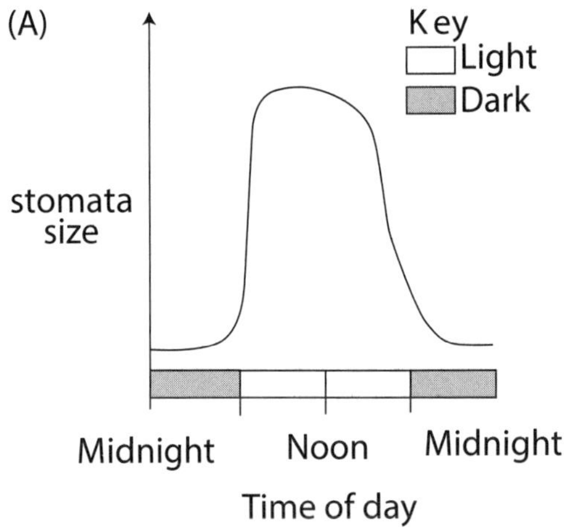

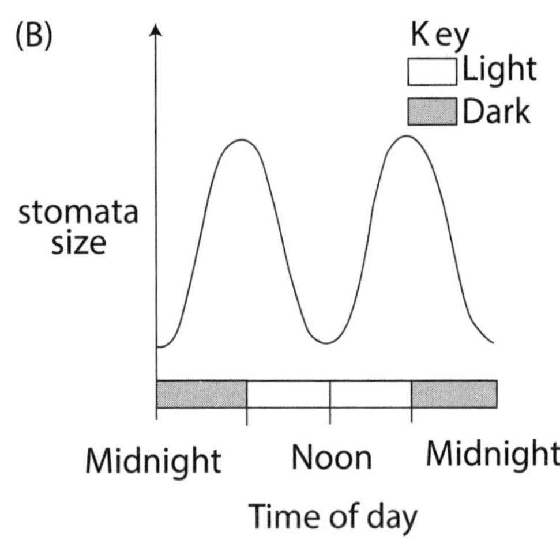

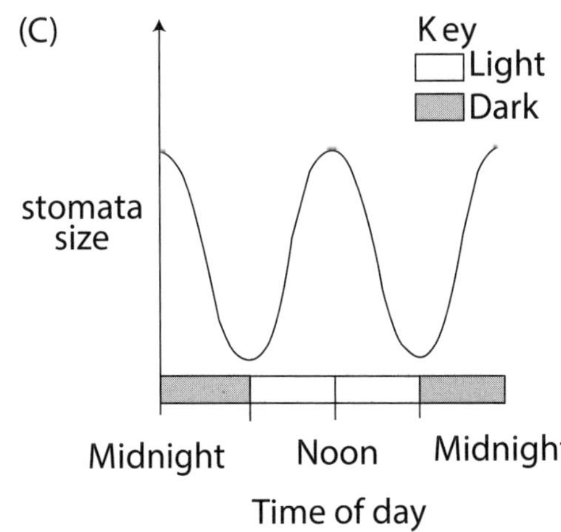

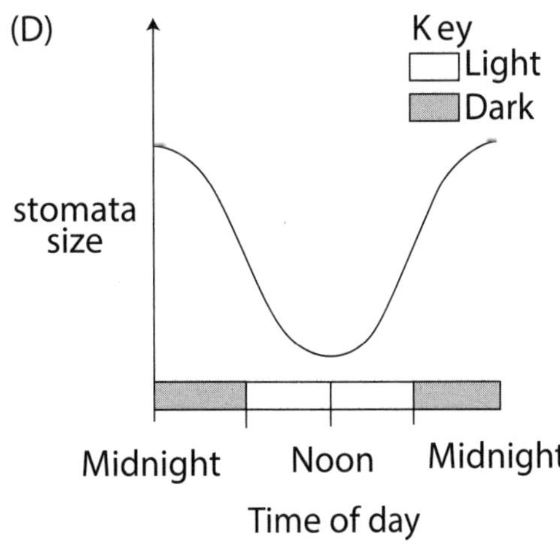

12 A student conducted an experiment where carbon dioxide was bubbled through water. The student recorded the pH meter reading every 5 seconds.

Time (s)	*pH meter reading*
5	7.5
10	7.2
15	7.0
20	6.8

What would be the most valid conclusion the student could draw from these results?

(A) The change in pH was too small to be significant.

(B) The water became less acidic as the amount of CO_2 in the water increased.

(C) The water became more acidic as the amount of CO_2 in the water increased.

(D) The change in pH might not be related to the CO_2 being bubbled through the water.

13 Why do organ transplant patients need anti-rejection medication?

(A) To minimise infection

(B) To prevent T-lymphocyte growth

(C) To stimulate the interaction of B- and T-lymphocytes

(D) To prevent the recipient's blood type changing and adopting the immune system of the donor

14 The diagram shows phenotypic changes in a population over a number of generations.

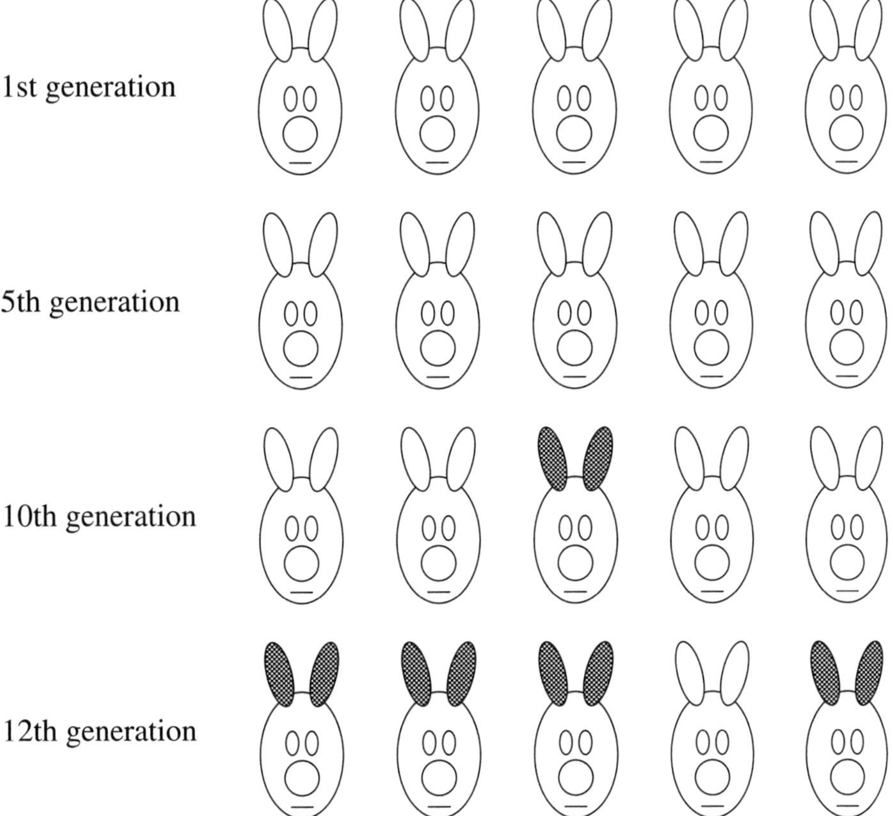

Which evolutionary concept is demonstrated in the diagram?

(A) Convergent evolution

(B) Divergent evolution

(C) Transitional forms

(D) Punctuated equilibrium

15 A patient was being treated for an infection using an antibiotic. At seven day intervals a swab was taken from the mouth and cultured onto a fresh agar plate.

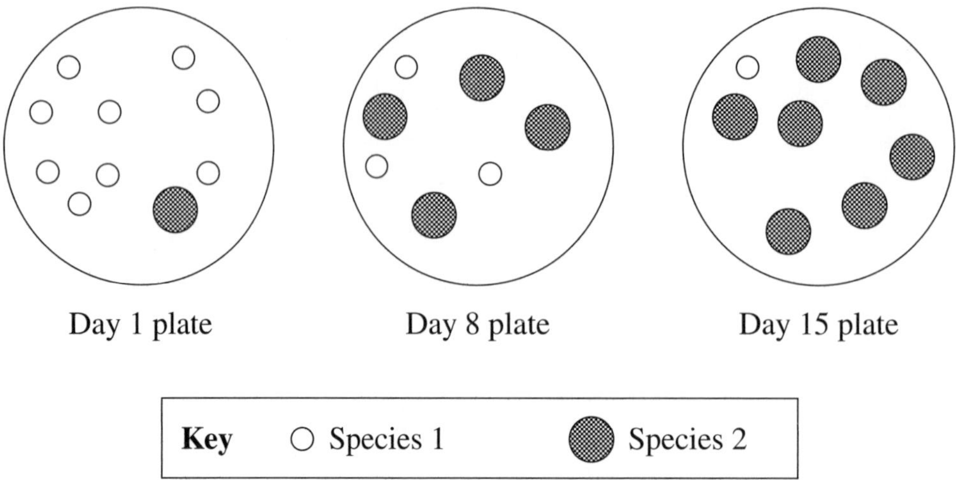

Key	○ Species 1	● Species 2

What is the most likely cause of the change in mouth microflora shown in these culture plates?

(A) Species 1 was a food source for Species 2 allowing more of Species 2 to grow.

(B) Species 1 and Species 2 are both fungi, but Species 1 is killed by the antibiotic while Species 2 uses the antibiotic as food.

(C) As the numbers of Species 2 increased, they changed the chemical conditions of the agar plate stopping the growth of Species 1.

(D) There is normally a balance between the numbers of each species but the removal of Species 1 by the antibiotic allowed more of Species 2 to grow.

2009 HIGHER SCHOOL CERTIFICATE EXAMINATION

Biology

Centre Number

Student Number

Section I (continued)

Part B – 60 marks
Attempt Questions 16–27
Allow about 1 hour and 45 minutes for this part

Answer the questions in the spaces provided.

Question 16 (3 marks)

Complete the following table. **3**

Pathogen	*Distinguishing characteristic of the pathogen*	*Disease caused by this type of pathogen*
Bacteria		
Fungi		
Protozoans		

Question 17 (6 marks)

As part of an independent research project, a student studied a genetic condition suffered by members of his family. The student wrote the following summary:

- I am male and I have the condition.
- My mother does not have the condition.
- My father and his brother have the condition.
- My father's sister and my father's mother do not have the condition.
- My father's father has the condition.

(a) Construct a pedigree of this family. **3**

(b) Why are diagrams, such as pedigrees, useful in analysing data? **1**

....

....

(c) The student made the following conclusion from his study. **2**

"As only males have the condition, it must be a sex-linked genetic condition."

Assess the validity of the student's conclusion, and provide support for your assessment.

....

....

....

....

....

....

Question 18 (5 marks)

On a ship at sea there is an outbreak of a disease affecting the human digestive system. **5**

Explain how cleanliness in food, water and personal hygiene practices on the ship could assist in controlling this disease.

....

....

....

....

....

....

....

....

....

....

Question 19 (6 marks)

Draw a diagram of a mammalian kidney. **6**

On your drawing, label THREE regions involved in the excretion of waste products AND indicate the main process that occurs in each of the three regions.

2009 HIGHER SCHOOL CERTIFICATE EXAMINATION

Biology

Centre Number

Section I – Part B (continued)

Student Number

Question 20 (5 marks)

(a) Outline the steps you would follow in a first-hand investigation of pathogens and insect pests in plant shoots and leaves. **3**

....

....

....

....

....

....

(b) Describe ONE possible risk in this investigation and ONE precaution needed to maintain safety. **2**

....

....

....

....

Question 21 (6 marks)

The data in the table shows the results from an investigation that measured growth in a species of wheat.

Temperature (°C)	–5	0	5	10	12.5	15	17.5	20	22.5	25	27.5	30	35	40
Relative growth (%)	0	0	0	20	40	60	80	100	75	50	25	0	0	0

(a) Use the data to draw an appropriate graph. **4**

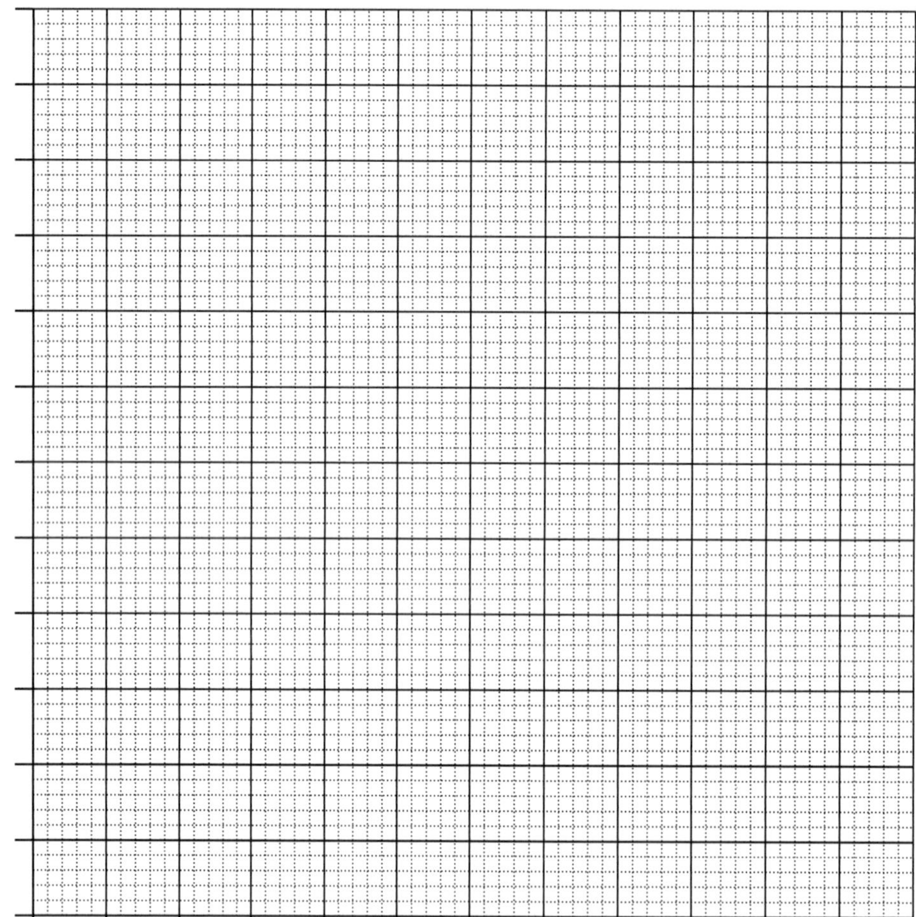

(b) From the graph, describe how the growth of wheat responds to temperature. **2**

....

....

....

....

Question 22 (3 marks)

Construct a table to identify THREE types of T-lymphocytes, and outline the role of each type in the immune response. 3

2009 HIGHER SCHOOL CERTIFICATE EXAMINATION

Biology

Centre Number

Section I – Part B (continued)

Student Number

Question 23 (5 marks)

(a)

> A woman recently conceived Britain's first baby guaranteed to be free from hereditary breast cancer. Doctors screened for an embryo that was free from a gene that can cause breast cancer.
>
> The screening was performed due to the long history of this form of cancer in the family and the fact that any daughter born with the gene would have a 50%–80% chance of developing breast cancer.

Explain the possible impact of this reproductive technology on the genetic composition of the population. **2**

....

....

....

....

(b) Describe the potential impact of the use of reproductive technologies on the path of evolution using ONE named plant or animal that has been genetically altered. **3**

....

....

....

....

....

....

Question 24 (6 marks)

An experiment was conducted to determine the effects on body function of increasing ambient temperature. Participants (100 individuals) were asked to lie down and remain as still as possible during the entire experiment. The following observations were made:

- The starting temperature of the room was 22°C.
- The mean body temperature of all participants at the start of the experiment was 37.1°C.
- As the room temperature increased, sweating and heart rate also increased.
- As the room temperature returned to 22°C, sweating and heart rate returned to normal.

In a control group of 100 participants who were lying down in a room where the ambient temperature was maintained at 22°C, no changes in sweating and heart rate were observed.

(a) Identify the dependent and independent variables in this experiment. **2**

....

....

....

(b) Identify the body system that monitors and responds to changes in external temperature. **1**

....

....

Question 24 continues

Question 24 (continued)

(c) Draw and label a model of a feedback mechanism that explains the observations made in the experiment. **3**

End of Question 24

Question 25 (4 marks)

A fisherman pricked his finger on a fish hook. Soon after he noticed that the injured finger was red and swollen. Some time later he felt a throbbing sensation in his arm. His doctor prescribed a course of antibiotics.

(a) Identify TWO defence adaptations used by the body in response to the injury. **2**

....

....

....

....

(b) Identify the role of the antibiotics in the management of this injury. **2**

....

....

....

....

2009 HIGHER SCHOOL CERTIFICATE EXAMINATION

Biology

Centre Number

Section I – Part B (continued)

Student Number

Question 26 (3 marks)

Describe how processes such as enantiostasis and homeostasis are used to maintain metabolic functions when salt concentration varies in plants living in estuarine environments. **3**

....

....

....

....

....

....

Question 27 (8 marks)

Most offspring resemble their parents in a number of characteristics, but there are often some characteristics in the offspring that are unexpected. **8**

Explain, using examples, how genetics and the environment can affect the phenotype of individuals.

2009 HIGHER SCHOOL CERTIFICATE EXAMINATION

Biology

Section II

25 marks
Attempt ONE question from Questions 28–32
Allow about 45 minutes for this section

Answer the question in a writing booklet. Extra writing booklets are available.

Question 28	Communication
Question 29	Biotechnology
Question 30	Genetics: The Code Broken?
Question 31	The Human Story
Question 32	Biochemistry (*Not included in this reproduction*)

Question 28 — Communication (25 marks)

(a) Draw a diagram of the human eye and label THREE sites of refraction. **3**

(b) (i) Outline ONE technology that can be used to treat or prevent blindness. **1**

(ii) What are the possible implications for society of the use of this technology? **3**

(c) The graph shows the voltage changes recorded in a neurone when it was stimulated at different intensities.

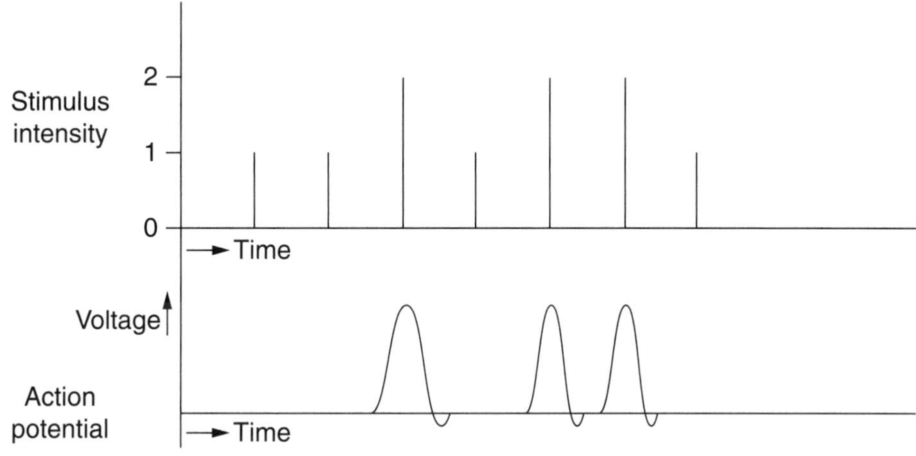

Explain why all the stimuli do not generate an action potential. **5**

(d) (i) Why is the process of accommodation important in visual communication? **2**

(ii) Describe how you would use lenses to model the process of accommodation. **4**

(e) Explain how an understanding of hearing mechanisms has been used to develop technologies to overcome hearing difficulties. **7**

Question 29 — Biotechnology (25 marks)

(a) (i) What is used to cut DNA in recombinant DNA techniques? **1**

(ii) Outline the role of ligases in recombinant DNA techniques. **2**

(b) Before the 18th century, fermentation was a 'cottage' industry carried out in homes or small shops in villages. Early in the 18th century, fermentation began to develop into a large scale industry.

(i) Identify ONE fermentation product made prior to the 18th century. **1**

(ii) Describe ONE scientific discovery that has led to the expansion of fermentation since the 18th century. **3**

(c) The diagram represents a model of protein synthesis.

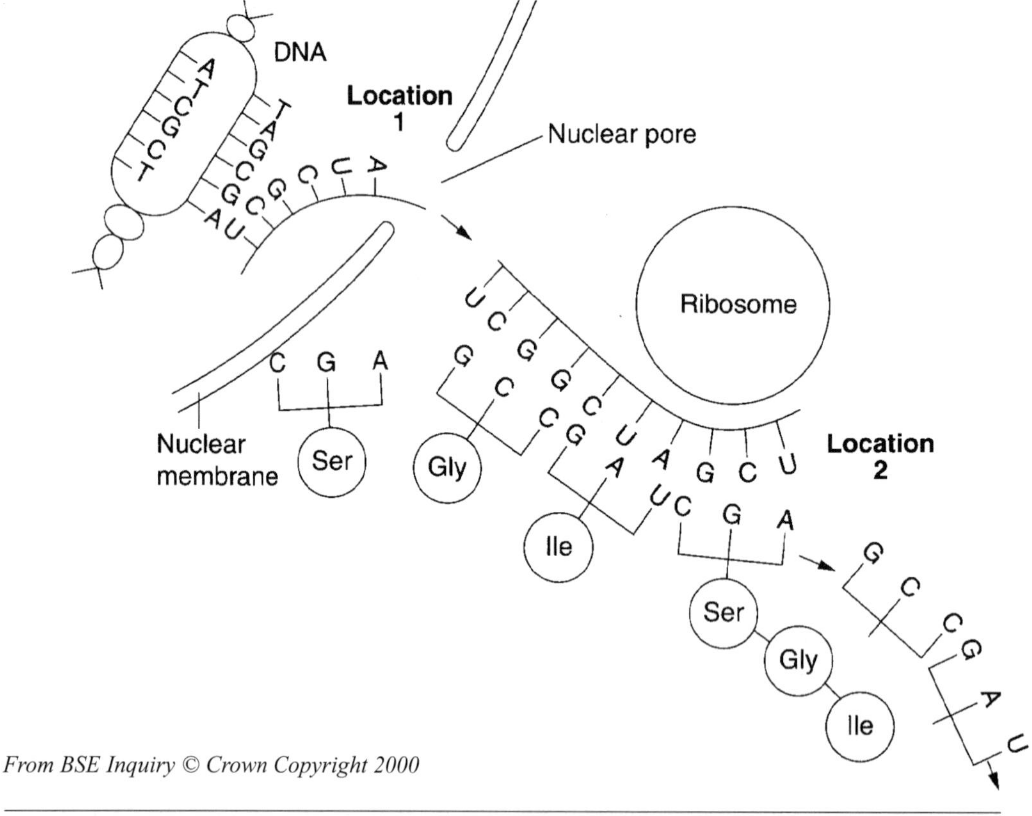

From BSE Inquiry © Crown Copyright 2000

(i) Name TWO types of RNA. **1**

(ii) Describe the role of RNA in the processes occurring at **locations 1** and **2** in the diagram. **4**

Question 29 continues

Question 29 (continued)

(d) (i) Describe how you would extract DNA from cells. **5**

(ii) Outline how you would identify the DNA you extracted. **1**

(e) Explain how applications of modern biotechnology have produced products that are useful to humans. **7**

End of Question 29

Question 30 — Genetics: The Code Broken? (25 marks)

(a) (i) What are the multiple alleles in the inheritance of ABO blood groups? **1**

(ii) Using a diagram, show how parents with blood groups A and B could have a child with blood group O. **2**

(b) (i) Identify ONE example of cloning an animal or plant. **1**

(ii) Describe ONE possible benefit for society of the use of this cloning. **3**

(c) The diagram represents a model of protein synthesis.

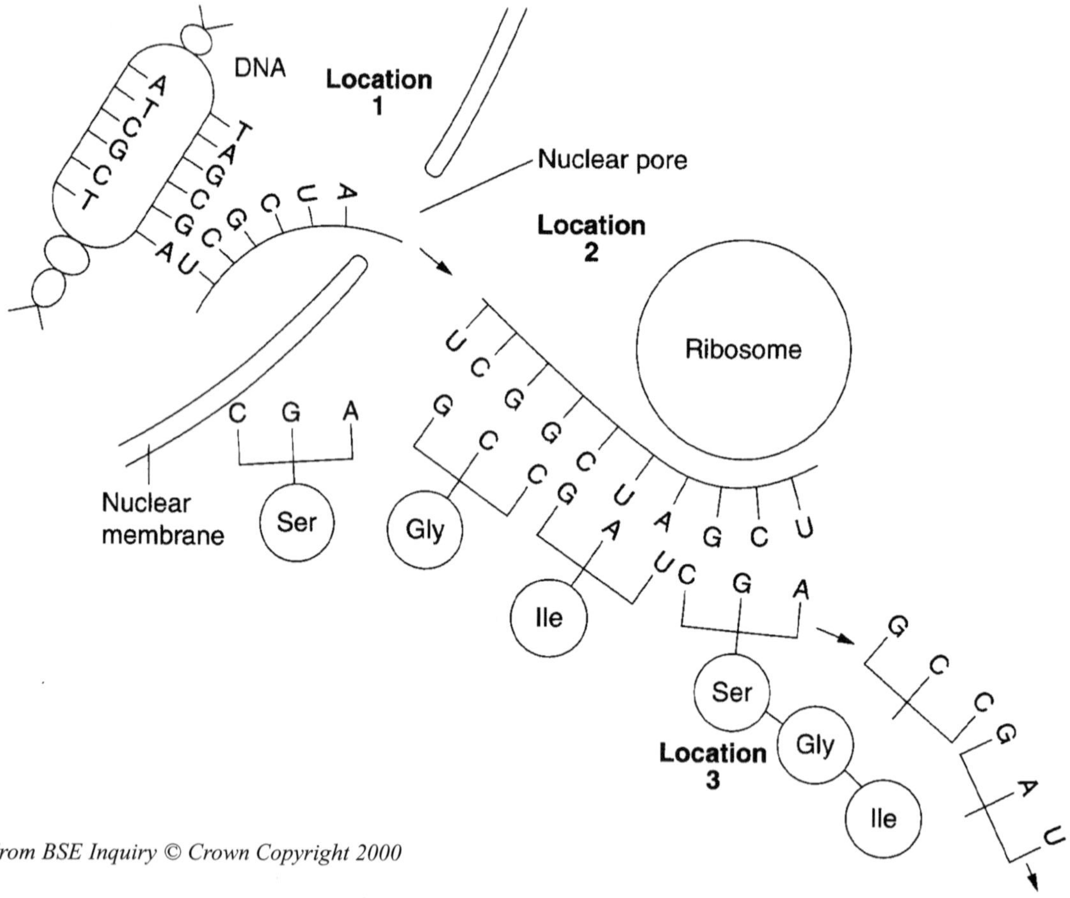

From BSE Inquiry © Crown Copyright 2000

Describe the processes occurring at each of the **locations 1**, **2** and **3** in the diagram. **5**

Question 30 continues

Question 30 (continued)

(d) The inheritance patterns of two genes indicate that they are linked. **6**

Describe how you would use a model of linkage to explain this.

(e) Explain how our understanding of genetic mutation could lead to the management of some diseases using gene therapy. **7**

End of Question 30

Question 31 — The Human Story (25 marks)

(a) Copy and complete the following table in your writing booklet. **3**

Classification level	*Human classification*	*Identifying feature*
	Primate	
	Hominid	
	Homo	

(b) (i) Identify ONE method used to date fossils. **1**

(ii) Outline the problems associated with relying only on the fossil record to interpret the past. **3**

(c) The diagram shows two different models of human migration and evolution.

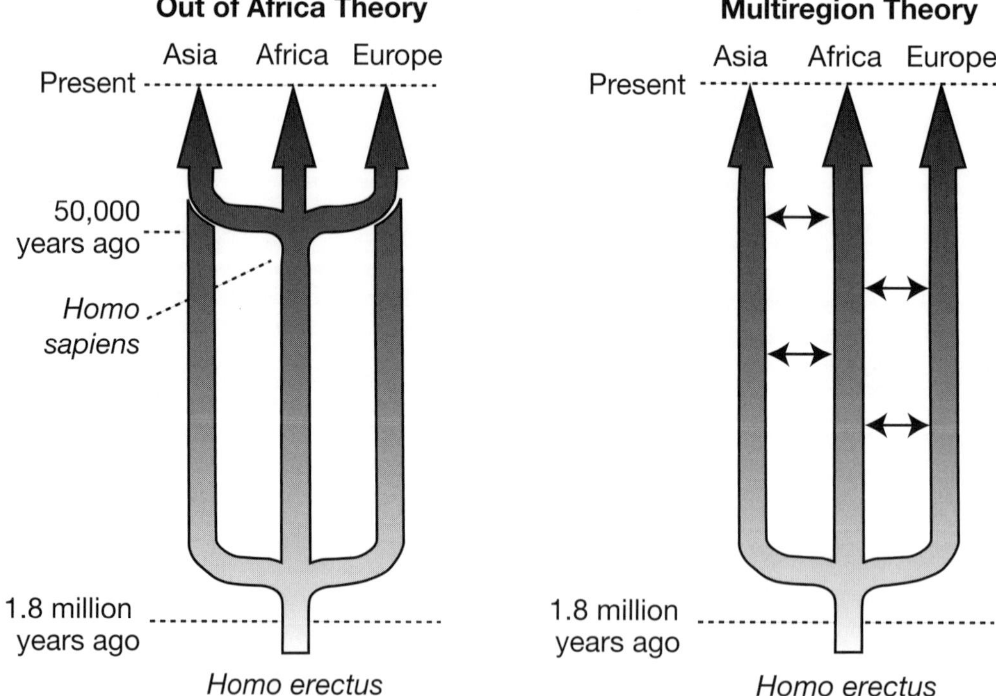

Describe the evidence that can be used to support each of the theories shown in the diagram. **5**

Question 31 continues

Question 31 (continued)

(d) (i) Name TWO species (other than *Homo erectus* and *Homo sapiens*) known from fossil evidence that have been used to trace human evolutionary relationships. **1**

(ii) Outline the similarities and differences you would look for when studying fossils of these species. **5**

(e) Explain the biological evidence, other than fossils, that can be used to improve our understanding of the evolution of humans. **7**

End of Question 31

End of paper

2009 HSC Examination Paper

Sample Answers

Section I Part A

1 C New alleles are created by mutations whereas crossing over creates new combinations of alleles and hence variation.

2 B Watson and Wilkins are associated with DNA discovery, Burnet predominantly with vaccination and Pasteur with causes of infectious diseases.

3 B The skin and chemical barriers are associated with 'first line of defence' only and hence with prevention of entry of pathogens into the body.

4 A This option removes the healthy animals and minimises contact with the diseased animals. Vaccination will further assist in decreasing the chance of disease spread more effectively than any other option.

5 C Mendel did not distribute his work widely and as such was overlooked for many years.

6 C The primary role of the kidney is to excrete nitrogenous waste and to regulate water levels in the body.

7 A If 30% of bases are adenine, then another 30% are cytosine, leaving 40% to be equally divided between guanine and thymine, resulting in 20% each respectively.

8 B At least one of the parents must be brown-eyed if the boy has brown eye colour as brown is a dominant characteristic, so the father must be brown-eyed and heterozygous for blue, resulting in a blue-eyed daughter, as the mother is blue-eyed.

9 C Oxygen is attached to the haemoglobin in the lungs to be transported to the rest of the body. In the muscles, respiration takes place, resulting in carbon dioxide, which is dissolved in the blood and carried as bicarbonate ions.

10 A The enzyme will function until it denatures. This is the only option directly related to enzyme function.

11 D This is the only graph to show consistent decrease in stomata size as temperature increases during daylight.

12 C As the carbon dioxide dissolves in the water, the pH decreases over time.

13 B Transplant recipients need to prevent their body from rejecting the newly introduced tissue.

14 D The sudden manifestation of the characteristic over so few generations demonstrates punctuated equilibrium rather than divergent evolution, which would occur over a longer period of time (greater number of generations).

15 D The antibiotic is effective against Species 1. The removal of this species has enabled the population of Species 2 to increase.

Section I Part B

16 Sample answers could include:

Pathogen	*Distinguishing characteristic of the pathogen*	*Disease caused by this type of pathogen*
Bacteria	Moneran, single-celled, procaryotic	Meningiccocal, tetanus, pneumonia, tuberculosis
Fungi	Parasitic, microscopic tubular filaments/threads, hyphae, spores	Ringworm, athletes' foot, rusts, moulds, blight
Protozoans	Single-celled, eucaryotic	Malaria, giardia, dysentry

17 (a)

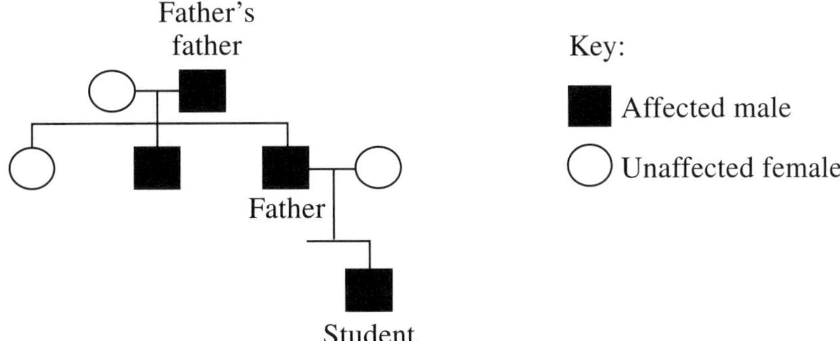

(b) Diagrams such as pedigrees are useful in analysing data because they summarise information, make it clearer/easier to interpret and show relationships between data, better than a long descriptive paragraph.

(c) The assessment of the condition is correct: it is a sex-linked genetic condition. The males in this pedigree are the affected ones and the mother and father's mother are carriers for the condition, not showing any symptoms of it. The males each inherit an affected X chromosome from an unaffected female and one from their father. The father's sister, who has not had any children, would also have to be a carrier as her father had the condition and would have passed on an affected X chromsome to her.

18 Transmission of a disease on the ship could be mitigated through the use of control measures such as general hygiene and cleanliness, which will destroy the pathogen, prevent it from being communicated and thereby reducing the risk of disease transmission. Control measures include:

- general cleanliness, such as washing hands prior to eating/preparing food, using gloves when handling food, cleaning food preparation and serving areas/ equipment
- correct food preparation (i.e. cooking thoroughly, storage at correct temperatures) and disposal
- isolating sick or affected people

- covering of mouths/noses when sneezing/coughing
- regular washing and laundering of clothing/towels
- chlorination of pool water
- ensuring the purity of drinking water by boiling it
- appropriate treatment and elimination of sewage from the ship.

19

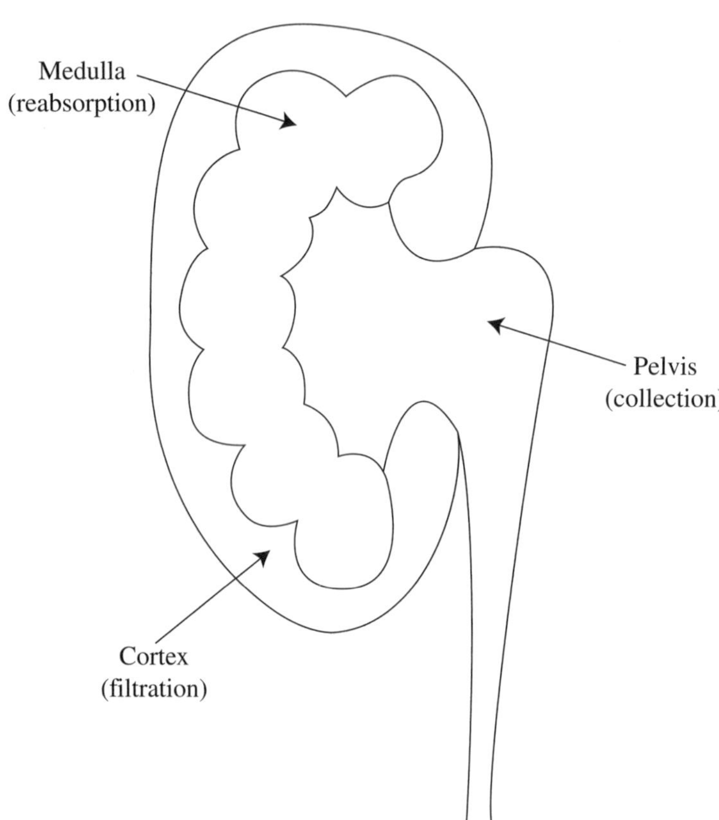

20 (a) Sample answer could include the following.

Aim: To carry out an investigation of pathogens and insect pests in plant shoots and leaves.

Method:

1 Using gloves, collect a variety of plant specimens from the gardens around the school or local park.

2 Using a hand lens or binocular microscope, examine the specimens for evidence of pests and pathogens. Look for things like bites, mould, rust, blights, galls, mildew and macroscopic organisms such as aphids, scale insect, borers, lerps, mites.

3 Tabulate the observations, including a description of the findings and a labelled sketch and possible identification. Alternatively, take photographs and use a key to identify the possible cause of the pest/pathogen evidence.

(b) Possible risks in this activity could include sunburn, bites, scratches or falls while collecting specimens. Safety precautions include wearing hat, sunscreen and gloves to protect hands. In the laboratory, correct laboratory procedures need to be followed when using hand lenses/microscopes or scalpels when investigating the specimens.

21 (a)

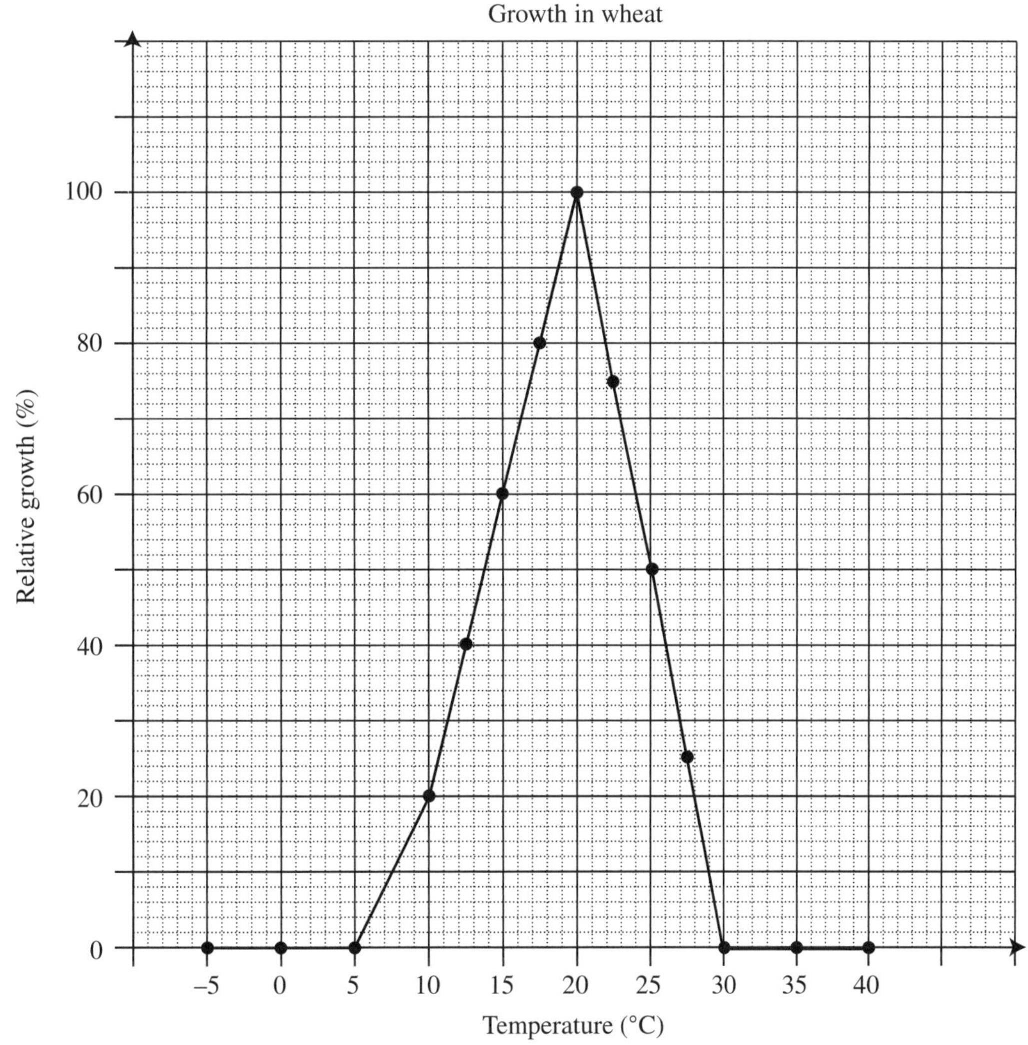

(b) The growth of wheat is best at temperatures of 17.5–22.5°C. Outside of these temperatures the growth is not as significant. Below 5°C and above 30°C there is no growth.

22 Answers could include any of the following cell types.

Roles of three types of T-lymphocytes		
Cytotoxic T cells	*Memory T cells*	*Helper T cells*
Destroy any cells that carry foreign antigens, removing foreign protein such as bacteria or transplant tissue	Recognise any antigen when it reappears, enabling antibody production for the particular antigen	Secrete interleukens that regulate the function of both *cytotoxic T cells* and *B cell* function

Other T-lymphocytes could include: *inducer T cells*, *natural killer cells* and *suppressor T cells*.

23 (a) Such reproductive technology may have the ability to decrease the occurrence or even eliminate such harmful genes from a population. Screening for the breast cancer gene prior to gestation can reduce the chance of this disease developing in an individual and prevent the gene from being passed on from one generation to the next. This will reduce the burden on society for treatment and increase individual quality of life.

(b) A potential impact of using reproductive technology on agricultural animals such as sheep or cows can be to create genetic changes that can quite quickly become widespread within a population. Artificial insemination, which involves the injection/transfer of male semen into a female animal, can be used to promote particular or favourable characteristics, such as features that make the products of these animals more economically viable, e.g. greater butter fat in milk, better quality/grade of wool, etc. This can have an impact on the natural evolutionary trend of the animals, as these features may not have developed otherwise and may not necessarily be of survival advantage to the animals outside of the farm.

24 (a) Dependent variables: sweating and heart rate.

Independent variable: room temperature.

(b) The hypothalamus in the Central Nervous System.

(c)

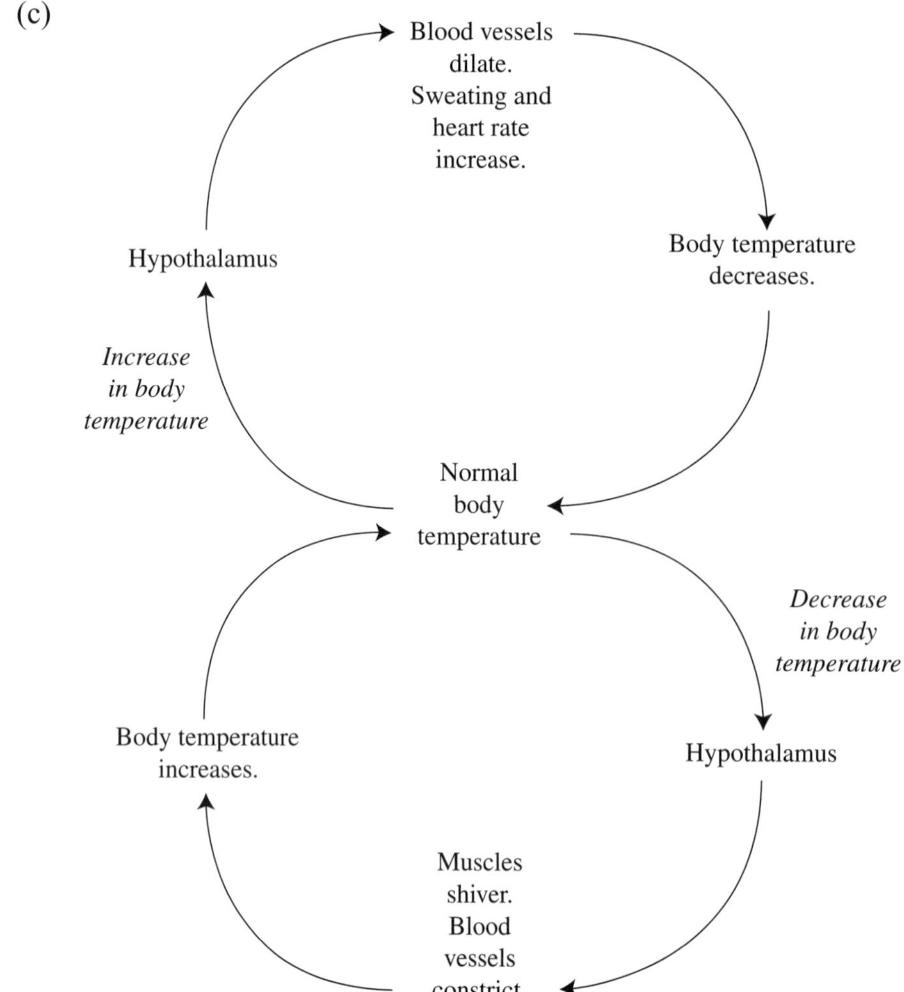

25 (a) Two defence adaptations used by the body in response to the injury are part of the inflammatory response. The injured area swells and reddens due to increased blood flow to the area causing the blood vessels in the area to dilate, confining the pathogen. Increased numbers of white blood cells in the area assist in destroying the pathogens introduced by the hook. Phagocytes can then actively ingest and destroy foreign material.

(b) Antibiotics can be used to manage the injury by fighting and destroying any disease-causing bacteria that may have been introduced by the hook and preventing their multiplication.

26 Enantiostasis is used to maintain metabolic and physiological function in response to variations in an organism's environment. Plants living in estuarine environments need to control the amount of salt entering them from their surroundings. This is done by plants such as Grey mangroves excluding salts from entering their roots through filtration. Other species excrete salts by concentrating salts in glands in the leaves. The released salt is then washed away by rain. Alternatively, salt can be accumulated in the leaves, then the leaves shed.

27 Genetics and the environment interact to determine the phenotype of an individual. While all characteristics are inherited from an offspring's parents, genetic variations can arise due to a number of factors: firstly, only a haploid number of chromosomes are present in any gamete, which is random; independent assortment of genes in gamete formation; mutation; crossing over—all of which result in a unique individual.

The expression of these genetic combinations (phenotype) can be affected by the environment the organism lives in. If conditions are not suitable or varied, then the genes may not be expressed correctly or revealed fully. For example, malnutrition can affect the gene for height, resulting in stunted growth. Insufficient water or nutrients in plants can affect the size of a plant or even cause 'stress flowering', where the reproductive cycle is artificially accelerated due to lack of water and the need to have pollination occur for species survival. The colour of hydrangea will vary depending on the soil composition, such as the concentration of lime or aluminium, or soil pH. Studies have shown that identical twins that have been separated at birth may develop their identical genotype differently due to differences in health or nutrition, resulting in different IQ, weight and height.

Hence phenotypic variation within an organism is the direct result of genetics and environmental interactions. So offspring do not necessarily have identical phenotypes to that of their parents, even though they may have similar genotypes.

Section II—Options

Question 28 — Communication

(a)

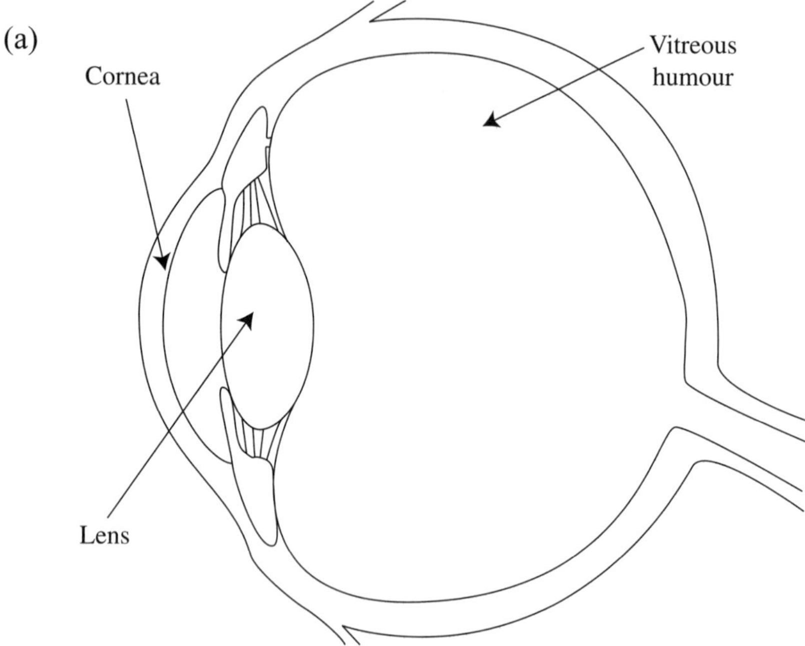

(b) (i) One technology that can be used to treat or prevent blindness is *cataract surgery*, whereby protein build-up in the lens is removed to improve eyesight, or the lens is replaced to restore vision.

(ii) The use of such technology would result in improved vision in society, particularly for the elderly, who are more prone to cataracts. This would enable this part of the population to have greater quality of life, remain in the workforce longer and decrease the burden to society, as they would be less reliant on social security or care from others.

(c) Environmental changes/stimuli for a neurone will affect the plasma membrane's permeability to ions, thereby changing the membrane's potential. Membrane potential will change as a response to stimulation, causing depolarisation if it is greater than –70 mV. Not all stimuli generate an action potential in the graph because the depolarisation is not strong enough (minimum change required is 15 mV) to generate an action potential. The threshold or amount of positive change in the membrane potential that is required before action potential is produced is not reached in a number of the stimuli in the graph.

(d) (i) Accommodation is important in visual communication to ensure that the lens in the eye can be adjusted such that the image produced is focused directly on the retina. If the light rays do not meet on the retina, then the image may form either before or after, resulting in a blurred image. This means that communication is not achieved as the message cannot be 'seen' or interpreted correctly by the brain.

(ii) The process of accommodation can be modelled through the use of a variety of convex/concave lenses and a ray box.

1 Place a lens of a particular size and shape a known position/distance away from a ray box with a four-slit/beam slide attached. Trace the beam produced by the ray box and the resulting pattern (focus point) as it passes through the lens, recording on a white sheet of paper.

2 Repeat using different sized and shaped lenses, one at a time, placing them the same position/distance away from the ray box. Record the pattern each time.

3 Repeat again using multiple lenses and record the patterns produced.

4 Compare these patterns by measuring the distance of the focus point in each situation and correlate to the sizes and shapes of the lens/es used. This will provide a model for how the refractive structures within the eye work together to produce a focused image on the retina.

(e) Understanding of hearing mechanisms has been used to develop technologies to overcome hearing difficulties.

- Firstly, the eardrum or tympanic membrane and ossicles are used to amplify sounds through the oval window and are easily damaged by loud sounds. This vibration increases the pressure waves that promote transmission through the fluid of the cochlea. If there is damage to these structures, hearing aids can be used to assist with the physical transmission of the vibration through the middle ear to the inner ear.
- Secondly, in the inner ear the cochlear contains the organ of corti, which has tiny hair cells that attach to the auditory nerve. Stimulation of these hair cells results in the transmission of sounds to the brain for interpretation of messages via the auditory nerve. If there is damage to this part of the ear, cochlear implants have been developed to augment the conversion of the mechanical vibrations into electrochemical impulses for transmission to the brain. This artificially determines the frequency and duration of the vibration and intensity of sound.

Hence, understanding hearing mechanisms has enabled the development of two very different technologies that can assist in the improvement of hearing, thereby increasing the success of communication for individuals and improving their quality of life.

Question 29 — Biotechnology

(a) (i) A *restriction enzyme* is used to cut DNA in recombinant DNA techniques. Different restriction enzymes cut the DNA in different places.

(ii) Restriction enzymes cut the DNA at particular sequences leaving 'sticky ends'. These sticky ends rejoin to other pieces of DNA with the corresponding base sequence. Ligases are enzymes. They are involved in the process of reattaching the cut DNA fragments to form a new piece of DNA.

(b) (i) Bread (or alcohol, yoghurt, cheese) was made prior to the 18th century.

(ii) One scientific discovery that has led to the expansion of the fermentation industry since the 18th century is *strain isolation*. This involves isolating the particular strain of a microorganism that facilitates the fermentation process and growing it in pure cultures. The technique can be used with bacteria, virus or fungi. Once identified as useful to a particular process, the microorganism can be cultured in the laboratory and its qualities 'enhanced' through experimentation. This could involve changing a particular variable, such as acidity, by progressive increments until it kills some of the microorganisms while resistant strains survive. The survivors will be cultured in increasing acid strength environments and periodically checked to ensure that they are still able to function for their primary purpose (e.g. fermentation). Over time, the strain becomes progressively more resistant to acidic conditions.

Strain isolation has also been used to increase the growth rate, and hence the quantities, of useful byproducts, such as is the case with a strain of penicillin. Fleming's initial cultures demonstrated the value of penicillin as a cure for infectious disease. However, it was not commercially viable because it couldn't be produced in sufficient quantities. Florey, Chain and several other microbiologists worked on producing a strain of penicillin using strain isolation techniques, e.g. bombardment with X-rays, or aerating in deep fermentation tanks, which greatly increased the output and availability of penicillin.

(c) (i) Messenger RNA (mRNA) and transfer RNA (tRNA).

(ii) *Location 1* is the site of *transcription*. Messenger RNA (mRNA) is involved in this process. mRNA is built up into a long single-strand molecule where each base of the RNA is complementary to the base in the corresponding DNA. The mRNA is small enough to pass through the nuclear pores and hence leave the nucleus. It will take its message through the cytoplasm and bond to the ribosome.

Location 2 is the site of *translation* in the ribosome. The mRNA is made of a sequence of triplets or codons (three bases) which are translated by the transfer RNA (tRNA). The three-base sequence on the tRNA is complementary to the mRNA (an anticodon). Once bonded to the mRNA, the tRNA then transfers its specific amino acid (which it has collected in the cytoplasm) to the end of the growing polypeptide chain. An enzyme is involved in linking the amino acids together so that each amino acid is detached from its parent tRNA and the tRNA leaves the ribosome to pick up another amino acid from the cytoplasm. The process of polypeptide synthesis continues until a 'stop code' is encountered on the mRNA strand. Once this occurs, the ribosome releases the amino acid chain.

(d) (i) The following method describes the process of extracting DNA from *onion cells*. (Other living cells may be used and the methods will differ slightly.)

1 First we need to break down the onion into its constituent cells so that we can extract the nuclear material. To do this, chop the onion into small pieces and place the pieces in a blender or food processor with half a cup of salty water. Strain the mixture into a beaker.

2 To dissolve some of the chemicals in the cell membranes, add some dishwashing liquid to the beaker and stir. Allow 5 minutes for the breakdown of the cell membranes. Stir periodically.

3 To separate the nuclear material from some of the larger chemicals in the cells, filter the mixture through a very fine filter, e.g. a coffee filter.

4 To further separate the DNA, add an alcohol, e.g. methylated spirits to the mixture. This should be done carefully to produce a separate alcohol layer that sits above the water layer. The DNA will not dissolve in alcohol, so it will precipitate into the interface between the water and the alcohol. It will appear as a white fluffy layer, which may be picked up using a toothpick or probe. This is onion DNA.

(ii) The DNA, once extracted from the living cells, looks like thick, whitish strands. These may be carefully lifted from the interface to reveal the characteristic long DNA strands.

(e) *Note:* A number of products that are useful to humans have been produced via applications of modern biotechnology. Several will be described below. When a plural (e.g. 'applications', 'products') is used in a question, it is important to ensure that you have more than one example in your answer.

Modern biotechnology has produced a number of beneficial products in the areas of forensic science, medicine, animal and plant biotechnology and aquaculture. In medicine, examples include the application of tissue engineering in organ transplants and skin grafts, gene delivery via nasal spray, and the synthetic production of insulin. Plant and animal biotechnology has produced monoclonal antibodies and recombinant vaccines, while aquaculture has been responsible for the production of pharmaceuticals and marine farming.

Tissue engineering is the process of producing functioning tissue from natural or synthetic sources. There is a significant shortage of organs available for transplant surgery and there are also problems associated with rejection and the need for immune-suppressants following transplant surgery. Tissue engineering can address both of these concerns. A scaffold is used, often made from a synthetic polymer, which is shaped into the particular desired organ, e.g. skin. The scaffold is then seeded with cells taken from the intended transplant recipient and growth factors are added. As the cells multiply they fill the scaffold, taking on the ultimate shape of the organ. Once the three-dimensional structure has been filled with cells, the structure can be transplanted into the recipient where the cells will take on their ultimate function. Blood vessels attach to the new tissue and supply the new cells while the synthetic scaffold breaks down and dissolves. The presence of the recipient's own cells reduces the risk of rejection. Different types of scaffolds have been used for different types of organs.

Another useful biotechnology application that has been useful to humans involves the production of *monoclonal antibodies*. Antibodies are a component of the immune response, the third line of defence against infectious, disease-causing organisms. Antibodies are specific, and confer an ongoing protection against the same antigens as a result of memory cells. As a result, antibodies have played a significant role in the effectiveness of vaccines. Antibodies can also be used to detect the presence of bacterial or viral products. However, the challenge to the effective use of antibodies is the production of large quantities of pure antibodies. Desired antibodies used to be collected from a laboratory host animal that had been injected with the particular antigen. Its blood would only yield small quantities of the desired antibodies as well as other undesirable substances.

A biotechnology solution to the problem is to identify natural cells that produce the desired antibodies and another type of cell that grows continuously (e.g. tumour cells). Biotechnology allows us to form a hybrid of these two cells, called a hybridoma. Effectively, we have a cell that grows and divides continually while retaining its antibody-producing function. It is an antibody production factory. These cells are called monoclonal as they originated from a single fused cell. An example of monoclonal antibody technology is the fusing of bone marrow tumour cells (myeloma), which can be grown in pure cell culture, with mammalian spleen cells (producing a specific antibody). The antibodies produced by these monoclonal cells are much purer and more effective than traditional drugs in combating disease. The antibodies attack only the specific antigens, reducing the risk of side effects.

Question 30 — Genetics: The Code Broken?

(a) (i) I^A, I^B and i.

(ii)

Group A IA i	Parents	Group B IB i

	IB	i
IA	IA IB	IA i
i	IB i	i i ← Group O child

(b) (i) Growing potatoes from potato tubers is an example of cloning. This gives genetically identical plants.

(ii) The offspring are all genetically identical. This can be of benefit to society if the parent plant chosen is of very good quality and well suited to the conditions of that area. All offspring can be expected to be of a similar high quality, ensuring high productivity for the farmer. If the farmer used seeds, the offspring, which are products of sexual reproduction, would all be different.

(c) *Location 1:* Transcription. The DNA molecule unzips and RNA nucleotides line up on their complementary bases on the DNA strand (A-T, C-G). These nucleotides then join to make a molecule of mRNA.

Location 2: The ribosome attaches to the end of a mRNA molecule. As it moves along it 'activates' a triplet of bases, which attract a tRNA molecule with a complementary triplet of bases. Each tRNA molecule carries an amino acid that is specific to its triplet of bases.

Location 3: The amino acids brought to the ribosome by the tRNA molecules combine via peptide bonds to make a polypeptide. The sequence of amino acids in this polypeptide is precisely determined by the sequence of bases in the original mRNA molecule.

(d) Linked genes are genes that are located on the same chromosome. For this reason their alleles do not 'sort out' independently of each other at gamete formation as predicted by Mendel's Law of Independent Assortment (which only applies to genes on different chromosomes).

Inheritance patterns of linked genes are different from those of unlinked genes, and this can be modelled using pipecleaners.

Example: For an organism of genotype Aa Bb, the pipecleaners can be marked to show alleles as follows, to represent chromosomes in the diploid parent.

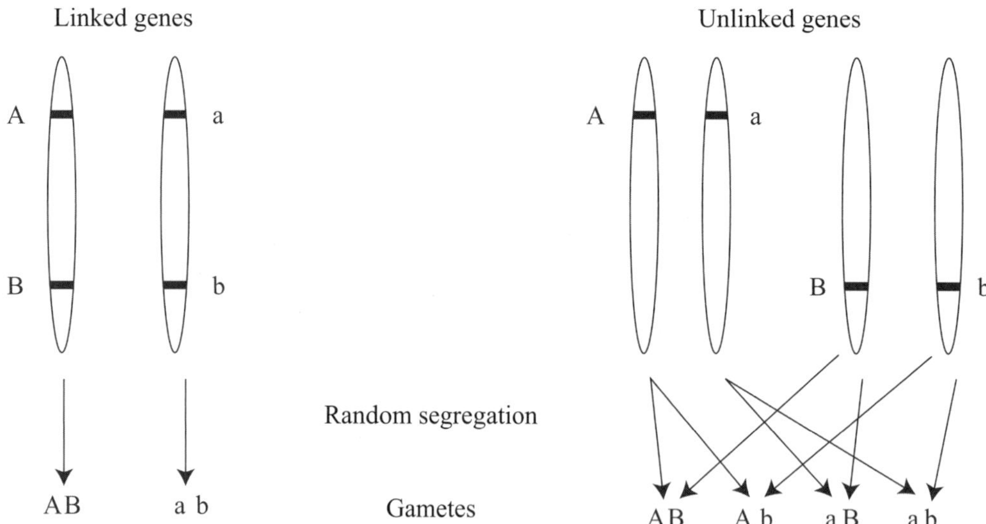

In the case of linked genes, modelling random segregation in meiosis (assuming no crossing over) would produce gametes, half of which would have the genotype AB, and the other half ab. In a cross between two individuals of the same genotype, a phenotype ratio of 3 : 1 would result.

In the case of the unlinked genes, modelling random segregation would give four different gamete genotypes as shown in the diagram. A cross between two parents of the same genotype would give a phenotype ratio of 9 : 3 : 3 : 1.

(e) A genetic mutation is a change to the DNA at the level of the gene (as opposed to the level of the chromosome). It can involve a base substitution, where a base is swapped, or a frame shift, where a base is deleted.

Since the precise base sequence of a gene determines the precise sequence of the amino acids in the polypeptide or protein produced, any change can lead to a change in that protein. If the protein is an enzyme, then that might affect body function, leading to diseases such as Tay-Sach's disease or cystic fibrosis. If it is a structural protein, then it can lead to a disease resulting from the compromising of that structure, such as spherocytosis, a condition involving misshapen red blood cells.

Knowledge of the cause of this type of disease allows us to devise treatments and management regimes that address the cause, rather than just treating the symptoms.

For example, traditional treatments for *cystic fibrosis* have involved massaging the chests of sufferers and having them inhale from nebulisers. This treats the symptom of mucus build-up in the lungs. A gene therapy treatment would involve them inhaling droplets containing the unmutated allele of the gene, carried by a suitable vector such as an adenovirus. This would be taken up the cells lining the lungs and allow those cells to produce the enzyme that they were previously unable to produce. Only through a precise knowledge of the genetic cause of the disease is such a management regime possible.

Another disease that could potentially be managed by gene therapy is the immunodeficiency disease *SCID*. It results from a genetic mutation that leads to the production of an enzyme. Knowledge of its genetic cause has allowed doctors to manage it by culturing stem cells that have taken up the unmutated allele and reintroducing them to the bone marrow of sufferers, allowing them to produce the enzyme and therefore produce functioning immune cells.

Question 31 — The Human Story

(a)

Classification level	*Human classification*	*Identifying feature*
Order	Primate	– Opposable thumb or toe
Family	Hominid	– Bipedalism
Genus	*Homo*	– Large brain (relative to body size) – Prominent chin

(b) (i) Examples: Radioactive decay dating (C^{14} dating) and stratigraphy.

(ii) The fossil record can give large amounts of knowledge but it can be misleading or incomplete.

The fossil record can be incomplete due to a paucity of fossils. Not all organisms that die are fossilised. Organisms without hard body parts are not fossilised. The conditions for fossilisation are not always present when an organism dies. Often fossils are disturbed due to predator activity or natural disasters such as earthquakes. The sediments containing fossils may be eroded away and lost. These all lead to gaps in the fossil record.

Different scientists have different ideas, which can affect their interpretations of the fossils. The same evidence can be interpreted in different ways because of preconceived ideas or religious or cultural beliefs resulting in conflicting theories about the course of evolution.

(c) The *Out of Africa Theory* proposes an initial migration of *Homo erectus* from Africa, followed by a more recent second wave of hominids migrated from Africa replacing species like *Homo neanderthalensis*. This culminated in the evolution of the *Homo sapiens* species. The evidence for this theory comes from fossil and genetic sources.

The oldest fossil of *Homo sapiens* has been found in Africa. Ancient skeletal remains of *Homo sapiens sapiens* have been found in South Africa and dated to between 74 000 and 115 000 years old. In Africa there are also transitional fossils classified as *Homo sapiens idaltu* linking pre-modern (archaic) and modern humans. This suggests that humans evolved in Africa.

Genetic evidence, specifically by comparing mtDNA of modern human populations and using the consistent rate of mutations, has led to a model of a ‘molecular clock’. This evidence suggests that through maternal inheritance of mtDNA the lineage can be traced back to Africa and a single female ancestor, ‘mitochondrial Eve’. This again suggests that the evolution of modern humans began in Africa.

The *Multiregion Theory* states that there was local evolution in different regions, from *Homo erectus* to *Homo sapiens*. This occurred simultaneously many times all over the world and in each region.

Evidence for this theory comes from fossils found outside Africa and dated before the proposed second wave out of Africa. Examples of these fossils include ‘Mungo Man’ found in southern NSW and dated at 60 000 years old and ‘Java Man’ found in Indonesia.

Genetic evidence from mtDNA studies of 'Mungo Man' showed that its mtDNA lineage has no descendants alive today. This suggests that this fossil, essentially anatomically identical to modern *Homo sapiens*, is not descended from 'mitochondrial Eve'. It supports the idea that evolution from *Homo erectus* to *Homo sapiens* occurred simultaneously in regions outside Africa.

(d) (i) Examples: *Australopithecus africanus* and *Homo habilis*.

(ii) *Similarities:*

- Bipedal due to the position of the foramen magnum and an S-shaped spine
- Opposable thumb resulting in a precision grip
- Flattening of the muzzle area

Differences:

- Cranial *capacity of Australopithecus africanus* (420–490 cc) smaller than that of *Homo habilis* (500–800 cc)
- Age of *Australopithecus africanus* fossils (3.2 to 2.5 million years) greater than age of *Homo habilis* fossils (2.2 to 1.6 million years)
- Front teeth of *Australopithecus africanus* relatively large with the canine teeth not projecting beyond the others; reduced tooth size in *Homo habilis*.

(e) Biological evidence such as DNA–DNA hybridisation and karyotype analysis can be used to improve our understanding of the evolution of humans.

In *DNA–DNA hybridisation* the following steps are used to collect the evidence:

1 Double-stranded DNA from human and primate samples is denatured using temperatures around 85°C.

2 Denatured DNA from two different species is then mixed. This is then cooled.

3 As it cools the DNA spontaneously forms double strands, many of which will be hybrids composed of one strand from each species.

4 These hybrids are then reheated, and the temperature at which this DNA denatures is compared with that of the original sample to identify the degree of similarity. The greater the temperature difference, the less similar the DNA strands.

Using this technology, chimpanzee–human hybrid DNA has been shown to denature at only 1°C lower than that of pure human DNA, and this has been shown to correspond to a 97.6% similarity in base sequence. This has led to the conclusion that chimpanzees and humans shared an ancestor more recently than other apes.

Karyotype analysis technology involves growing cells, outside the body, stopping them during metaphase or prophase, then staining and photographing them. Chromosomes are compared between species for number, position of banding, size, and structure of the banding patterns. If the patterns of the stained chromosomes of two or more organisms appear identical, then it is assumed that these chromosomes are homologous and have been retained from one or more common ancestors.

Karyotype analysis has revealed that the chromosomes of chimpanzees, gorillas and humans are very closely related to each other and that those of the orangutan are the most different. Such evidence leads to the proposal that chimpanzees, gorillas and humans shared an ancestor more recently than orangutans.

CHAPTER 9

BOARD OF STUDIES
NEW SOUTH WALES

2010
HIGHER SCHOOL CERTIFICATE EXAMINATION

Biology

General Instructions

- Reading time – 5 minutes
- Working time – 3 hours
- Write using black or blue pen
- Draw diagrams using pencil
- Board-approved calculators may be used
- Write your Centre Number and Student Number where required

Total marks – 100

Section I

75 marks

This section has two parts, Part A and Part B

Part A – 20 marks

- Attempt Questions 1–20
- Allow about 35 minutes for this part

Part B – 55 marks

- Attempt Questions 21–30
- Allow about 1 hour and 40 minutes for this part

Section II

25 marks

- Attempt ONE question from Questions 31–35
- Allow about 45 minutes for this section

Section I
75 marks

Part A – 20 marks
Attempt Questions 1–20
Allow about 35 minutes for this part

Use the multiple-choice answer sheet for Questions 1–20.

---

1 Which of the following provides evidence that supports the theory of evolution?

(A) Clones

(B) Punnet squares

(C) The fossil record

(D) The DNA double helix

2 Which method did Rosalind Franklin use during her investigations of the structure of DNA?

(A) Protein sequencing

(B) Monohybrid crosses

(C) Computer modelling

(D) X-ray crystallography

3 Cilia prevent the entry of pathogens into the human body by

(A) providing a protective body covering.

(B) producing secretions toxic to pathogens.

(C) moving trapped pathogens to the mouth.

(D) increasing circulation of blood to the infected area.

4 Which of the following correctly compares the urine of marine and freshwater fish?

(A) Similar volumes of urine are produced by freshwater and marine fish.

(B) Smaller volumes of urine are produced by freshwater fish than marine fish.

(C) The urine produced by freshwater fish is less concentrated than that of marine fish.

(D) The urine produced by both freshwater and marine fish is similar in concentration.

5 The gamete plays an important role in sexual reproduction because it carries

(A) genetic information from both parents.

(B) half the genetic information of the parent.

(C) all of the genetic information of the parent.

(D) double the genetic information of the parent.

6 An increase or decrease in the salt concentration in interstitial fluids triggers a response so that the concentration returns to a set value.

What is this mechanism called?

(A) Diffusion

(B) Homeostasis

(C) Enantiostasis

(D) Positive feedback

7 Koch contributed to an understanding of disease by developing

(A) a method to link a particular pathogen to the cause of a disease.

(B) an experiment that disproved the theory of spontaneous generation.

(C) a method of killing microbes by heating and thus preventing decay.

(D) an immunisation program based on a knowledge of immune responses.

8

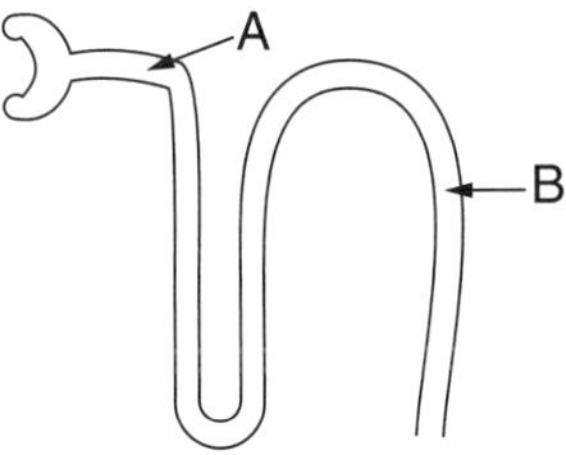

A diagram of a nephron

What happens to the concentrations of glucose, urea and protein as fluid moves from A to B in the nephron?

	Glucose Concentration	*Urea Concentration*	*Protein Concentration*
(A)	Unchanged	Decreases	Decreases
(B)	Unchanged	Decreases	Increases
(C)	Decreases	Increases	Increases
(D)	Decreases	Increases	Unchanged

9 Experiments were carried out to show the effects of pH and temperature on enzyme activity. The experiments also tested the effects of a chemical called an inhibitor. The results are shown in the graphs.

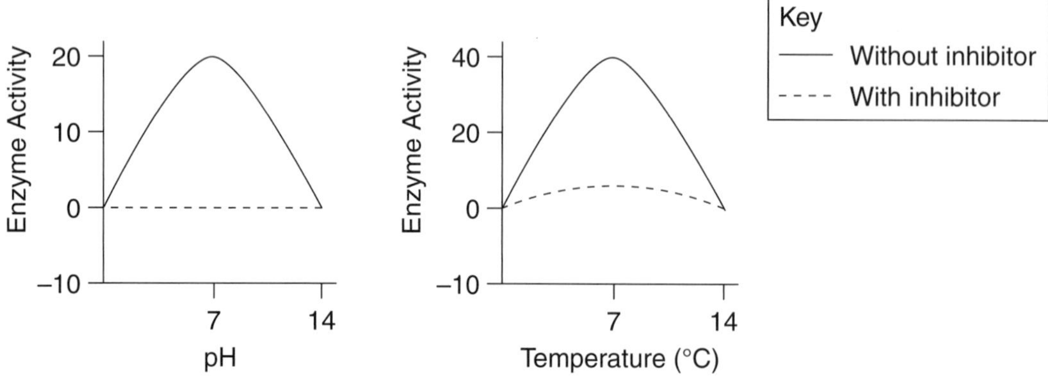

The best conclusion that can be drawn from these results is that the inhibitor affects

(A) pH.

(B) temperature.

(C) enzyme activity.

(D) enzyme concentration.

10 An experiment was conducted to test the effect of dissolved carbon dioxide (CO_2) on the pH of water. It was found that the pH of water decreased as CO_2 was added.

How do these findings relate to the acidity of blood as it circulates in the body?

(A) Blood in the veins of muscles becomes less acidic.

(B) Blood in the arteries of muscles becomes less acidic.

(C) Blood in the capillaries of the brain becomes less acidic.

(D) Blood in the capillaries of the lungs becomes less acidic.

11 How can widespread use of artificial insemination alter the genetic composition of a population?

(A) It results in many genetically identical individuals.

(B) It makes certain alleles more common in a population.

(C) It decreases the number of chromosomes in some individuals.

(D) It ensures that only the genetic composition of the males is altered.

12 How are lipids transported in mammalian blood?

(A) They are attached to proteins.

(B) They are dissolved in the blood.

(C) They are part of the membrane of red blood cells.

(D) They are attached to haemoglobin in red blood cells.

13 Which of the following shows DNA replication in the correct order?

(A) Two DNA double helices → strands separate → matching bases pair up → DNA double helix

(B) DNA double helix → strands separate → matching bases pair up → two DNA double helices

(C) Strands separate → two DNA double helices → matching bases pair up → DNA double helix

(D) DNA double helix → strands separate → two DNA double helices → matching bases pair up

14 A pedigree is shown.

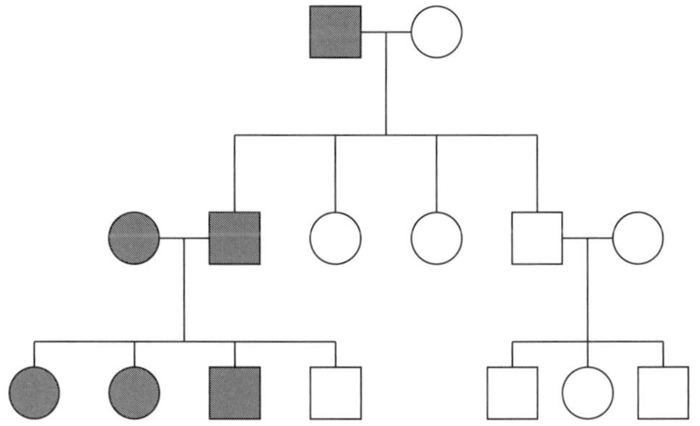

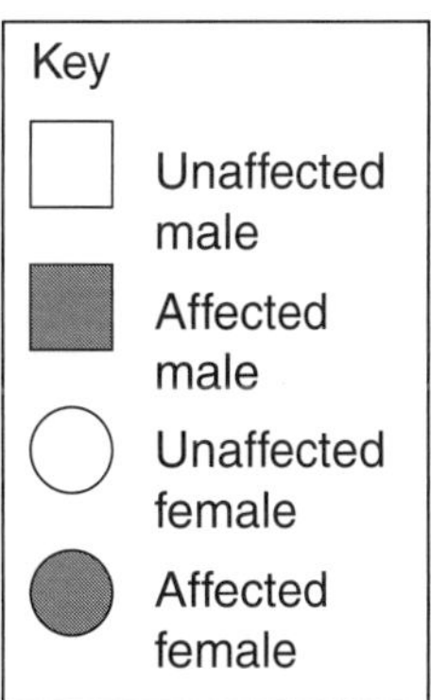

What type of inheritance is shown in the pedigree?

(A) Sex-linked recessive

(B) Sex-linked dominant

(C) Non sex-linked recessive

(D) Non sex-linked dominant

15 A rainforest vine grows up a tree to reach the top. Near the ground, the leaves are small and oval. At the top, the leaves are large and round.

This change in leaf size is an example of

(A) variation in a species.

(B) a commensal relationship.

(C) a co-dominant gene relationship.

(D) environment affecting phenotype.

16 The following events occur after DNA is subjected to radiation. The events are listed in no specific order.

P: Change in protein structure
Q: Change in polypeptide sequence
R: Change in cell activity
S: Mutation

What is the correct sequence of steps?

(A) S, P, Q, R

(B) S, Q, P, R

(C) R, Q, S, P

(D) R, S, Q, P

17 What feature of prions distinguishes them from all other types of pathogens?

(A) Prions are not cells.

(B) Prions do not contain DNA.

(C) Prions do not contain nucleic acids.

(D) Prions cannot reproduce outside a host cell.

18 A model of a virus is shown.

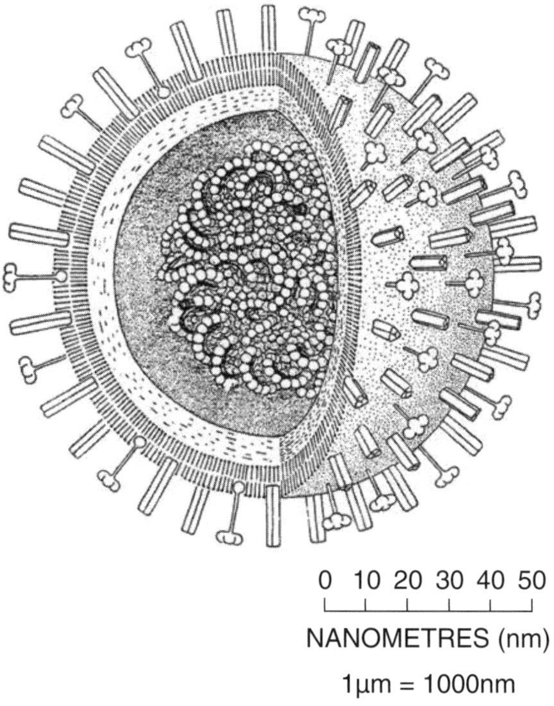

What is the approximate diameter of this virus?

(A) 13 cm

(B) 13 nm

(C) 130 μm

(D) 0.130 μm

19 Which of the following observations helped develop a model for the transmission of malaria?

(A) Immunisation against the pathogen prevented transmission of malaria.

(B) Use of antibiotics decreased the incidence of malaria in the community.

(C) After transmission, B cells in the infected individual produced antibodies against malaria.

(D) After swamps were drained there was a major decline in the numbers of individuals catching malaria.

20 The diagram illustrates an immune response.

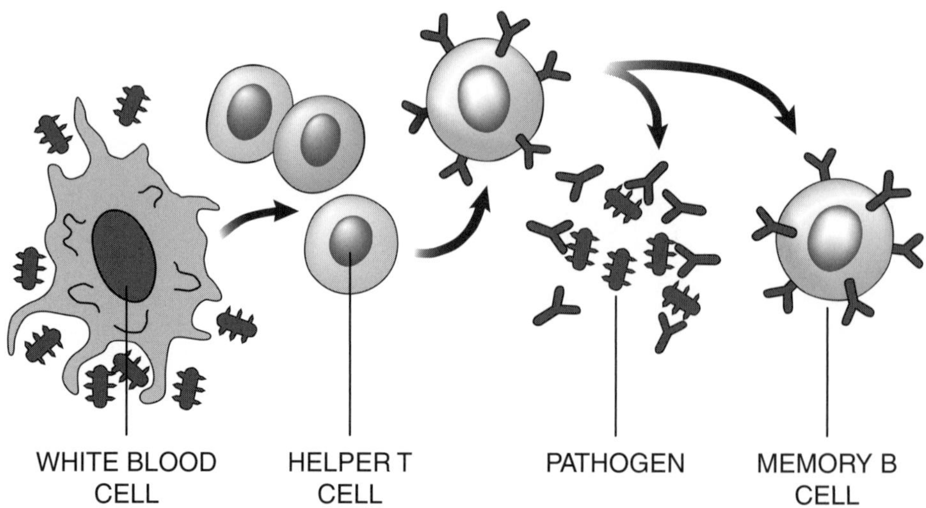

Below is a list of statements, each describing a step in the immune response.

1. Antibodies are produced to immobilise the pathogens.
2. B cell is activated by a helper T cell.
3. Helper T cells are activated by the white blood cell.
4. Memory B cell is ready to respond to further infections.

What is the correct sequence of events?

(A) 2, 3, 1, 4

(B) 2, 3, 4, 1

(C) 3, 2, 1, 4

(D) 3, 4, 2, 1

2010 HIGHER SCHOOL CERTIFICATE EXAMINATION

Biology

Centre Number

Section I (continued)

Student Number

Part B – 55 marks
Attempt Questions 21–30
Allow about 1 hour and 40 minutes for this part

Answer the questions in the spaces provided. These spaces provide guidance for the expected length of response.

Question 21 (2 marks)

Gregor Mendel and Thomas Morgan both used breeding experiments to deduce fundamental principles of genetics.

Complete the four blank boxes in the table. **2**

	Mendel's Monohybrid Cross	Morgan's Fruit Fly Experiments
First Cross Parents Phenotype	tall × short	red eyed female × white eyed male
First Cross (F_1) Parents Genotype		$X^R X^R \times X^r Y$
First Cross Punnet Square	<table><tr><td></td><td>t</td><td>t</td></tr><tr><td>T</td><td>Tt</td><td>Tt</td></tr><tr><td>T</td><td>Tt</td><td>Tt</td></tr></table>	
F_1 Phenotype		

Question 22 (6 marks)

The following data were recorded about the effectiveness of antimalarial drugs for treating malaria.

	Effectiveness of drug (%)	
Year	*Mefloquinine*	*Quinine*
1976	100	90
1978	100	85
1980	100	80
1984	100	72
1988	90	64
1992	70	58

(a) Graph the data on the grid. **3**

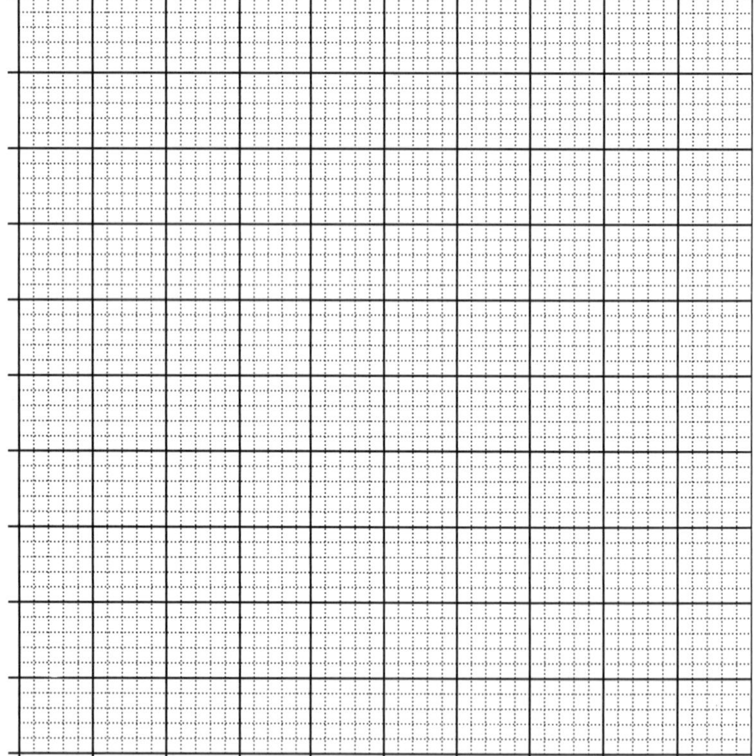

Question 22 continues

Question 22 (continued)

(b) Use these data to explain the impact of human processes on biodiversity. **3**

....

....

....

....

....

....

End of Question 22

2010 HIGHER SCHOOL CERTIFICATE EXAMINATION

Biology

Section I – Part B (continued)

Centre Number

Student Number

Question 23 (7 marks)

(a) Use an example to explain why hybridisation within a species is carried out. **2**

....

....

....

....

(b) Use an example of a named transgenic species to discuss the social and environmental impact of this technology. **5**

....

....

....

....

....

....

....

....

....

....

....

Question 24 (5 marks)

Design an experiment that tests the effect of opening a window on the blood oxygen saturation level of the people in a room. 5

Independent variable	....
Control	....
Variables to be kept constant and justification for keeping each constant.	Variable 1: .... Justification: ....
	Variable 2: .... Justification: ....
Technology used to measure oxygen saturation in blood	....

2010 HIGHER SCHOOL CERTIFICATE EXAMINATION

Biology

Section I – Part B (continued)

Centre Number

Student Number

Question 25 (5 marks)

(a) Justify your choice of equipment or resources to perform a first-hand investigation to draw a longitudinal section of xylem tissue. **2**

....

....

....

....

(b) Draw a diagram of a longitudinal section of xylem tissue and label ONE characteristic feature. **3**

Question 26 (5 marks)

An epidemiological study of lung cancer, cigarettes and death rates produced the following data.

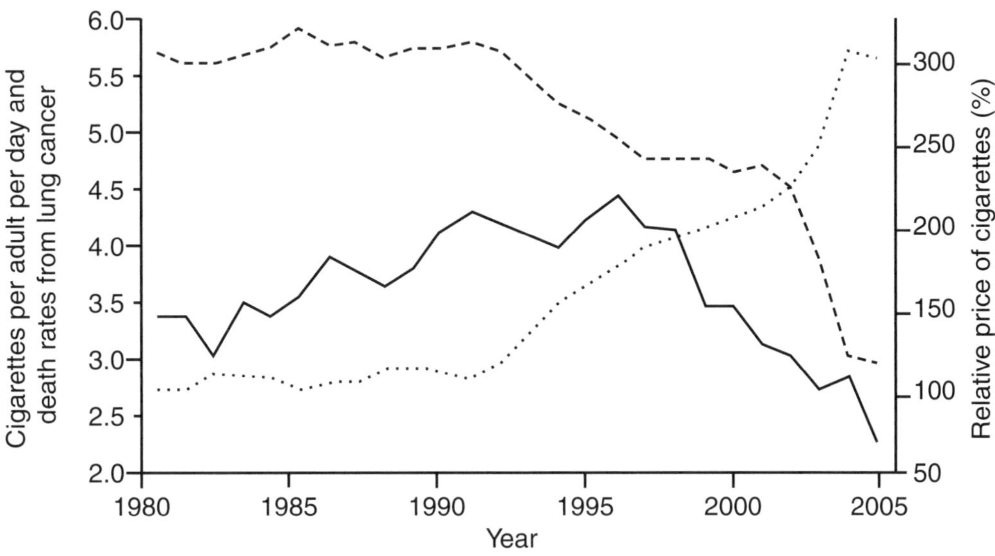

- - - - - - Cigarettes per adult per day
——— Lung cancer death rates per 100,000 (divided by 4): men aged 35–44
·········· Price of cigarettes relative to 1980 (%)

Using these data alone, you are asked to report on the likelihood of reducing lung cancer by increasing the price of cigarettes. Present your evaluation including limitations of the information provided. **5**

2010 HIGHER SCHOOL CERTIFICATE EXAMINATION

Biology

Centre Number

Section I – Part B (continued)

Student Number

Question 27 (3 marks)

(a) Outline examples of effective quarantine regulations in Australia that protect animal and plant health. Give one example for each.

(i) animal health **1**

....

....

....

(ii) plant health **1**

....

....

....

(b) Using one of your examples, explain why this method is effective. **1**

....

....

....

Question 28 (8 marks)

Organ transplants may trigger an immune response which can lead to organ rejection.

The diagram below represents a model of two heart cells, one from a transplant recipient and one from a donor.

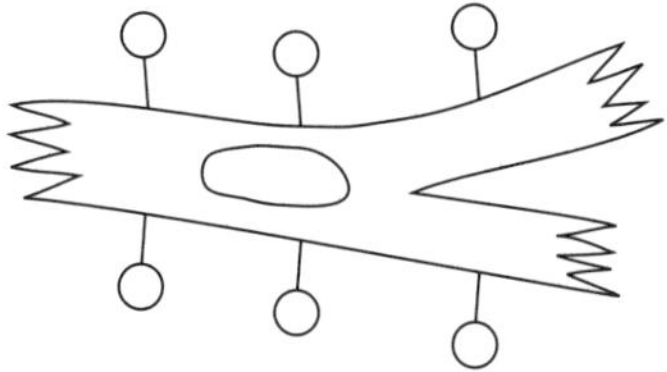

Recipient heart cell

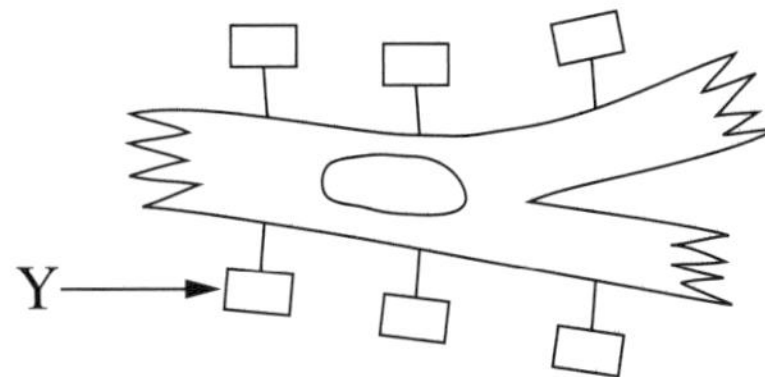

Donor heart cell

(a) What does Y represent? **1**

....

(b) Assess the effectiveness of the model in explaining the cause of organ rejection in a transplant recipient. **3**

....

....

....

....

....

....

(c) Name and outline the role of TWO types of T lymphocytes in organ rejection. **4**

....

....

....

....

....

....

Question 29 (7 marks)

You have been asked to write a report on the responses of plants to temperature change.

You find three sources of information.

Article 1 from **Science News** – a web based site reporting on current science

> Plants are highly sensitive to changing temperatures and can coordinate an appropriate response for variations as little as 1 °C, according to a new finding that can help explain how plants will respond to climate change. Plants not only 'feel' the temperature rise, but also respond by activating hundreds of genes and deactivating others. This offers new leads in the quest to create crop plants better able to withstand high temperature stress.
>
> Plants can sense, formulate reaction to temperature change. By IANS 11 January 2010 http://blog.taragana.com/science/2010/01/11/plants-can-sense-formulate-reaction-to-temperature-change-3207/

Article 2 from **Climate Research** – a scientific journal

> We analysed the flowering times of a violet and a tree at similar latitudes in the UK and Poland over 26 years. Careful analysis showed that both species in both locations showed significant responses to temperature variation although plants in the UK were more responsive than those in Poland. We conclude that locally adapted species may differ in their projected flowering times under future climate warming.
>
> Modified abstract from Tryjanowski P, Panek M, Sparks T (2006). Phenological response of plants to temperature varies at the same latitude: case study of dog violet and horse chestnut in England and Poland. *Climate Research*, 32 (1): 89-93.

Article 3 from **Wikipedia** – an online user-modifiable encyclopedia

> Increases in CO_2 concentration affect how plants photosynthesise[1]. Increased CO_2 can also lead to increased carbon to nitrogen ratios in the leaves of plants, possibly changing herbivore nutrition[2].
>
> 1. Steffen, W & Canadell, P (2005). 'Carbon Dioxide Fertilisation and Climate Change Policy.' 33 pp. Australian Greenhouse Office, Department of Environment and Heritage: Canberra
>
> 2. Gleadow RM, et al (1998). Enhanced CO_2 alters the relationship between photosynthesis and defence in cyanogenic *Eucalyptus cladocalyx F. Muell. Plant Cell Environ.* 21: 12–22.
>
> http://en.wikipedia.org/wiki/Effect_of_climate_change_on_plant_biodiversity

Question 29 continues

Question 29 (continued)

(a) From these articles, identify two responses of plants to temperature change. **2**

....

....

....

....

(b) Evaluate the relevance of the information to your report, and the reliability of each of the sources given. **5**

....

....

....

....

....

....

....

....

....

....

....

....

....

....

End of Question 29

Question 30 (7 marks)

Geological and biological history of New Zealand

Event	*Time*
Australia and New Zealand separated	85–65 million years ago
New Zealand drifted east and subsided, its land mostly under seawater (most fossils are marine)	85–22 million years ago
Mammals became abundant worldwide	60 million years ago
Earliest migratory bird fossils	55 million years ago
New land created by volcanoes in New Zealand	22 million years ago to present
Many new, unique species of birds appear in the fossil record	20 million years ago to present
Islands completely devoid of mammals. Birds occupied niches that were usually occupied by mammals	700 years ago

Use this information and other relevant knowledge to demonstrate how the practice of biology has led to the validation of current theories of evolution. **7**

....

....

....

....

....

....

....

....

....

....

....

Question 30 continues

Question 30 (continued)

End of Question 30

2010 HIGHER SCHOOL CERTIFICATE EXAMINATION

Biology

Section II

25 marks
Attempt ONE question from Questions 31–35
Allow about 45 minutes for this section

Answer parts (a)–(c) of the question in a writing booklet. Answer parts (d)–(e) of the question in a SEPARATE writing booklet. Extra writing booklets are available.

Question 31	Communication
Question 32	Biotechnology
Question 33	Genetics: The Code Broken?
Question 34	The Human Story
Question 35	Biochemistry *(Not included in this reproduction)*

Question 31 — Communication (25 marks)

Answer parts (a)–(c) in a writing booklet.

(a) Construct a table to identify the structures used by insects, fish and mammals to detect vibrations. **3**

(b) The vocal folds are different when a person sings a high pitched note and a low pitched note. Draw TWO labeled diagrams to illustrate this difference. **4**

(c) On the cross-section of the eye and the graph, corresponding retina locations are indicated according to their angle from the fovea.

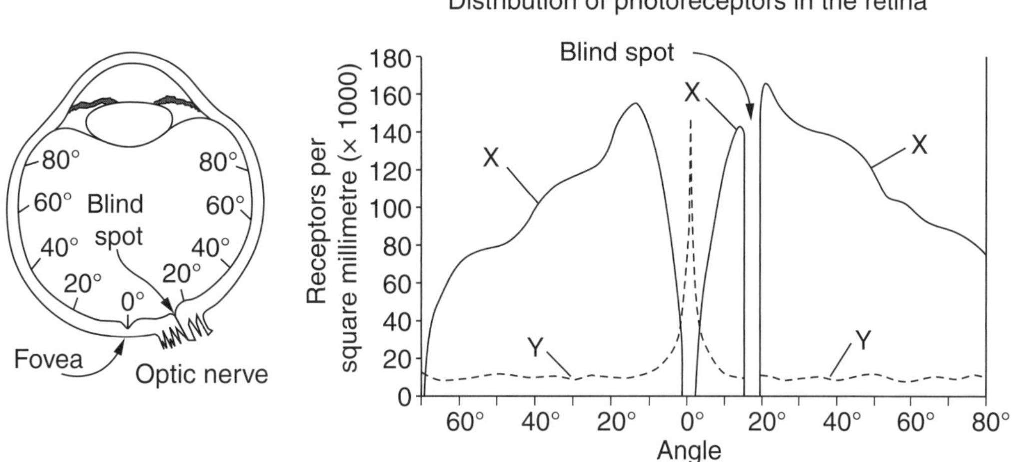

(i) What label should be given to the line Y? **1**

(ii) Explain why the structure of cones varies depending on their location in the retina. **2**

(iii) Outline the role of rhodopsin in rods. **2**

Question 31 continues

Question 31 (continued)

Answer parts (d)–(e) in a SEPARATE writing booklet.

(d) In your course, you did a first-hand investigation on a mammalian brain similar to the one shown.

The brain shown is from a mammal that was brought into a vet surgery after surviving a fall. Brain testing showed no action potentials occurring in region X.

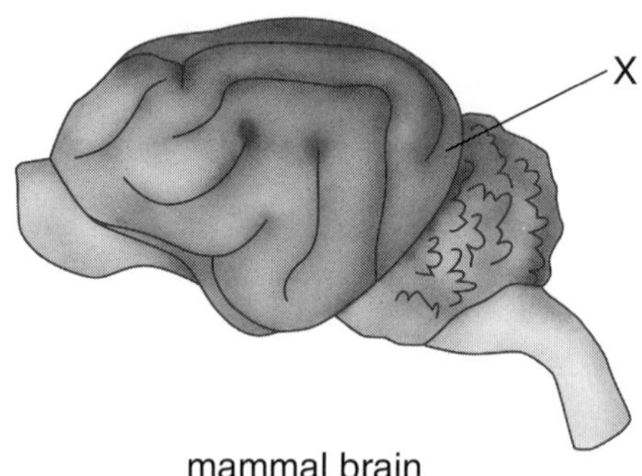

mammal brain

(i) Explain TWO possible causes for the lack of action potentials in region X. **4**

(ii) Outline how this condition could change the behaviour of the mammal. **2**

(e) The following article was found in a newspaper.

> **Movie experience just gets better and better –**
> **3D movies, 3D glasses and surround sound systems**
>
> In a 3D movie, two different images are projected onto the same screen. 3D glasses are worn by the audience to ensure that one image is seen by one eye and the other image is seen by the other eye.
>
> Surround sound systems allow sounds to be produced in different areas of a cinema so that members of the audience feel like they are at the centre of the action that is displayed on the screen.
>
> These new technologies use our knowledge of depth perception, sound shadows and how sights and sounds are received, transmitted and then interpreted by the brain to improve our movie experience.

Evaluate how our understanding of the eye and the ear has led to the development of technologies such as those above. **7**

End of Question 31

Question 32 — Biotechnology (25 marks)

Answer parts (a)–(c) in a writing booklet.

(a) Construct a table to identify an industrial fermentation process, the microorganism used, and a product of the process. **3**

(b) Draw a flowchart showing the sequence of events that results in the formation of recombinant DNA. **4**

(c) A farmer raised some animals. Offspring from each generation were chosen for crosses for the next generation. The farmer collected data which are shown in the table.

Generation number	*Breeding males*	*No. of offspring*	*No. of early deaths*	*Number of male animals surviving to breeding age*
1	non-brown	53	27	2 non-brown 8 brown
2	brown	45	12	3 non-brown 15 brown
3	brown	62	10	1 non-brown 27 brown

(i) Outline the form of biotechnology used by the farmer. **1**

(ii) Using the farmer's data, assess the effectiveness of this use of biotechnology. **4**

Question 32 continues

Question 32 (continued)

Answer parts (d)–(e) in a SEPARATE writing booklet.

(d) Fire and Mello discovered a cellular process now known as RNA interference. In this process, small nucleotides bind to specific mRNA sequences causing their destruction. This discovery is revolutionising molecular therapeutics for devastating diseases.

(i) Explain how protein synthesis is affected by RNA interference. **2**

(ii) Describe the roles of DNA and RNA in an application of biotechnology. **4**

(e) Assess how the development of new biotechnologies has led to ethical and social issues. **7**

End of Question 32

Question 33 — Genetics: The Code Broken? (25 marks)

Answer parts (a)–(c) in a writing booklet.

(a) Construct a table to identify how each of the following mutations affects chromosome number in an organism: **3**

Trisomy, Polyploidy, Base Substitution.

(b) A somatic cell has a diploid number of 4. **4**

Draw diagrams that show the similarities and differences between the chromosomes in:

- this diploid cell and
- a haploid cell which could result from meiotic division of this cell.

(c) The pedigrees show the inheritance of two genetic disorders in the same family. Person 8 is not a carrier of the vision defect.

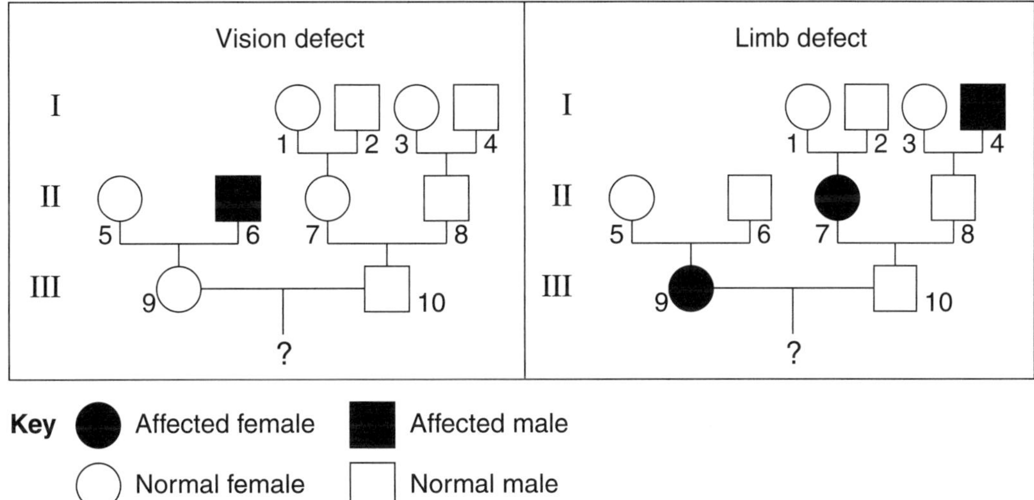

(i) Predict whether genes for each of these defects are dominant or recessive. **1**

(ii) Individuals 9 and 10 have requested genetic counselling. **4**

Predict the possible ratios of the phenotypes of their children if the genes were linked. Predict the ratios if they were not linked. Show working.

Question 33 continues

Question 33 (continued)

Answer parts (d)–(e) in a SEPARATE writing booklet.

(d) (i) Explain how data can be collected and analysed to identify the relative position of linked genes. **3**

(ii) Give THREE reasons why the human genome project could not be achieved by studying linkage maps. **3**

(e)

New Artifical Life Form Produced!

Recently, knowledge of gene cloning and gene cascades has been used to produce a whole artificial chromosome. This chromosome was inserted into a bacterium and resulted in surviving, reproducing bacteria.

Evaluate how our understanding of gene cloning and gene cascades has led to the development of new applications for technologies such as the one shown above. **7**

End of Question 33

Question 34 — The Human Story (25 marks)

Answer parts (a)–(c) in a writing booklet.

(a) Construct a table to identify a feature of humans which classifies them as chordates, mammals and animals respectively. **3**

(b) Draw a diagram of a prosimian hand/foot and a human foot. Label two corresponding features of each hand/foot that demonstrate the differences between these primates. **4**

(c) The diagram compares the inheritance of nuclear DNA with mitochondrial DNA.

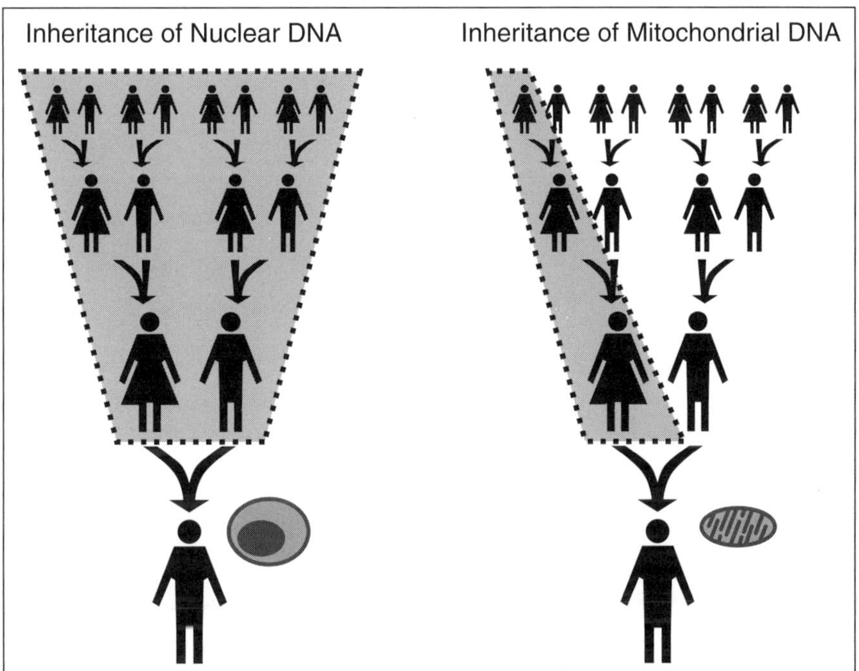

(i) Compare the pattern of inheritance of nuclear DNA with mitochondrial DNA. **1**

(ii) Assess whether the analysis of nuclear DNA or mitochondrial DNA is more useful to evolutionary biology. **4**

Question 34 continues

Question 34 (continued)

Answer parts (d)–(e) in a SEPARATE writing booklet.

(d) The graph below shows the results of an investigation into cranial capacity of fossil hominins.

- Data were obtained from measurements of hominin cranial capacity published in scientific journals
- Each data point on the graph represents the average cranial capacity of the adult skull at a single archaeological dig
- The results from 215 archaeological digs are represented

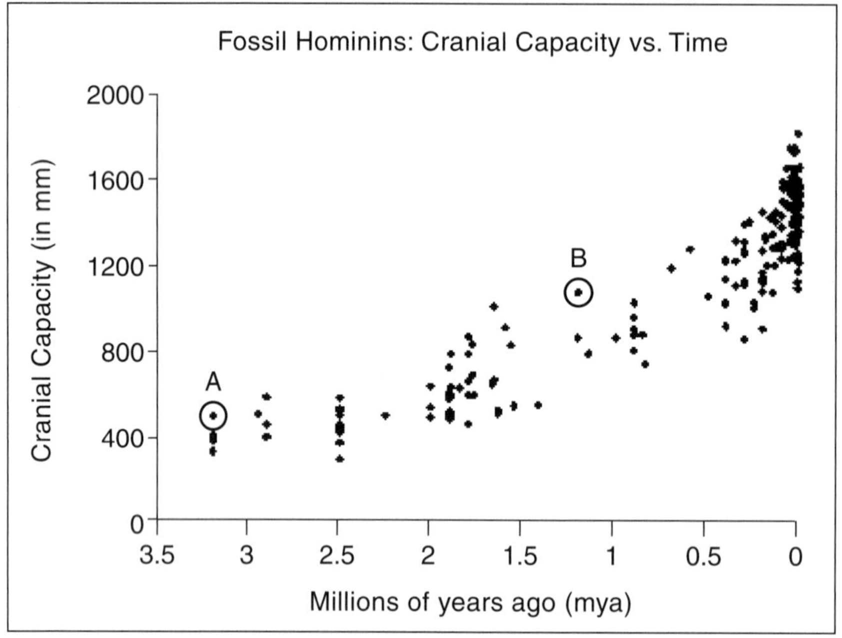

(i) Justify the methods used to collect and present the data. **4**

(ii) For either data point A or B propose the name of the hominin species and its regional location. **2**

(e) Assess the significance for future evolution of the similarities and differences in cultural development between humans and other primates. **7**

End of paper

2010 HSC Examination Paper

Sample Answers

Section I Part A (Total 20 marks)

1 C The fossil record provides evidence of development from simple to more complex organisms.

2 D X-ray crystallography was the technique used by Rosalind Franklin to identify the helical shape of DNA.

3 C Cilia form part of the first line of defence for preventing the entry of foreign pathogens into the body by removing the pathogens from the trachea to the mouth.

4 C Urine of freshwater fish is more dilute as they tend to gain water by osmosis from their surroundings and as such have an excess of water in their body. Marine fish tend to lose water to their surroundings and as such have concentrated urine.

5 B A gamete has half the number of chromosomes of each parent so that when male and female gametes unite in fertilisation a full complement is restored.

6 B Homeostasis is the term used to describe the process whereby internal conditions are kept constant.

7 A Koch contributed a method known as 'Koch's postulates' that enabled systematic identification of a pathogen.

8 D The concentrations of proteins will remain unchanged. Glucose will be reabsorbed and the urea will increase as the blood passes through a nephron.

9 C The inhibitor has significant effects on enzyme activity with both factors represented (i.e. pH and temperature).

10 D As acidic blood travels to the lungs in arteries (from the muscles to the heart in veins and then in arteries to the lungs), it loses its carbon dioxide to the air in the capillaries of the lungs and so becomes less acidic.

11 B Artificial insemination will make desired characteristics more prevalent in a population as greater numbers of stock will possess those characteristics and can pass them on.

12 A Lipids are transported in blood as lipoproteins.

13 B This is the only correct sequence indicating unzipping and complementary base pairing resulting in identical copies of DNA.

14 D Non sex-linked as only one male offspring in the first generation possesses the characteristic. Dominant because one male offspring in the second generation does not possess the characteristic even though both his parents are affected.

15 D Change in leaf size is an example of environment affecting phenotype as the leaves higher up are more readily able to photosynthesise because of proximity to light.

16 B This is the only correct sequence indicating mutation and change as a result of radiation and consequently change to cell activity.

17 C Nucleic acids make up DNA, so C is the better option.

18 D The diameter of the virus measures 130 nm or 0.130 μm.

19 D This observation led to a link between the breeding of mosquitoes and stagnant water.

20 C This option gives the correct sequence for the action of antibody mediated response.

Section I Part B

21

	Mendel's Monohybrid Cross	Morgan's Fruit Fly Experiments
First Cross Parents Phenotype	tall × short	red eyed female × white eyed male
First Cross (F1) Parents Genotype	TT × tt	$X^R X^R \times X^r Y$
First Cross Punnet Square	(t, t / T: Tt, Tt / T: Tt, Tt)	(X^r, Y / X^R: $X^R X^r$, $X^R Y$ / X^R: $X^R X^r$, $X^R Y$)
F1 Phenotype	100% tall	100% red eyed

Mendel's Monohybrid Cross — First Cross Punnet Square:

	t	t
T	Tt	Tt
T	Tt	Tt

Morgan's Fruit Fly Experiments — First Cross Punnet Square:

	X^r	Y
X^R	$X^R X^r$	$X^R Y$
X^R	$X^R X^r$	$X^R Y$

(2 marks)

22 (a)

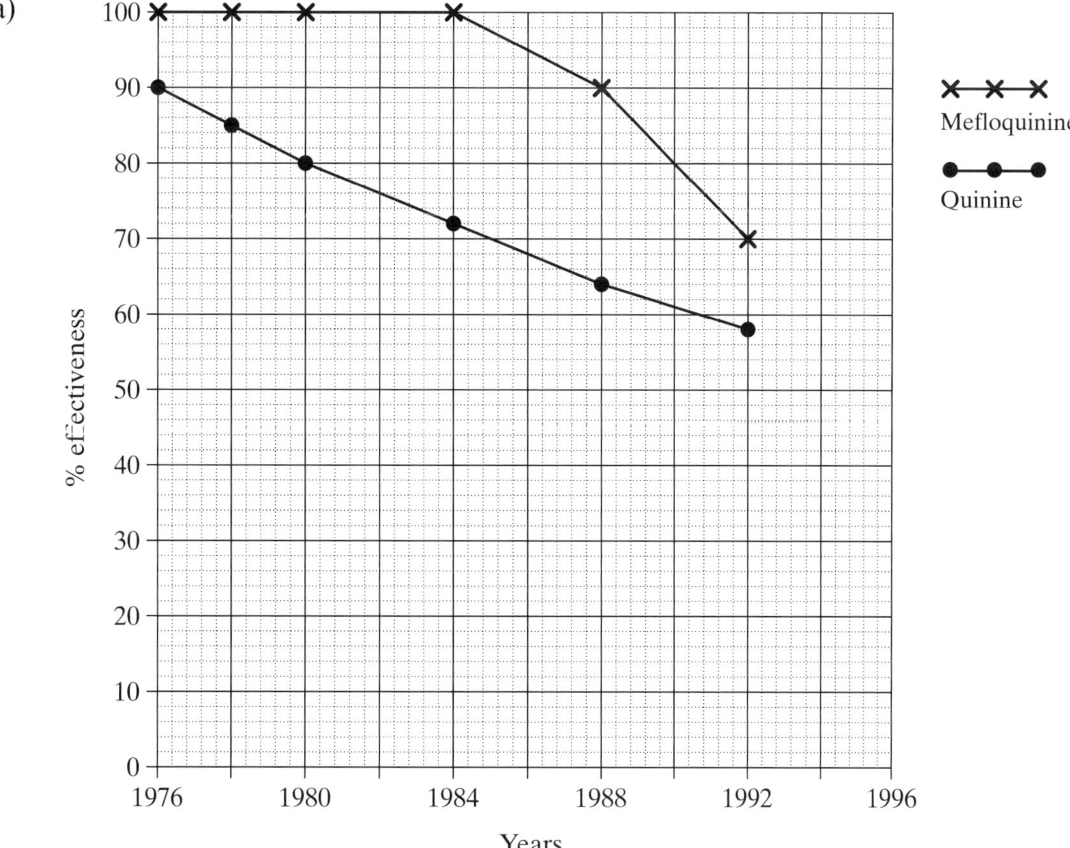

(3 marks)

(b) Human processes can create selection pressures affecting biodiversity. These anti-malarial drugs show significant effectiveness against malaria-causing agents. These drugs eventually become less effective as natural selection enables drug-resistant malarial protozoa to increase, thereby reducing the effectiveness of the drugs over time and increasing biodiversity in the long term. *(3 marks)*

23 (a) Hybridisation within a species increases genetic variation in the gene pool as a result of different combinations of characteristics. For example, in agriculture higher yielding crops, such as barley with increased fibre content, have been developed, as have seedless/juicier mandarins that are easier to peel. *(2 marks)*

(b) Transgenic species are created when desirable genetic characteristics are introduced into the DNA of a species and can then be passed on to the offspring. For example, frost-free strawberries carry a gene from salmon to enable them to grow in areas subject to frost, which would normally kill the plants; sheep have been successfully modified to carry extra growth hormone, resulting in increased size; and Bt cotton has been modified to make it disease resistant.

The advantage of using a transgenic species is that it can increase revenue for the farmer through higher sales and reduction of production costs, increasing the overall profit margin. The disadvantages are the scepticism of various community sectors about the ethics of 'tampering' with nature and genetic material, and the unknown factors that may arise in the modified organism in the long term. Over time biodiversity may be affected, as some genetic variations may be eliminated from the gene pool in certain populations. *(5 marks)*

24 Sample answer. Student answers will vary.

Independent variable	*room with open window*
Control	*room with closed window*
Variables to be kept constant and justification for keeping each constant.	Variable 1: *size of rooms* Justification: *both rooms have the same amount of oxygen/air to start with*
	Variable 2: *number of people in the rooms and their activity* Justification: *need to have similar amount of air being used in each sample/test*
Technology used to measure oxygen saturation in blood	*pulse oximeter*

(5 marks)

25 (a) To produce a longitudinal section of xylem tissue you would require a cutting board and a scalpel or knife to cut the stem material of a plant such as celery, which had been soaked for an hour in red food colouring to provide contrast in the vascular tissue. Glass slides and cover slips are required for mounting the thin slices of specimen, which are viewed under a microscope to magnify the cells. *(2 marks)*

(b)

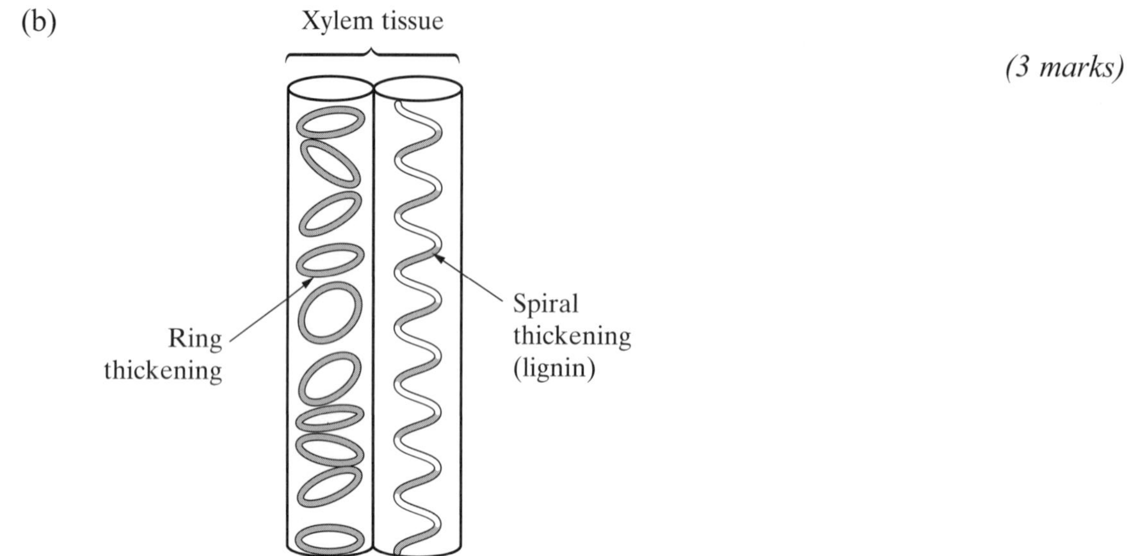

(3 marks)

26 The data indicates that overall the number of deaths decreases in proportion to the rising price of cigarettes. The average cost of cigarettes over the same period has increased threefold. In the period between 1980 and 2005 death rates have dropped from about 3.4 to 2.25 and the number of cigarettes smoked per day has dramatically reduced from approximately 5.75 to 3 per day. The evidence strongly supports the notion that increasing the price of cigarettes will reduce the number of cigarettes smoked and hence have a significant impact on death rates. The data does have some limitations, however, as it only takes into consideration a narrow section of the community—that is men aged 35–44—rather than both genders over a wider age group. *(5 marks)*

27 (a) (i) Animals coming into Australia are inspected and quarantined. *(1 mark)*

(ii) Plant health is protected in Australia by quarantine measures such as the ban on transport of fruit interstate to prevent the spread of disease. All fruit is confiscated at airports, and border inspection stations are used to check cargo. *(1 mark)*

(b) The screening and inspection of live animals at border entry has been very effective in keeping the Australian livestock industry free from many diseases, including mad cow and foot and mouth, by preventing the entry of diseased animals that could potentially infect the local stock. *(1 mark)*

28 (a) antigen *(1 mark)*

(b) This model can easily be used to explain organ rejection as it shows the difference between the recipient and donor tissues. The donor cells have different antigens and they will trigger the T lymphocytes in the recipient to commence rejection by destroying the foreign tissue cells. *(3 marks)*

(c) T cells have surface receptors that recognise specific antigens. They activate the helper T cells to produce clones of cytotoxic (killer) T cells and memory T cells specific to that antigen and these destroy it by secreting cytokinins. *(4 marks)*

29 (a) Responses of plants to temperature change include the activation or deactivation of genes, variation of flowering times and variation in the ratio of carbon to nitrogen in leaves. *(2 marks)*

(b) The relevance of the information used in this report could be evaluated by looking at the currency of the reference material used and a comparison of the content published. Do the articles address specifically the main issues associated with plant responses to temperature change? The reliability of the articles used can be checked by ascertaining the source of the information.

Article 1 comes from a blog, and while it looks like it might be from a 'science' site, it is essentially of 'unknown' origin. It could be scientifically accurate, but additional information, such as the author's name, qualifications and authority on the subject, would be required to validate the information provided. It contains information that is relevant to plants and responses to temperature. Article 2 is an extract from a journal article on climate research and so could be relatively reliable, although it is detailing species native only to the northern hemisphere. The article does make reference to plant response to temperature change. Article 3 has two references, one from a government department in Canberra and one from a plant cell journal, both of which could be reliable, although the article itself is from Wikipedia, which could be less than accurate. The information provided does not directly relate to plant responses to temperature change but rather to plant responses to carbon dioxide concentrations, photosynthesis and herbivore nutrition, and so it is not really relevant for the report. *(5 marks)*

30 The two current theories of evolution are the Darwin/Wallace theory of natural selection and the theory of punctuated equilibrium.

The Darwin/Wallace theory of natural selection is a theory of the mechanism by which evolution occurs. It states that all species show variation, in any generation more offspring are produced than can survive and that consequently there is a struggle for existence. Individuals with characteristics that allow them to survive and reproduce more successfully do so, and they pass these characteristics on to the next generation. These characteristics therefore become widespread in subsequent generations.

Punctuated equilibrium is evolution that is proposed to have occurred in short, rapid bursts followed by long periods of stasis (little change or evolution of species). This is different from Darwin's theory, since he suggested a gradual sequence of evolution over extended periods of time.

The biological practices used to support the theories are investigation of fossil records and biogeography (which can be seen from the table) as well as biochemistry, comparative anatomy and comparative embryology.

Biological practices such as the use of fossil or palaeontological evidence are commonly used to support the theory of evolution. Prior to the data in the table, the two landmasses, Australia and New Zealand, were connected in a supercontinent called Gondwana. During this time, mammals and birds lived on both continents. However, when the landmasses separated, the species became isolated. Over time, different conditions led to different selective pressures, and the organisms that survived best reproduced and passed on their characteristics (natural selection). Thus, the mammal and bird fossils look different in New Zealand than in Australia, despite having common ancestors. This is a biological practice that provides evidence validating the theory of natural selection, as these birds and mammals have evolved characteristics that suit them only to the very specialised environment in which they find themselves.

Another biological practice that provides evidence for natural selection is the observation of the similarities of biochemistry (e.g. DNA hybridisation, haemoglobin or cytochrome C). Using DNA hybridisation, where a single strand of DNA from one organism (e.g. a human) is added to a single stand of DNA from another organism (e.g. a chimp) and the number of matches between bases is measured, we can see how similar or different two species are. If there is a large number of matches, then the organisms shared a common ancestry recently, while if there are fewer matches, then they have diverged much earlier and different selective pressures have led to different genes becoming common in their species.

The many new unique species of birds appearing in the fossil record at 20 mya in the NZ record could also be used to support punctuated equilibrium. It shows that in a period of 2 million years new land was formed and new species of birds appeared. This could be attributed to the new selective pressures that existed on the newly formed NZ landmass, leading to rapid changes in the species. These changes can be seen by using the biological practice of palaeontology (studying the fossils of these organisms).

Thus, the use of biological practices such as palaeontology (as seen in the table) and biochemistry support the evolutionary theories of natural section and punctuated equilibrium. *(7 marks)*

Section II—Options

Question 31 — Communication

(a)

Animal	*Method used to detect vibrations*
Insects	Hairs on antennae in mosquito; tympanum on legs or abdomen of grasshoppers, cicadas, moths and butterflies
Fish	Lateral line consisting of mechano-receptors along the side of the fish; inner ear with semi-circular canals to detect vibrations from swim bladder
Mammals	Ears with tympanic membrane and ossicles to transmit vibrations to the inner ear where a cochlea with organ of corti, which has sensitive hair cells detecting vibration, is located

(3 marks)

(b) (Sample diagrams)

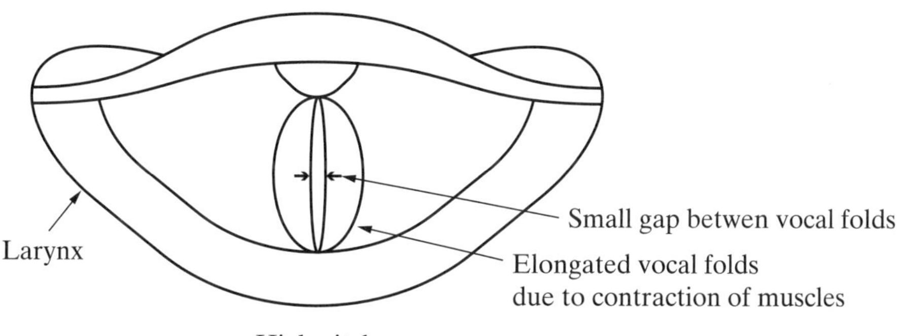

High pitch

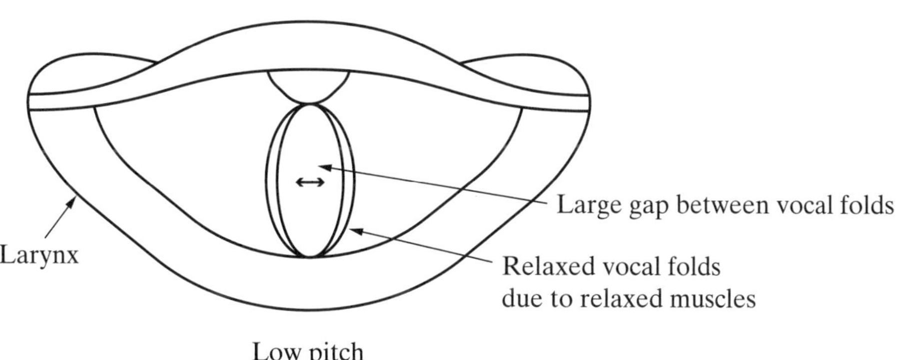

Low pitch

(4 marks)

(c) (i) Line Y represents the cones. *(1 mark)*

(ii) The structure of cones varies in that the three types of cones contain differing amounts of photopsins. The cones are thinner in shape within the fovea, where they are arranged in a hexagonal lattice and are most numerous. As you move away towards the periphery, the cones thicken. They fall away to a constant level of 10–15 degrees and are interspersed with rods. The blind spot has no receptors at all. *(2 marks)*

(iii) Rods are most numerous around the perimeter of the retina. The role of rhodopsin in rods is that it is sensitive to blue-green light. When light hits a rod a change occurs in the shape, activating the rhodopsin molecule and causing a change in the electrical potential. This triggers a nerve impulse that passes through to the optic nerve. The rods work in low light levels, enabling the detection of shape and movement but no colour detail as they have only the one pigment type. *(2 marks)*

(d) (i) The area depicted as being damaged is the cerebrum. The absence of action potential in the region after the fall may be a result of the threshold for the action potential not being reached and hence stimulation of the area will not result in a response (the all or nothing effect). Alternatively, the blood vessels may have been damaged in the fall, resulting in an inadequate supply of oxygen and subsequent inability of the tissue in the area to allow the passage of the nerve impulse, as the sodium/potassium pump cannot operate with reduced blood flow. *(4 marks)*

(ii) Changes to the animal's behaviour because of this damage could result from memory loss; sensory interpretation, especially hearing and sight, could be affected; or possibly voluntary muscular control could be affected. *(2 marks)*

(e) The understanding of the eye and ear enables us to use these (eye/ear) receptors to make the most of technologies such as 3D movies and surround sound systems. Forward-facing eyes permit an overlap in the viewing field and so film-makers use twin cameras to film sequences that are then overlaid to enhance depth perception. This is amplified by the wearing of 3D glasses to ensure that each eye detects a different image, giving rise to a stereoscopic effect. This means objects in 3D movies appear to 'leap out' at you.

The understanding of how sound shadows work has led to the use of multiple speaker systems to ensure that the head/objects does not prevent sound waves reaching the brain. This allows the richness of sounds that are detected with surround sound systems, as opposed to the poorer quality of sounds received from stand-alone, isolated speakers from which sound waves could be blocked.

The simultaneous detection of sight and sound enhances the audience experience and cinemas using these technologies have become very popular, compared with those using the traditional image with a speaker either side of the screen.

Together with an understanding of the transmission of these stimuli from the receptor rods/cones and organ of corti, via the optic and auditory nerves respectively, to the cerebrum, the technology described in the stimulus can be effectively utilised for our entertainment. *(7 marks)*

Question 32 — Biotechnology

(a) Answers will vary, but one possible answer is:

Industrial fermentation process	*Microorganism*	*Product*
Production of lactic acid. Carbohydrates such as sugars and molasses are fermented in large vessels containing nutrients, stirrers and pH/ temperature controls, together with aeration devices. These 'bioreactors' speed up the fermentation process in which lactic acid bacteria convert carbohydrate to glucose and then to pyruvic acid and the final product, lactic acid.	Lactic acid bacteria include *Streptococcus thermophilus, Streptococcus bovis, Lactobacillius bulgaricus* and *Lactobacillius lactis*.	Lactose is converted firstly to glucose and then through glycolysis to produce lactic acid via pyruvic acid.

(3 marks)

(b) Answers will vary but the following is a basic example.

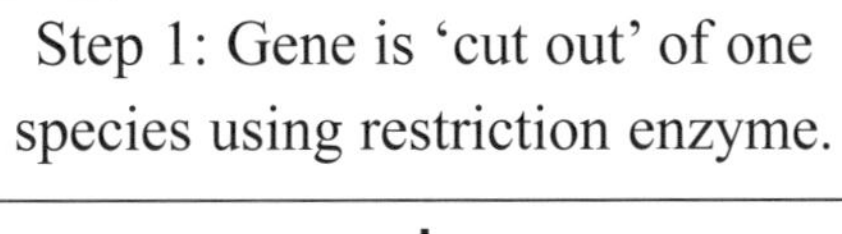

↓

Step 2: Circular piece of DNA (plasmid) is removed from a bacterial cell and cut open with restriction enzyme (e.g. from *E. coli*).

↓

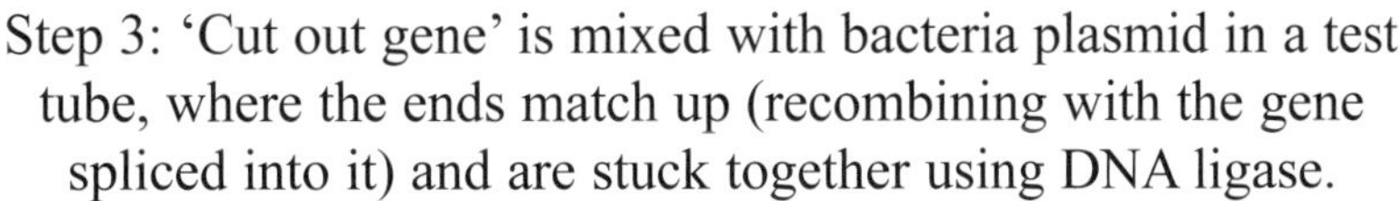

↓

Step 4: 'Transformation into bacteria'. Plasmid containing recombinant DNA is re-inserted into the bacteria by mixing it with calcium chloride or the use of electroporation.

↓

Step 5: Introduced gene will now be expressed by the bacterial cell.

(4 marks)

(c) (i) Artificial selection is the term attributed to a practice used by agriculturalists in order to select the most favourable characteristic in a species and facilitate the breeding of the organisms that possess those desirable characteristics. This results in a population of organisms that have the greatest chance of exhibiting those specific phenotypes. *(1 mark)*

(ii) In the example given the cattle have been artificially bred for the phenotype brown and an increased number of offspring being produced and surviving. The outcome of the three generations of crosses of the desired characteristics has been relatively successful. Generation 1, where non-brown breeding males were used to inseminate the cows, resulted in 53 offspring of which approximately 50 per cent died early with a 1:4 ratio of non-brown to brown males surviving to breeding age. This is compared to generation 3 where 62 offspring are produced with only 10 deaths, a significant reduction, and a 27:1 ratio of brown to non-brown breeding males surviving. So in this example the selective breeding has been successful in creating a reliably brown, high-yielding stock. *(4 marks)*

(d) (i) Each codon provides information for one amino acid, which is synthesised into polypeptides at the ribosomes. The genetic code determines the order of the amino acid in the polypeptide or protein, which acts as functional molecules such as hormones or enzymes. Genetic information is first synthesised from a DNA template (transcription) and the protein is then synthesised from a RNA template, called translation. If interference occurs during the translation process, where complementary mRNA is formed, this will result in a different polypeptide or protein, having a significant impact on the function of the cell. These changes can be deleterious to a cell, where the cell causes faulty function, or beneficial, resulting in differences that promote function or survival. This is central to natural selection. RNA interference has an important role in defending cells against parasitic genes and in directing development, as well as gene expression in general. *(2 marks)*

(ii) New biotechnology applications in the treatment of diabetic patients have been developed, including the production of insulin through gene technology, a process that utilises the human insulin gene and inserts it into the patient's own cells. The gene is inserted into liver cells rather than damaged pancreatic tissue, as the liver cells are similar to the beta cells that produce insulin. The new gene or section of DNA (human rather than from another animal) is inserted into the patient's liver cell nuclei. The translation of this section of DNA into complementary bases then involves reverse transcription, which will create a copy of RNA that can then produce the new polypeptide or protein. This, in turn, will complete the new strand and form the insulin. *(4 marks)*

(e) New biotechnologies that have led to significant ethical and social issues are genetically modified organisms (GMOs) and animal and gene cloning. When considering GMOs the major concerns are interfering with nature; access to the benefits; loss of biodiversity; inability to control the spread of transgenic species through natural mechanisms such as wind pollination, thus affecting natural balances in the environment when the genetically engineered species compete with native species for food and shelter; labelling of consumable products made with GMOs; and possible effects down the track that have not been identified as yet. Concerns about animal and gene cloning once again include interfering with nature and evolutionary processes. For example, cloning of stem cells is considered unethical as applications involving humans may see some characteristics favoured over others, and cloning of transgenics inside bacteria can result in new genomes, altering natural evolutionary processes. Other concerns are a possible decline in biodiversity and the possibility that large groups of organisms will have weaknesses or be vulnerable to diseases. This could be potentially devastating, especially if they were accidentally released into the environment. How patents on this technology and the species derived will be controlled is unknown as yet and the potential for individuals/groups to have ultimate control of the technology is of paramount concern, as the results could be misused in things such as weapon manufacture, which would not be a desirable outcome. *(7 marks)*

Question 33 — Genetics: The Code Broken?

(a) n = haploid number

$2n$ = diploid number

Mutation type	*Trisomy*	*Polyploidy*	*Base substitution*
Chromosome number	$2n + 1$	xn (where $x > 2$)	$2n$

(3 marks)

(b)

(4 marks)

Diploid cell 4 chromosomes: 2 chromosome no. 1 2 chromosome no. 2	**Haploid cell** 2 chromosomes: 1 chromosome no. 1 1 chromosome no. 2

(c) (i) Not possible to say for the vision defect gene. The limb defect allele must be recessive to the normal allele as person 9 shows the trait, while neither of her parents does.

(1 mark)

(ii) Assuming vision defect is recessive and that 10 is homozygous for the normal vision allele:

Let V = normal vision and v = defective vision

Let L = normal limbs and l = defective limbs

If the genes are linked:

Individual 9 is $\frac{\text{Vl}}{\text{Vl}}$ and individual 10 is $\frac{\text{VL}}{\text{Vl}}$.

$\frac{\text{Vl}}{\text{vl}} \times \frac{\text{VL}}{\text{Vl}}$

	VL	Vl
Vl	$\frac{\text{VL}}{\text{Vl}}$	$\frac{\text{Vl}}{\text{Vl}}$
vl	$\frac{\text{VL}}{\text{vl}}$	$\frac{\text{Vl}}{\text{vl}}$

Phenotype ratio = 1 normal vision and normal limbs : 1 normal vision and defective limbs

If the genes are **not** linked:

Individual 9 is Vvll and individual 10 is VVLl.

Vvll × VVLl

	VL	Vl
Vl	VVLl	VVll
vl	VvLl	Vvll

Phenotype ratio = 1 normal vision and normal limbs : 1 normal vision and defective limbs

(Other answers are possible, depending on the assumptions made. In this case the ratios are the same. Had other assumptions been made, they would have been different.) *(4 marks)*

(d) (i) If genes are linked they are situated on the same chromosome and, therefore, their alleles should be inherited together. In other words, they don't obey Mendel's law of independent assortment: they don't sort out independently of each other at meiosis. Genes are, however, sometimes separated if crossing over occurs in between them. The farther apart they are, the more likely this is to happen.

In fast breeding species that produce many offspring it is possible to determine the relative frequencies at which linked genes are separated at crossing over and, therefore, their relative distances apart. The distance apart of two genes is expressed in centimorgans (the number of recombinations per 100 offspring). The higher the number in centimorgans, the farther apart the two genes are.

This is best investigated using a fast breeding species with large numbers of offspring, such as fruit flies. *(3 marks)*

(ii) Three reasons are:

1. Chromosome mapping using cross-breeding experiments requires large numbers of controlled crosses between individuals whose genotypes are known. Humans produce too few offspring to give statistically valid results.
2. The crosses have to be controlled so that specific individuals of known genotype are used in each cross. Such control is not possible, either practically or ethically, with humans.
3. The genotypes of the individuals breeding need to be known for the genes under investigation. This is only possible in a species where breeding can be controlled over a number of generations, giving pure breeding strains and known hybrids. Again, this is not feasible with humans. Humans are slow breeders, it would take many years, and again the degree of control necessary is not practical or ethical.

(3 marks)

(e) Gene cloning is the process by which identical copies of a gene, a section of DNA, are made. Essentially this involves removing the gene from its chromosome by using restriction enzymes, which cut the DNA at specific base sequences, leaving characteristic 'sticky ends'. This gene is then spliced into a bacterial plasmid and reintroduced to a bacterium. As the bacterium reproduces, new copies of the plasmid, and hence the spliced gene, are produced. In this way many copies of the gene can be produced, or cloned.

This technique is critical for the development of the technology mentioned in the article. Without the development of techniques that use restriction enzymes to cut DNA molecules, and ligase enzymes to recombine them, it would not be possible to assemble an 'artificial chromosome'. Many copies of this chromosome, or fragments of it, would need to be produced in the course of producing this 'artificial chromosome'.

Gene cascades are assemblages of genes that act in sequence during the process of development of the organism. Gene A produces a protein that not only performs a specific function but also triggers the expression of gene B, and so on. The precise sequence of genetic expression is critical to the correct sequential development of the features of an embryo.

Knowledge of the functioning of gene cascades would be an important part of developing an 'artificial chromosome'. These cascades are evolutionarily conservative assemblages. They are found throughout, for example, the animal kingdom and are very similar in related species. In producing an artificial chromosome it would be important to realise that some of the genes require other genes to be present to trigger their activity.

(7 marks)

Question 34 — The Human Story

(a)

Classification level	*Identifying feature*
Animals	• lack rigid cell walls; heterotrophic
Chordates	• possess notochord or spine
Mammals	• possess mammary glands; four-chambered heart

(3 marks)

(b)

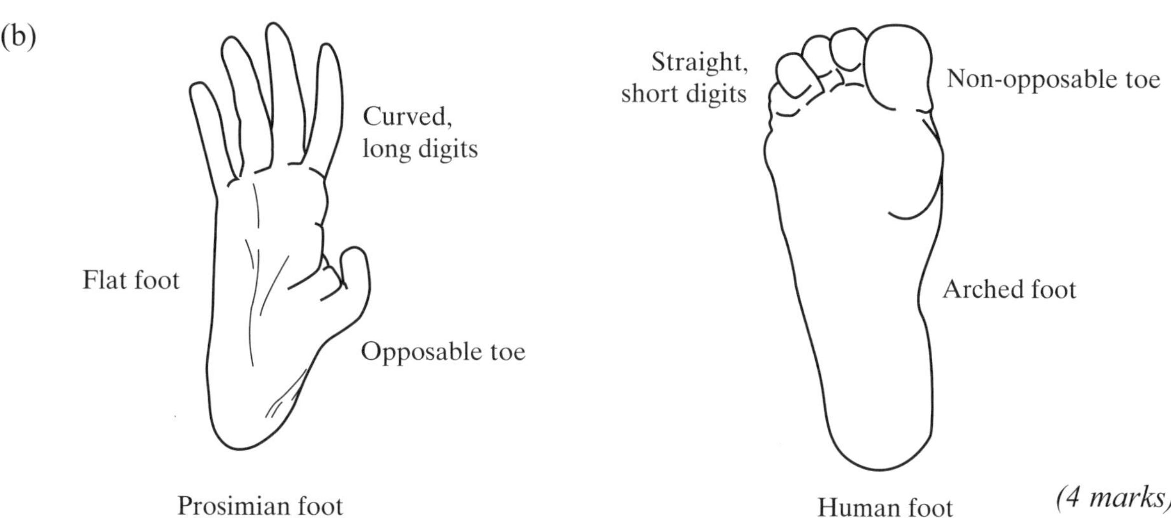

(4 marks)

(c) (i) Nuclear DNA is inherited by the offspring from both parents (mother and father) whereas mitochondrial DNA is inherited by the offspring from the mother (maternal inheritance). *(1 mark)*

(ii) Mitochondrial DNA is maternally inherited. Analysis of mtDNA is considered to be more useful to evolutionary biology and is commonly used to construct 'evolutionary trees'.

mtDNA is more useful in evolutionary biology because it is present in large numbers and fewer samples are required to collect data. It has a higher mutation rate, making it easier to track the lineage of humans, and there is no recombination of genes, making it easier to trace, through the maternal line, a direct genetic lineage. This can be compared with nuclear DNA, where frequent recombinations and crossings over during meiosis result in an enormous range of possibilities of genetic lineage.

mtDNA analysis should be supplemented with nuclear DNA studies where relationships can be confirmed by comparing the degree of similarity of nuclear DNA between samples. *(4 marks)*

(d) (i) Data from scientific journals are reliable because of the peer review process and the reputation of these journals for publishing valid and reliable information. The reliability of the collected data was enhanced by choosing only adult skulls and by taking the average of a large number of skulls collected from archaeological digs.

A scatter plot was used to present the data. This provided a clear visual representation of the data so that the relationship between cranial size and the age of the fossil could be deduced. *(4 marks)*

(ii) A: *Australopithecus afarensis*, Africa OR B: *Homo erectus*, Africa *(2 marks)*

(e) Cultural development is the process of acquiring knowledge by direct teaching or imitation. Cultural development of primates and humans can be compared in the areas of technology, medicine, genetic engineering and communication.

Feature	*Humans*	*Other primates*	*Impact on future evolution*
Technology	• Sophisticated use of technology to make work easier (e.g. computers) • Specialised use of tools (e.g. machines that reduce the effort of work)	• Primitive use of tools (e.g. chimps using sticks to collect termites) • Stone tools and weapons used by early primates • Little specialisation in use of tools	• Humans are able to manipulate their environment by using machines and tools (e.g. air conditioners) and thus increase their survival in all environments. • Evolution for humans is now not dependent on normal selection pressures such as temperature and availability of food and water as it was in the past and is for primates.
Modern medicine	• Effective drugs to fight infections (e.g. antibiotics) • Prenatal care, birth control • Assisted reproduction (IVF) • Vaccines, surgery • Treatments for diseases (e.g. insulin for diabetes, kidney dialysis, chemotherapy)	• No development of cultural processes involving medicine • Survival is dependent on natural immunity	• Humans have reduced mortality and increased life expectancy. • Harmful genes are no longer removed and are being maintained in the human gene pool. • Humans have reduced ability to respond to changes in the environment and thus are more likely to be wiped out by epidemics such as 'swine flu'.

Continued

Feature	*Humans*	*Other primates*	*Impact on future evolution*
Genetic engineering	• Genetically modified plants and animals in agriculture (including transgenics) and selective breeding are used to treat disease and develop new products to improve the quality of life and survival. • Cloning to improve agricultural products or to replace diseased organs	• This technology not developed by primates	• There is less diversity in the human gene pool, reducing the possibility of survival in future climate changes or exposure to a 'superbug'. • Monocultures in crop agriculture could mean starvation for humans if disease strikes the crop. • Harmful genes are maintained in the gene pool.
Communication systems	• Sophisticated use of language to improve survival by warning of danger and for continuing cultural development and survival	• Visual, auditory and olfactory communication; vocalisation limited but can be used to increase survival	• By changing the selection pressures, humans are able to improve their survival and thus are not subject to traditional environmental selective pressures. • Use of communication to teach and long distance communication also improves survival (e.g. tsunami warnings).
Population mobility	• Isolation reduced by development of efficient transport and thus more interbreeding • Successful genes often not chosen for survival reasons but for fashion or religious/ethnic bias	• Mobility still constrained by the local environment • Population isolation possible if major environmental catastrophes (e.g. volcanic eruptions and floods) occur	• There is increased mixing of the human gene pool. • There is less variability in the gene pool when selective pressures change (e.g. spread of diseases such as SARS and swine flu). • 'Superbugs' may be powerful selective agents because they can be rapidly spread by mobile populations.

To sum up, evolution for humans in the future is no longer being shaped by the selective pressures of climate, food availability, predators, prey and availability of a mate. The use of technology, modern medicine, genetic engineering and advanced communication processes means that the gene pool includes many genes that would previously have been eliminated when sick individuals died. Selective pressures are not the major force on the evolution of humans as they are for primates or were for humans in the past. *(7 marks)*

CHAPTER 10

BOARD OF STUDIES
NEW SOUTH WALES

2011
HIGHER SCHOOL CERTIFICATE EXAMINATION

Biology

General Instructions

- Reading time – 5 minutes
- Working time – 3 hours
- Write using black or blue pen
  Black pen is preferred
- Draw diagrams using pencil
- Board-approved calculators may be used
- Write your Centre Number and Student Number where required

Total marks – 100

Section I

75 marks

This section has two parts, Part A and Part B

Part A – 20 marks
- Attempt Questions 1–20
- Allow about 35 minutes for this part

Part B – 55 marks
- Attempt Questions 21–31
- Allow about 1 hour and 40 minutes for this part

Section II

25 marks
- Attempt ONE question from Questions 32–36
- Allow about 45 minutes for this section

Section I
75 marks

Part A – 20 marks
Attempt Questions 1–20
Allow about 35 minutes for this part

Use the multiple-choice answer sheet for Questions 1–20.

---

1 The DNA in plants and animals is composed of the same chemical components.

What is the biological significance of this statement?

(A) Plants and animals can interbreed.

(B) Plants and animals share common ancestry.

(C) Plants and animals are genetically identical.

(D) Plants and animals are composed of identical proteins.

2 Which substance is mainly bound to proteins when it is carried in mammalian blood?

(A) Nitrogenous waste

(B) Carbon dioxide

(C) Lipid

(D) Salt

3 In organisms, the maintenance of a constant internal environment is

(A) necessary because organisms must have a constant body temperature.

(B) necessary because enzyme activity is highest at specific temperatures.

(C) unnecessary because organisms are found in environments with a broad range of temperatures.

(D) unnecessary because the nervous system detects and responds to changes in ambient temperature.

4 Which of the following alternatives lists ALL of the chemical components of a chromosome?

(A) Sugar, phosphate and bases

(B) Lipids, DNA and protein

(C) DNA and protein

(D) Genes and DNA

5 The presence of freckles is a dominant characteristic. A child's mother has no freckles and its father is heterozygous for freckles.

What is the probability that this child will have freckles?

(A) 25%

(B) 50%

(C) 75%

(D) 100%

6 What was Macfarlane Burnet's contribution to our understanding of human disease?

(A) He determined that mosquitoes transmit malaria.

(B) He disproved the theory of spontaneous generation.

(C) He developed a method for establishing that a particular microbe causes a disease.

(D) He discovered how the body distinguishes between its own tissues and foreign cells.

7 How do cleanliness in food preparation and cleanliness in personal hygiene assist in the control of disease?

(A) They reduce the microflora.

(B) They reduce transmission of pathogens.

(C) They stop the transfer of dirt between people.

(D) They stop food that contains bacteria from being eaten.

8 Which type of cells destroy virally infected cells?

(A) B cells

(B) Killer T cells

(C) Helper T cells

(D) Cytotoxic T cells

9 Which list shows pathogens in order of increasing size?

(A) Bacterium, prion, virus, protozoan, macroparasite

(B) Macroparasite, prion, protozoan, bacterium, virus

(C) Prion, virus, bacterium, protozoan, macroparasite

(D) Virus, prion, bacterium, macroparasite, protozoan

10 Which of the following correctly identifies a source of variation in both asexual and sexual reproduction?

	Asexual	*Sexual*
(A)	Cloning	Natural selection
(B)	Crossing over	Cell division
(C)	Spontaneous generation	Fertilisation
(D)	Mutation	Mutation

11 The diagram represents one current theory for the movement of materials in phloem.

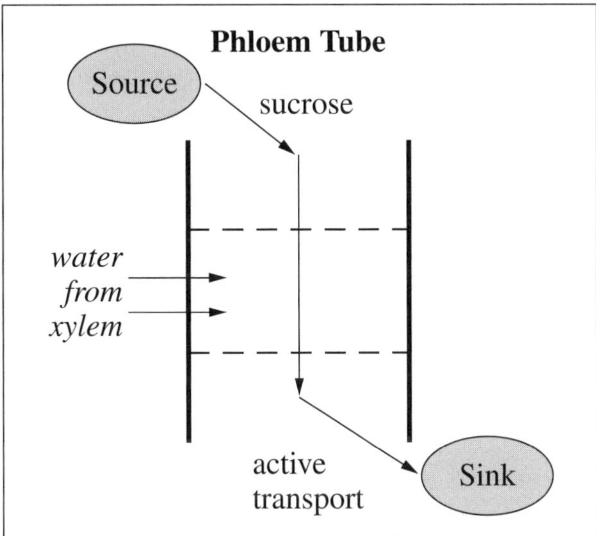

In the process illustrated in the diagram, water from xylem

(A) causes a build up of pressure.

(B) allows adhesion for capillarity.

(C) provides water for photosynthesis.

(D) dilutes sucrose for active transport.

12 An Australian insect produces uric acid and no other form of nitrogenous waste.

What is the purpose of this adaptation?

(A) To increase salt loss

(B) To reduce water loss

(C) To increase its toxicity to predators

(D) To reduce the chance of attracting predators

13 Which diagram best explains changes to the composition of blood in the lungs?

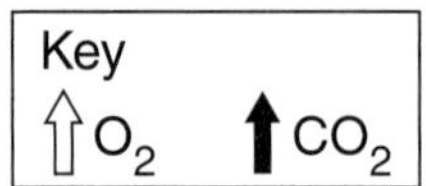

(A)

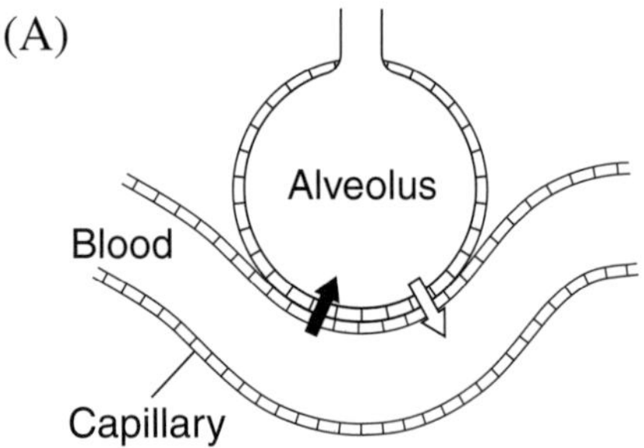

(B)

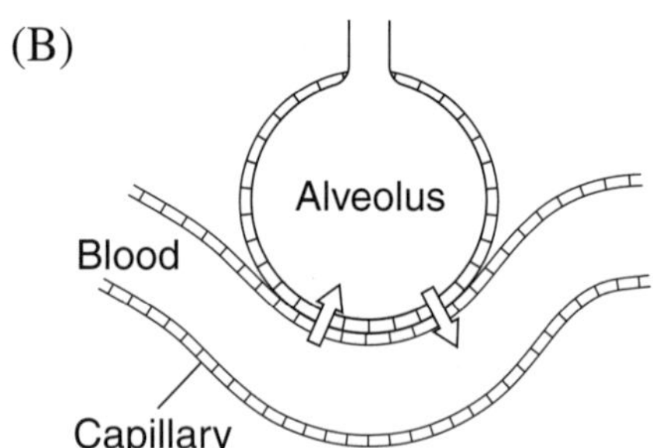

(C)

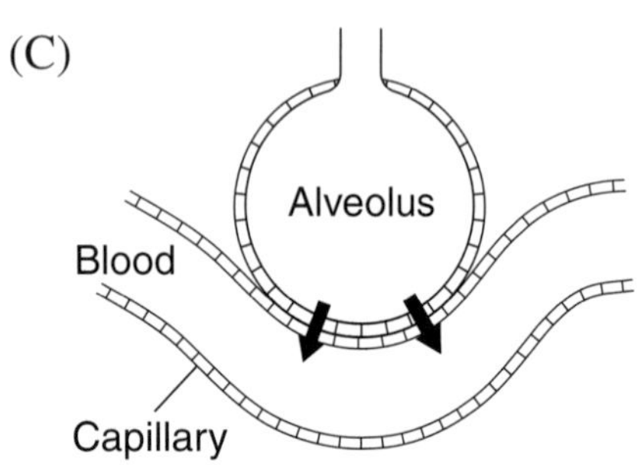

(D)

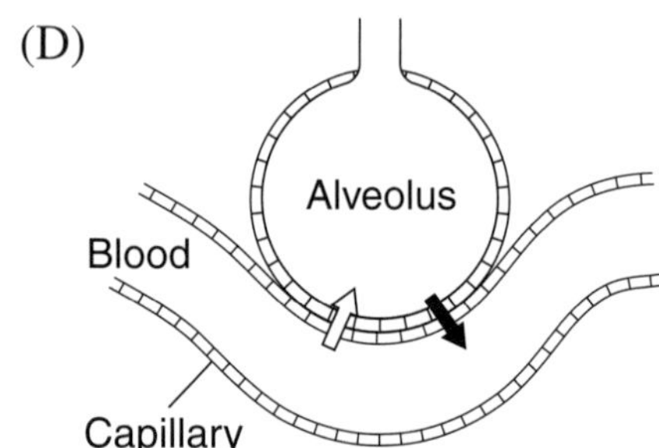

14 What role do mucous membranes and cilia play in the body's defence against pathogens?

(A) They cause cell death to seal off pathogens.

(B) They form part of a barrier preventing entry of pathogens.

(C) They remove pathogens as part of the inflammatory response.

(D) They recognise pathogens passing through the lymphatic system.

15 African ostriches, South American rheas and Australian emus are all large extant flightless birds. These observations provide evidence for the theory of evolution.

This evidence for evolution belongs to

(A) biochemistry.

(B) biogeography.

(C) paleontology.

(D) comparative embryology.

16 Which of the following shows the correct sequence of steps in the interaction between T cells (lymphocytes) and B cells (lymphocytes)?

	Step 1	*Step 2*	*Step 3*
(A)	T cells interact with antigens presented by macrophages	Activated T cells differentiate into helper T cells	Helper T cells produce chemicals that cause B cells to multiply
(B)	T cells interact with antigens presented by macrophages	Activated T cells differentiate into B cells	B cells produce chemicals that cause T suppressor cells to multiply
(C)	B cells interact with antigens presented by macrophages	Activated B cells differentiate into plasma cells	Plasma cells produce chemicals that cause T cells to multiply
(D)	B cells interact with antigens presented by macrophages	Activated B cells differentiate into B memory cells	B memory cells produce chemicals that cause macrophages to multiply

17 The pedigree shows the inheritance of a trait controlled by a pair of alleles.

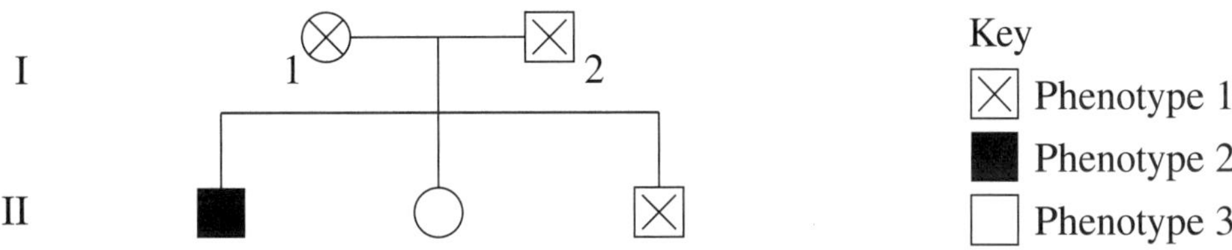

Which Punnett square correctly represents the cross between the parents in generation I?

(A)

	B	b
A	AB	Ab
a	aB	ab

(B)

	B	B
A	AB	AB
A	AB	AB

(C)

	B	b
B	BB	Bb
b	Bb	bb

(D)

	A	B
A	AA	AB
B	AB	BB

18 Diagrams of blood vessels are shown (not to scale).

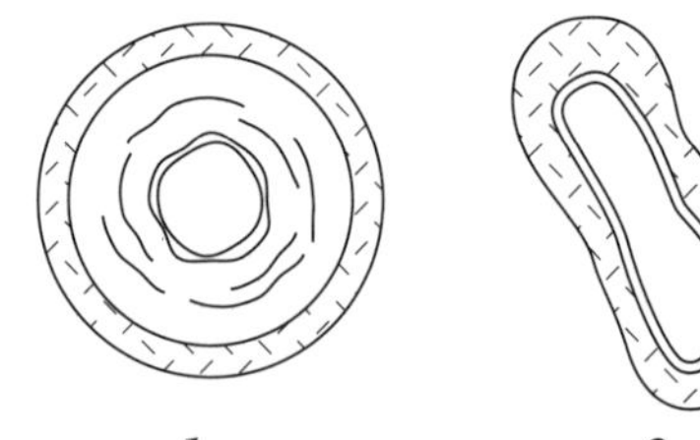
1

2

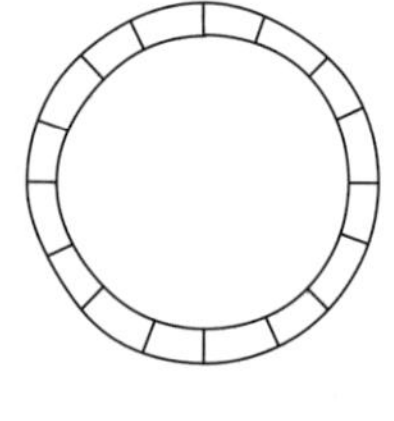
3

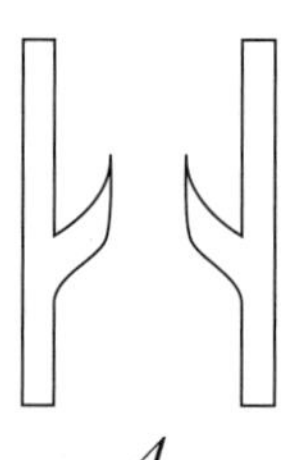
4

Which of the following correctly lists the names of the blood vessels shown?

	1	*2*	*3*	*4*
(A)	Artery	Vein	Capillary	Vein
(B)	Capillary	Artery	Artery	Vein
(C)	Artery	Vein	Vein	Capillary
(D)	Vein	Artery	Capillary	Capillary

19 Why are antibiotics ineffective in treating malaria?

(A) Malaria is not caused by bacteria.

(B) Malaria is not an infectious disease.

(C) Malaria is transmitted by mosquitoes.

(D) Malaria is associated with wet environments.

20 How does the fossil record provide evidence to support the concept of punctuated equilibrium?

(A) The fossil record is incomplete.

(B) The fossil record shows that some organisms have become extinct.

(C) The fossil record shows that there are short periods of rapid change in fossil forms.

(D) The fossil record shows that some organisms change gradually over geologic time.

2011 HIGHER SCHOOL CERTIFICATE EXAMINATION

Biology

Centre Number

Student Number

Section I (continued)

Part B – 55 marks
Attempt Questions 21–31
Allow about 1 hour and 40 minutes for this part

Answer the questions in the spaces provided. These spaces provide guidance for the expected length of response.

Question 21 (4 marks)

(a) Identify TWO methods used to treat drinking water to reduce the risk of infection. **2**

....

....

(b) Outline how the use of each of these methods reduces the risk of infection. **2**

....

....

....

....

Question 22 (4 marks)

Six students performed a trial experiment on enzyme activity. The enzyme they were studying acts on a cloudy suspension, breaking it down into a soluble form. The lesson ended and students were asked to stop their experiment. They then recorded the time the experiment had run and their observations. These data are collated in the table.

Student	*Volume of enzyme solution* (mL)	*Temperature* (°C)	*pH*	*Volume of cloudy suspension* (mL)	*Time* (minutes)	*Observation at end of lesson*
1	5	37	2	10	9	Cloudy
2	5	37	2	10	20	Clear
3	5	37	4	10	35	Clear
4	5	37	4	10	20	Cloudy
5	5	37	6	10	8	Cloudy
6	5	37	6	10	40	Cloudy

(a) What is the purpose of this experiment? **1**

....

....

(b) Describe TWO changes that would improve the validity of the data collected in the experiment. **2**

....

....

....

....

Question 22 continues

Question 22 (continued)

(c) The experiment was repeated with all trials running for 40 minutes. Complete the table to predict the results. **1**

Student	*Volume of enzyme solution* (mL)	*Temperature* (°C)	*pH*	*Volume of cloudy suspension* (mL)	*Time* (minutes)	*Observation at end of trial*
1	5	37	2	10	40	
2	5	37	2	10	40	
3	5	37	4	10	40	
4	5	37	4	10	40	
5	5	37	6	10	40	
6	5	37	6	10	40	

End of Question 22

2011 HIGHER SCHOOL CERTIFICATE EXAMINATION

Biology

Section I – Part B (continued)

Centre Number

Student Number

Question 23 (5 marks)

Several scientists were involved in determining the structure of DNA. **5**

To what extent did the quality of collaboration and communication between these scientists impact on their scientific research?

Question 24 (5 marks)

(a) Name a viral disease that is controlled by the use of a vaccine. **1**

....

(b) The flowchart shows the steps involved in the preparation of a vaccine to prevent a viral disease.

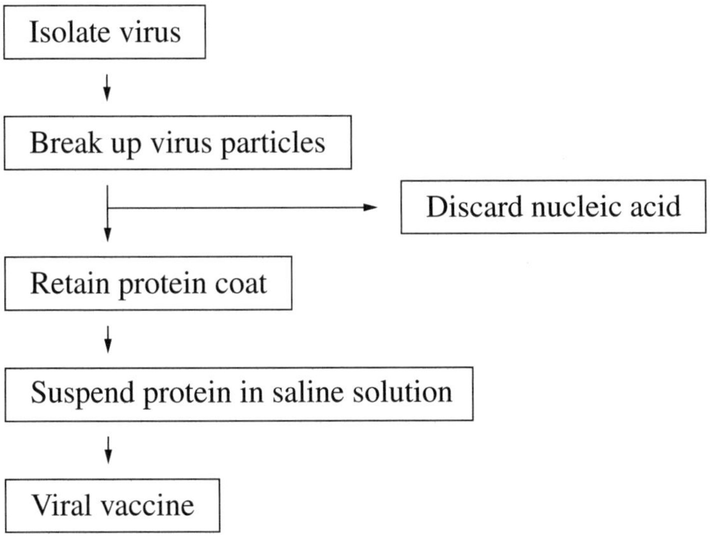

Isolate virus
↓
Break up virus particles
↓ → Discard nucleic acid
Retain protein coat
↓
Suspend protein in saline solution
↓
Viral vaccine

(i) Outline how a vaccine that is prepared in this way will produce immunity to a specific viral disease. **2**

....

....

....

....

(ii) Explain ONE procedure in the flowchart that ensures the safety of the prepared vaccine. **2**

....

....

....

....

2011 HIGHER SCHOOL CERTIFICATE EXAMINATION

Biology

Section I – Part B (continued)

Centre Number

Student Number

Question 25 (5 marks)

A diagram of a nephron is shown.

Processes that occur in a nephron

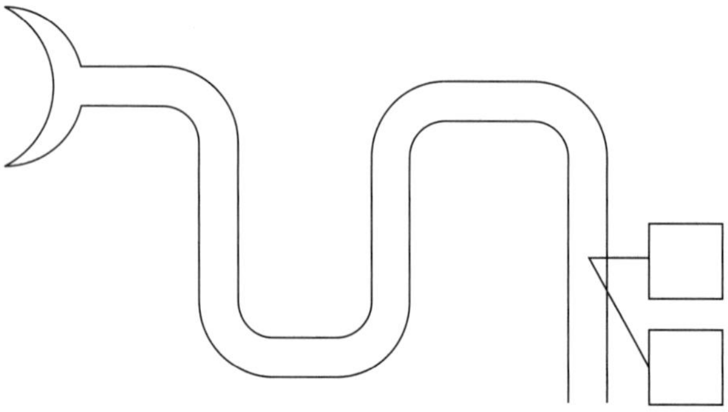

Key:
A Filtration takes place
B Water reabsorption takes place
C Glucose reabsorption takes place
D Antidiuretic hormone (ADH) has its effect

(a) Label each of the two boxes on the diagram using A, B, C or D to identify the processes that take place at this location. **2**

(b) Which one of the above processes (A, B, C or D) occurs due to active transport? **1**

....

(c) Outline the effect of aldosterone on the control of body fluids. **2**

....

....

....

....

....

Question 26 (6 marks)

In assessing the effectiveness of the global polio vaccination program started in 1988, a student collected Graphs A and B from two reputable sources.

Graph A: Reported annual global polio cases from 1988–1996

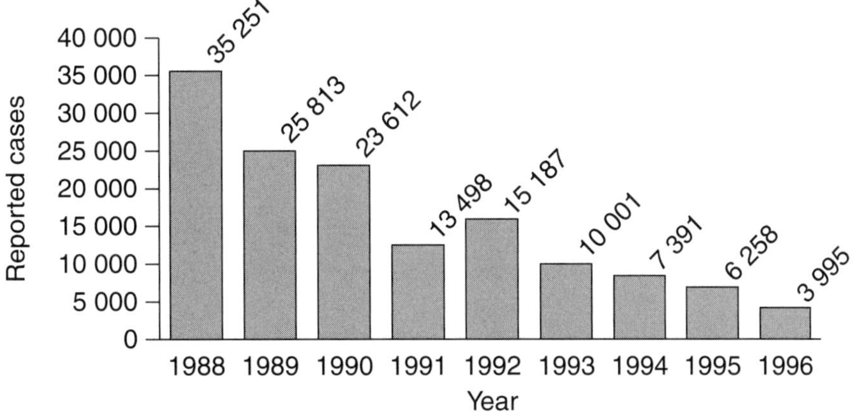

Graph B: Estimated annual global polio cases from 1988–1996

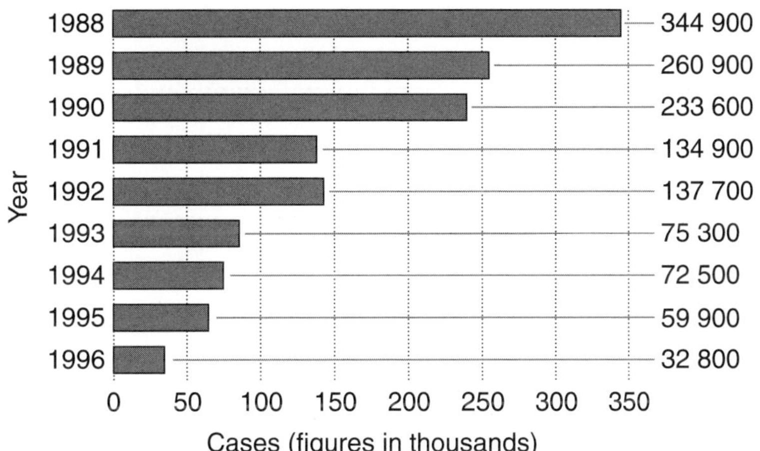

Question 26 continues

Question 26 (continued)

(a) Identify TWO similarities and TWO differences between the information contained in Graphs A and B. **4**

....

....

....

....

....

....

....

....

(b) Recommend TWO additional types of data that could be used for evaluating the effectiveness of the global polio vaccination program. **2**

....

....

....

....

End of Question 26

2011 HIGHER SCHOOL CERTIFICATE EXAMINATION

Biology

Centre Number

Section I – Part B (continued)

Student Number

Question 27 (4 marks)

A new non-infectious disease has been discovered in people in some aged care homes. **4**

Outline at least FOUR features of an epidemiological study that could be used to identify the cause of this disease.

....

....

....

....

....

....

....

....

....

....

Question 28 (5 marks)

Stem cells can develop into any type of body cell. When stimulated they can differentiate into specialised cells.

The skin contains stem cells as well as epidermal, blood vessel, fat and muscle cells.

New research suggests that stem cells obtained from the skin of a patient can be used to make red and white blood cells for that patient in the laboratory. Production of new improved artificial blood seems a likely outcome.

(a) Outline how gene expression in skin stem cells is linked to the maintenance of health after the skin is broken. **2**

....

....

....

....

(b) Explain features of the new artificial blood made from skin stem cells that are likely to be an improvement over an existing form of artificial blood. **3**

....

....

....

....

....

....

....

....

2011 HIGHER SCHOOL CERTIFICATE EXAMINATION

Biology

Section I – Part B (continued)

Centre Number

Student Number

Question 29 (5 marks)

You are required to plan and perform a first-hand investigation to identify microbes in food or in water. **5**

Complete the following table for your investigation.

Dependent variable	
Independent variable	
Control	
Safe work practices to be followed	

Question 30 (5 marks)

Scientists recorded the body temperature of 50 reptiles of the same species on the same day. They were kept in small cages in the shade in a hot desert habitat. Technology *X* was used to measure skin temperature. This was linked to technology *Y* and then to a computer as shown in the diagram. The graph shows the averaged data generated by the computer from this experiment.

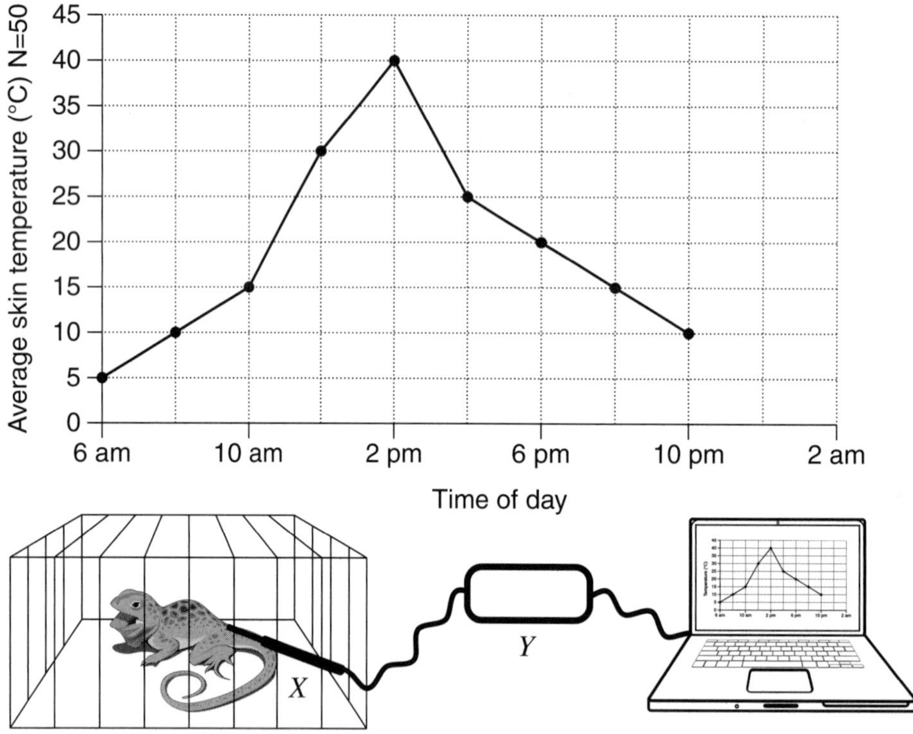

(a) Identify technology *X* and technology *Y*. **2**

....

....

(b) The scientists concluded that the body temperature of this species of reptile is only controlled by the ambient temperature. **3**

Construct an argument against this conclusion, based on the information provided.

....

....

....

....

....

....

2011 HIGHER SCHOOL CERTIFICATE EXAMINATION

Biology

Centre Number

Section I – Part B (continued)

Student Number

Question 31 (7 marks)

Fungi are a natural source of antibiotics. A scientist developed a new antibiotic by exposing a fungus to radiation. **7**

Information relevant to this antibiotic:

- It stops the activity of an enzyme in pathogenic bacteria.
- It has no effect on a similar enzyme in humans.
- The chemical composition of the enzyme in humans differs from the enzyme in the pathogenic bacteria by two amino acids.
- It is used to treat humans infected with the pathogenic bacteria.

Using this example and other relevant knowledge, describe how advances in our understanding of biology have implications for society.

....

....

....

....

....

....

....

....

....

....

....

....

Question 31 continues

Question 31 (continued)

End of Question 31

2011 HIGHER SCHOOL CERTIFICATE EXAMINATION

Biology

Section II

25 marks
Attempt ONE question from Questions 32–36
Allow about 45 minutes for this section

Answer parts (a)–(c) of the question in Section II Answer Booklet 1.
Answer parts (d)–(e) of the question in Section II Answer Booklet 2.
Extra writing booklets are available.

Question 32	Communication
Question 33	Biotechnology
Question 34	Genetics: The Code Broken?
Question 35	The Human Story
Question 36	Biochemistry *(Not included in this reproduction)*

Question 32 — Communication (25 marks)

Answer parts (a)–(c) in Section II Answer Booklet 1.

(a) The diagram below shows a microscopic view of a neuron.

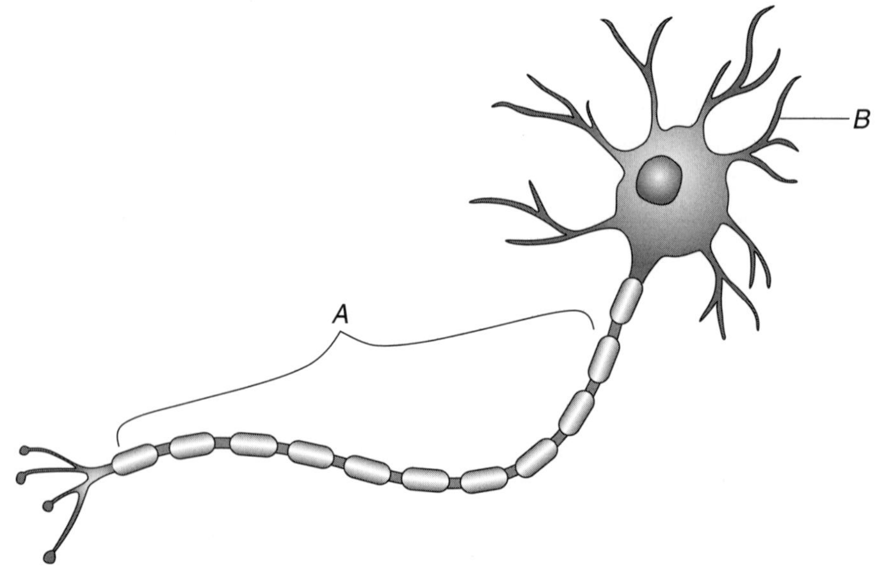

(i) What are the structures labelled *A* and *B*? Clearly indicate in your answer which is *A* and which is *B*. **2**

(ii) Identify the role of structure *B*. **1**

(b) Explain why some stimuli do not result in an action potential. Use a graph to support your answer. **4**

(c) Laser Assisted In-Situ Keratomileusis (LASIK) is a procedure that uses a laser beam to cut into the cornea to change its shape in a precise manner.

(i) Identify the functions of the cornea. **2**

(ii) Explain how LASIK could be used to treat myopia so that glasses or contact lenses are no longer required. **3**

Question 32 continues

Question 32 (continued)

Answer parts (d)–(e) in Section II Answer Booklet 2.

(d) Compare the hearing of THREE mammals, including humans, giving possible reasons for their differences in hearing. **6**

(e) Explain how the hearing aid and cochlear implant technologies assist hearing. Refer to the anatomy and function of the human ear and the role of the brain in coordinating responses to stimuli. **7**

End of Question 32

Question 33 — Biotechnology (25 marks)

Answer parts (a)–(c) in the Section II Answer Booklet 1.

(a) The diagram below shows a recombinant DNA technique.

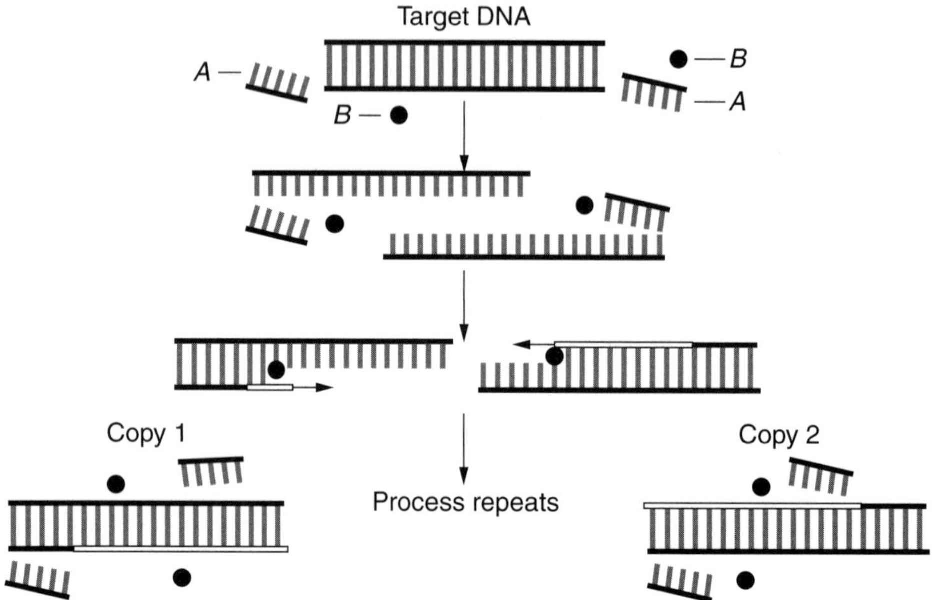

(i) What are represented by the components labelled *A* and *B*? Clearly indicate in your answer which is *A* and which is *B*. 2

(ii) Identify the process shown in the diagram. 1

(b) Explain why transcription and translation result in different products. Use a diagram to support your answer. 4

(c)

Ants cultivate grass for food

A species of ant has been observed to clear the surface of the ground for 5 metres around its colony and bury grass seeds there. The ants remove any other plants that may spring up among their crop. When the seeds are ripe, the ants harvest the seeds for food and retain some seeds for next year's crop. A ten-year study has shown that the average size of the grass seeds increased by 5%.

(i) Outline the difference between qualitative and quantitative observations, using examples from the text above. 2

(ii) Explain why the ant activity could be interpreted as biotechnology. 3

Question 33 continues

Question 33 (continued)

Answer parts (d)–(e) of the question in Section II Answer Booklet 2.

(d) Explain the risks and benefits to society of using recombinant DNA technologies. **6**

(e) Enzymes have been developed from bacteria that remove contaminating insecticides and herbicides from waterways. **7**

Explain how the technologies you have studied in this option could be applied to the development and production of such enzymes. Refer to techniques such as those used to extract enzymes, amplify genes and recombine DNA.

End of Question 33

Question 34 — Genetics: The Code Broken? (25 marks)

Answer parts (a)–(c) in Section II Answer Booklet 1.

(a) The table shows parts of DNA sequences.

	DNA Sequence
Original DNA	AAC TCG GTC AAT ATG
Mutation 1	AAC TCC GTC AAT ATG
Mutation 2	AAC TCG GTA ATA TGC

(i) What types of mutations are *Mutation 1* and *Mutation 2*? Clearly indicate in your answer which is *Mutation 1* and which is *Mutation 2*. **2**

(ii) Using the information provided in the table, what is the effect of *Mutation 1* in part (i)? **1**

	T	C	A	G
T	TTT Phe (F) TTC Phe (F) TTA Leu (L) TTG Leu (L)	TCT Ser (S) TCC Ser (S) TCA Ser (S) TCG Ser (S)	TAT Tyr (Y) TAC Tyr (Y) TAA STOP TAG STOP	TGT Cys (C) TGC Cys (C) TGA STOP TGG Trp (W)
C	CTT Leu (L) CTC Leu (L) CTA Leu (L) CTG Leu (L)	CCT Pro (P) CCC Pro (P) CCA Pro (P) CCG Pro (P)	CAT His (H) CAC His (H) CAA Gin (Q) CAG Gin (Q)	CGT Arg (R) CGC Arg (R) CGA Arg (R) CGG Arg (R)
A	ATT Ile (I) ATC Ile (I) ATA Ile (I) ATG Met (M) START	ACT Thr (T) ACC Thr (T) ACA Thr (T) ACG Thr (T)	AAT Asn (N) AAC Asn (N) AAA Lys (K) AAG Lys (K)	AGT Ser (S) AGC Ser (S) AGA Arg (R) AGG Arg (R)
G	GTT Val (V) GTC Val (V) GTA Val (V) GTG Val (V)	GCT Ala (A) GCC Ala (A) GCA Ala (A) GCG Ala (A)	GAT Asp (D) GAC Asp (D) GAA Glu (E) GAG Glu (E)	GGT Gly (G) GGC Gly (G) GGA Gly (G) GGG Gly (G)

Question 34 continues

Question 34 (continued)

(b) Explain how children with different ABO and Rhesus blood groups could be the offspring of the same parents. Use diagrams to support your answer. **4**

(c) Generally, chickens raised for meat are fed intensively to achieve fast growth. However, this also leads to the generation of a lot of body heat.

It has been found that appetite is reduced when animals are hot. A scientist, using the technology of selective breeding, has developed a featherless chicken.

The scientist claims that 'these chickens are not genetically modified as they come from a natural breed and were not cloned'.

(i) Outline the difference between selective breeding and animal cloning. **2**

(ii) Discuss the scientist's claim using your knowledge of selective breeding. **3**

Answer parts (d)–(e) in Section II Answer Booklet 2.

(d) Discuss the use of the Human Genome Project and traditional inheritance studies as ways to identify the location of harmful genes. **6**

(e) The origin of the dingo (an Australian native dog) is uncertain. It has commonly been classified as *Canis familiaris dingo*, a subspecies of the domestic dog, *Canis familiaris*. **7**

It has recently been suggested that the dingo should be reclassified as *Canis lupus dingo* to reflect a closer relationship to the Asian wolf, *Canis lupus*.

Explain how the technologies you have studied in this option could be applied to identifying the relationships between these species.

End of Question 34

Question 35 — The Human Story (25 marks)

Answer parts (a)–(c) in Section II Answer Booklet 1.

(a) The diagrams show labelled fossil strata that were found in different locations within a region.

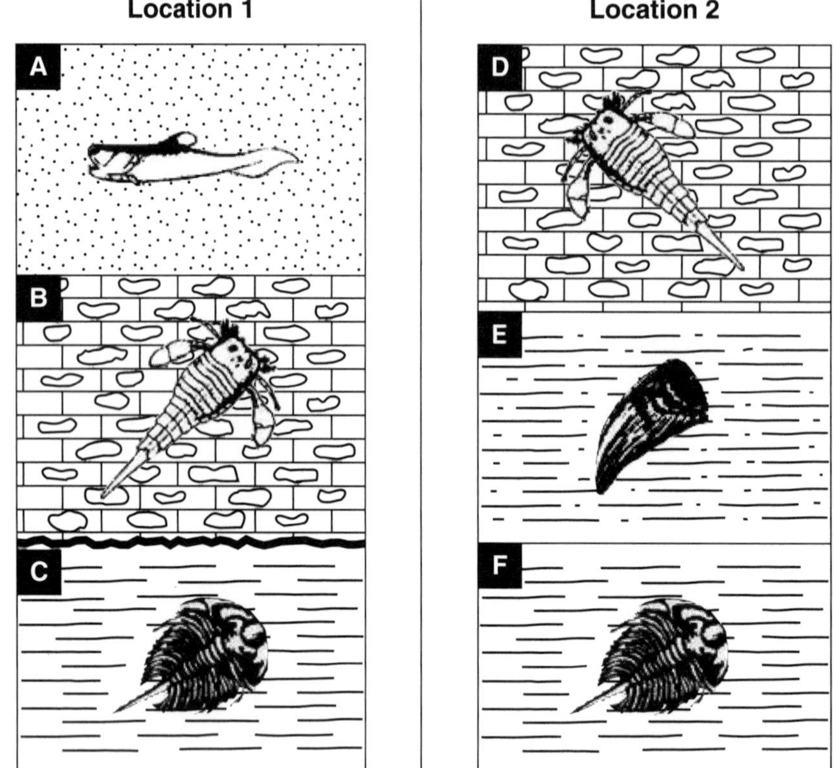

(i) Which rock layer(s) is/are the oldest and which is/are the youngest in this region? **2**

(ii) Name the relative dating technique used in your answer to part (i). **1**

(b) Explain TWO alternative views of the evolutionary relationships between FOUR identified hominids of the same genus. Use diagrams to support your answer. **4**

Question 35 continues

Question 35 (continued)

(c) In 2004, a new species, Homo floresiensis, was identified from fossils found in a cave on an Indonesian island. Dated to 18 000 years ago, these fossils were found in sediment containing stone tools and cooking hearths.

The adult skull of this upright bi-pedal fossil organism has a volume of 380 cm^3. The size of each adult organism was approximately the same as a three-year-old modern human.

Their arm to leg ratio was slightly larger than modern humans. They had hard, thicker eyebrow ridges than us, a sharply sloping forehead, and no chin.

The cooking hearths contained charred bones of game species, each estimated to weigh more than 1 000 kilograms. Stone tools included blades, spear heads and cutting and chopping utensils.

(i) Outline the fossil evidence from this description that suggests the advanced cultural development of *Homo floresiensis*. **2**

(ii) Discuss the validity of the classification of *Homo floresiensis* as a species separate from other known fossil hominids. **3**

Answer parts (d)–(e) in Section II Answer Booklet 2.

(d) (i) By explaining the mechanisms of biological evolution, compare early and current *Homo sapiens*. **4**

(ii) Explain why there is a greater genetic diversity between human populations in Africa than between human populations in other continents. **2**

(e) Insectivorous microbats and fruit-eating megabats are classified as members of the order Chiroptera. Lemurs are fruit and leaf eaters and are classified as members of the order Primates. **7**

Scientists have recently hypothesised that megabats are closer in evolutionary terms to lemurs than to microbats.

Explain how you could use technologies and the data they generate to test this hypothesis.

End of Question 35

End of paper

2011 HSC Examination Paper

Sample Answers

Section I Part A *(Total 20 marks)*

1 B If animals and plants contain the same chemicals in DNA, they must have come from the same ancestor.

2 C Lipids are carried as chylomicrons or lipoprotein particles.

3 B Enzyme function is dependent on optimum temperature being maintained constantly.

4 C Chromosomes are made up of 60% protein and 40% DNA.

5 B The mother's genotype is homozygous recessive and the father's is heterozygous for freckles.

6 D Ronald Ross determined mosquitos were carriers of malaria, Redi was responsible for discrediting spontaneous generation and Koch developed the method for establishing identification of particular microbes and the cause of a disease.

7 B Cleanliness in food preparation and hygiene reduce the transmission of pathogens but they cannot eliminate them or prevent them from being ingested.

8 B or D Cytotoxic T cells destroy virally infected cells. Both D and B were accepted by the examiner because of their similarity.

9 C Prions are approx. 10 nm, viruses 20–100 nm and bacteria 1–5 μm.

10 D Cell division, spontaneous generation and cloning do not result in variation.

11 A The difference in water concentration creates the osmotic pressure gradient.

12 B The insect needs to minimise water loss from waste elimination as it lives in a typically dry environment.

13 A Blood delivers high concentrations of carbon dioxide to lungs and leaves with high concentrations of oxygen.

14 B Mucous membranes and cilia are part of the primary or first line of defence, preventing the physical entry of pathogens into the body.

15 B The flightless birds are found on continents that were once joined together as the supercontinent of Gondwana.

16 A It is the helper T cells that cause multiplication of B cells, resulting in the interaction between the cell types.

17 D Parents in generation I are the same phenotype to start with and generation II has three phenotypes; therefore, the alleles are codominant.

18 A Veins contain valves (2 and 4 in LS) and are thinner than arteries (1) but thicker than capillaries (3).

19 A Antibiotics will treat only bacterial infections; malaria is caused by a plasmodium, a protozoan.

20 C Short periods of rapid change are the basis of punctuated equilibrium.

Section I Part B

21 (a) Two methods of treating drinking water are flocculation/filtration and chlorination. (2 marks)

(b) Filtration removes solid particles that may contain pathogens. Chlorination kills any microscopic pathogens that may not be filtered out. *(2 marks)*

22 (a) The purpose of the experiment was to determine the pH range that the enzyme functions best at. *(1 mark)*

(b) Include a sample without enzyme at the same temperature/pH as a control and check each sample at the same time intervals. *(2 marks)*

(c)

Student	*Time*	*Observation*
1	40	clear
2	40	clear
3	40	clear
4	40	clear
5	40	cloudy
6	40	cloudy

(1 mark)

23 The work of scientists that led to the determination of DNA structure was primarily independent rather than a true collaborative effort. Wilkins and Franklin worked in the same laboratory; however, their working relationship was strained as a result of a number of factors. These included the fact that Wilkins did not look upon Franklin as his equal and the fact that Franklin was unable to participate in collegial discussion in the male-dominated scientific community. Wilkins and Franklin did not work with Watson and Crick, who were independently working on the same project. However, Watson and Crick did work collaboratively with each other. It was only because of Wilkins' casual disclosure of Franklin's X-ray crystallography photographs of DNA molecules to Watson (without her knowledge or permission) that Watson and Crick were able to produce their now famous model. This does not demonstrate intentional collaboration on the part of those involved; if anything it highlights the competition between the scientists. However, had the four scientists collaborated, the DNA structure would have been discovered earlier. *(5 marks)*

24 (a) A named viral disease that is controlled by a vaccine could be, for example, influenza, measles or smallpox. *(1 mark)*

(b) (i) A vaccine prepared in this way will produce immunity to a specific viral disease, because it contains a weakened form of the virus that will not harm the body but will produce an immune response that will result in memory cells for the specific antigen. *(2 marks)*

(ii) The safety of the vaccine is ensured by the removal of the nucleic acid, thereby preventing the virus from having the ability to replicate within a host. *(2 marks)*

25 (a) Labels B and D *(2 marks)*

(b) C *(1 mark)*

(c) Aldosterone is produced by the adrenal glands and regulates the salt balance by increasing the permeability of the nephron to sodium. If a high blood volume and pressure is detected, the production of aldosterone is reduced, so that less salt and water is absorbed by the nephron tubules, eliminating the excess salt and water as urine. *(2 marks)*

26 (a) Two similarities between the graphs are, for example:

- the data shown is for the same time period, 1988–1996
- data is presented in multiples of thousands.

Two differences between the graphs are, for example:

- Graph A shows reported cases while Graph B shows estimated annual cases.
- Graph A includes the full figures on the axes, while Graph B simplifies the number of cases by showing the cases as multiples of thousands. *(4 marks)*

(b) Two additional types of data that could be used to evaluate the effectiveness of the global polio vaccination program could include the percentage of the population represented by these figures or the countries in which the cases occurred. *(2 marks)*

27 Features of an epidemiological study that could be used to identify the cause of the non-infectious disease include the age, gender, general health and socioeconomic background of all the people who have the disease, the diet/food source of the affected people versus the non-affected, the number of affected people and non-affected people in each of the aged care homes, the location of the homes and the timeframe of diagnosis. This data will provide important information about the patterns of the disease, the frequency of the disease, the population affected and the geographical location, all of which will provide vital clues about possible causes. *(4 marks)*

28 (a) Gene expression in skin stem cells is important for the maintenance of health after skin is broken because it enables the wound to heal. The stem cells will develop into epidermal cells to replace the damaged ones, as well as produce white blood cells that can prevent infection or red blood cells to carry oxygen to the area. *(2 marks)*

(b) Features of the new artificial blood made from stem cells that are likely to be an improvement over existing forms of artificial blood are that there would be no tissue rejection complications as the blood would be a perfect match. Because the blood contains red blood cells it could easily carry/supply oxygen to the body tissues, as current artificial blood does not. *(3 marks)*

29

Dependent variable	Number of microbes
Independent variable	Type of medium used to culture the microbes
Control	Temperature/volume of medium
Safe work practices to be followed	Keep culture vessels sealed and autoclave prior to disposal

(5 marks)

30 (a) X is a temperature probe; Y is a data logger. *(2 marks)*

(b) If the temperature of the animal is determined only by the ambient temperature, the animal would lose temperature at a similar rate to which it gained the temperature, rather than retaining the heat. *(3 marks)*

31 Advances in our understanding of biology have had significant impact on society over the years. As this example demonstrates it is possible for us to manipulate living things in our environment for our own advantage. In this case a fungus was used to develop an antibiotic that can be used to treat infection caused by a specific pathogenic bacterium. The discovery of antibiotics (by Fleming) has had major implications for society. Antibiotics work by killing bacteria (e.g. by perforating the bacterial cell walls). This advance has significantly reduced the incidence of bacterial diseases such as cholera, and this has increased the life span of people around the world. The work of Pasteur is also linked to the issue of disease control. His swan-neck experiment, in which broth exposed directly to the air showed signs of decay while broth in the flask with the swan neck stayed clear, disproved the idea of spontaneous generation and consequently led to new methods of preventing the spread of disease, such as by improved hygiene procedures like washing hands when handling food. *(7 marks)*

Section II—Options

Question 32—Communication

(a) (i) *A* axon, *B* dendrite *(2 marks)*

(ii) Dendrites are branching extensions of the cell body that conduct impulses involving chemical transmitter substances towards it or away from it. *(1 mark)*

(b) Some stimuli do not result in an action potential because a minimum threshold needs to be reached before an impulse is produced. Stimuli will cause the polarity of the axon membrane to be reversed. The polarity needs to be greater than the threshold potential to produce an action potential.

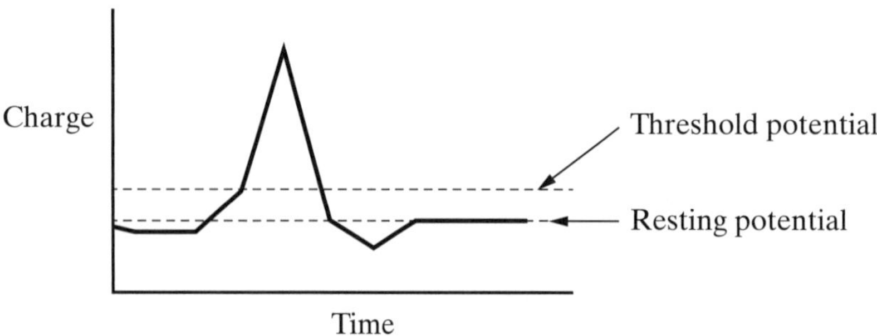

(4 marks)

(c) (i) The cornea is the thick, transparent front part of the eyeball. It refracts/bends light rays that pass through it. *(2 marks)*

(ii) LASIK can be used to treat myopia so that glasses or contact lenses are no longer required because the procedure makes very precise incisions into the cornea to flatten it, thereby changing its refractive ability. The image focal point is then increased so that it no longer falls short of the retina. *(3 marks)*

(d) Mammals are able to hear differently because of the ability to detect different frequency ranges. This is usually related to the type of organism, the environment it lives in and its habit. Humans are able to detect frequencies between 16 Hz and 20 000 Hz. Although children can typically hear up to about 25 000 Hz, the ability to hear the higher frequencies diminishes with age. Other mammals, such as dogs, can detect frequencies up to about 40 000 Hz, which is much higher than humans can. This enables dogs to be sensitive to sounds when hunting or locating prey, particularly in the dark. Some mammals are able to reflect very high frequencies or ultrasonics off objects to detect the size, motion or position of anything in their path. For example, bats when flying can produce frequencies up to 120 000 Hz for echolocation, used to navigate and locate food. *(6 marks)*

(e) Cochlear implants basically bypass the middle ear and enable sound vibrations to reach the inner ear. Hearing aids are used to amplify sounds from the outer ear through to the middle ear, so that they can be transmitted to the cochlea, where tiny sensitive hair cells in the organ of Corti detect the sound and transmit the signal along the auditory nerve to the brain for interpretation. This has enabled countless numbers of people to hear better if they have had damage to the outer ear (tympanic membrane) or middle ear (ossicles). Should there be severe or profound sensorineural hearing loss or damage to the inner ear, then the hearing aid is not going to be of assistance to the individual. A cochlear implant can be used to bypass the outer and middle ears as well as the cochlea. It comprises a microphone, speech processor and implanted receiver with electrode array that can stimulate the nerve endings in the cochlear, thereby transmitting the messages detected (vibrations) to the brain where the response is coordinated. Should the nerve endings be damaged, then this device will not assist either and the person would not be able to detect any sound whatsoever. The brain interprets the messages transmitted by the auditory nerve and coordinates responses to the stimuli by means of communication, or movements that are required to convey messages or keep the individual out of danger. *(7 marks)*

Question 33—Biotechnology

(a) (i) *A* DNA fragment, *B* restriction enzyme *(2 marks)*

(ii) Polymerase Chain Reaction (PCR) *(1 mark)*

(b) The two stages in polypeptide production are translation and transcription. Translation involves the synthesis of a polypeptide sequence from mRNA by translating genetic information from both the mRNA molecule and the ribosome. It takes place on the ribosome. Transcription takes place in the nucleus and involves the synthesis of RNA from a DNA template.

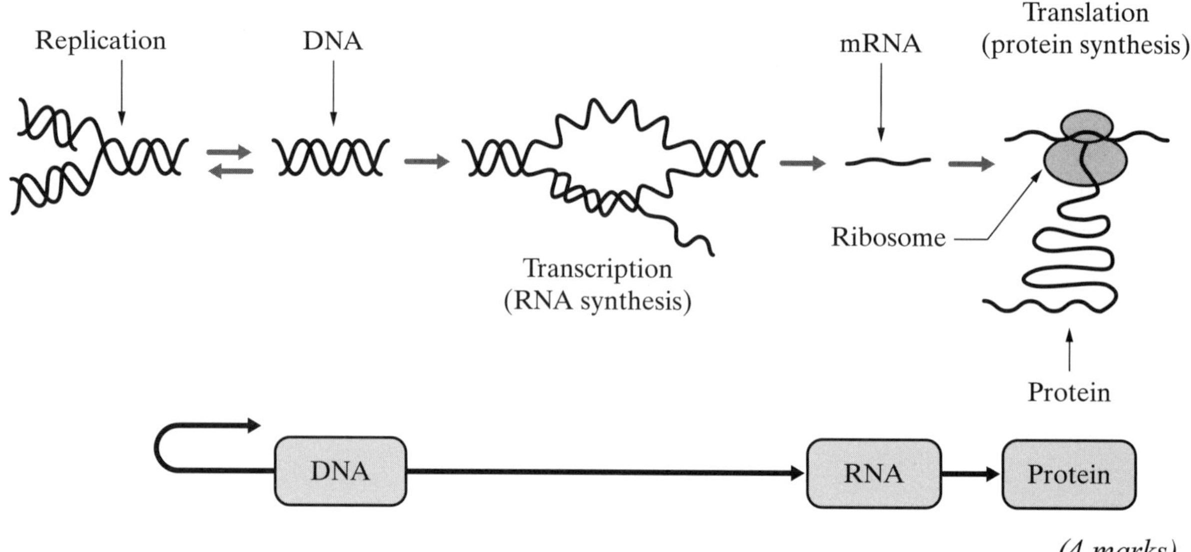

(4 marks)

(c) (i) A qualitative observation is one that describes the features of something, such as ‘ants remove any other plants that may spring up among their crop’. A quantitative observation is one that uses a measurement, for example ‘a ten-year study has shown that the average size of the grass seeds increased by 5%’. *(2 marks)*

(ii) The ant activity could be described as biotechnology because the ants ‘retain some seeds for next year’s crop’. This implies that they are actively selecting some seeds (the bigger ones) for planting the following year and, as a result, over a ten-year period the average size has increased. The ants have manipulated the seeds to their benefit. *(3 marks)*

(d) Recombinant DNA technologies have many benefits to society. Some of these include the ability to map genes on chromosomes through the use of genetic markers and the use of genetic engineering to modify the genetic blueprint of one organism by inserting genes from another. These technologies have resulted in the mapping of the human genome and creation of various transgenic species (e.g. BT cotton, BT corn, frost-free strawberries) that have advanced agriculture in particular. A large range of pharmaceutical products has also been produced utilising recombinant technology on microorganisms, such as antibiotics to treat infection and human insulin for diabetics. However, there may be risks associated with such advances, as they have not been used for long enough to determine any side effects that may arise and there is the possibility of the loss of biodiversity in the long run. *(6 marks)*

(e) Technologies that can be used to produce such enzymes could include:

- Genetic engineering can be used to alter genetic material in living cells so that they can carry out different functions or produce new substances. This can be done through the use of microinjection to transfer new DNA into the original cells.
- Recombinant plasmids can be used as vectors into plants, as can biolistics or electroporation.
- Recombinant DNA technology cuts the genes out of one species and inserts it into the DNA of another species in particular locations by the use of restriction enzymes to cut and enzyme DNA ligase to stick the ends back together. This is often done by using a bacterial plasmid (a circular self-replicating chromosome). Once transferred, the grown bacteria will produce the new substance according to the introduced gene. This has been used to extract particular enzymes that are then transferred to other species, creating transgenic organisms. Uses of such technology include the creation of bacteria that break down toxic waste from oil spills, the introduction of pest resistance into plants and the production of blood clotting protein and human growth hormone.
- Gene amplification is a process where a cell deliberately increases the number of copies of a particular gene more than the number of genes in the remainder of the genome. As such, it is not a form of duplication but increases the number of copies of a gene (proteins specifically coded for) available for transcription. This has been associated with the genes involved in resistance to some drugs. *(7 marks)*

Question 34—Genetics: The Code Broken?

(a) (i) *Mutation 1* represents a base substitution; *Mutation 2* represents a frame shift mutation. *(2 marks)*

(ii) In *Mutation 1* TCG is replaced by TCC. This will not have a significant effect, as both these code for the same amino acid, Serine or 'Ser', and the remaining reading frame is unchanged. *(1 mark)*

(b)

Possible blood type genotypes of parents	*Resulting blood group*
ii	O
$I^A I^B$	AB
$I^A I^A$, $I^A i$	A
$I^B I^B$, $I^B i$	B

The blood type can be determined by using the above blood type genotypes in a punnet square, for example: Parents $I^A I^B \times I^B I^B$.

Parent gametes	I^A	I^B
I^B	$I^B I^A$	$I^B I^B$
I^B	$I^B I^A$	$I^B I^B$

In this punnet the possible blood types of the offspring are 50% $I^B I^A$ (Type AB) and 50% $I^B I^B$ (Type B).

The rhesus blood groups can similarly be determined using a trihybrid cross, which is a grid of 8 × 8 squares, as there are three different pairs of alleles.

Genetic makeup	*Blood type*
++	Rh positive
+–	Rh positive
–	Rh negative

Parents' Rh types	*Possible allele combinations*	*Possible Rh in the children*	
Both +	++ and ++	++	positive
Both +	++ and +–	++ or +–	positive
Both +	+– and +–	++ or +– or –	positive or negative
Both –	– and –	–	negative
One + and one –	++ and –	+–	positive
One + and one –	+– and –	+– or –	positive or negative

- Rh negative children can have parents who are either Rh positive or Rh negative.
- If both parents are Rh negative, then the children will be Rh negative. *(4 marks)*

(c) (i) 'Selective breeding' is the identification of desirable characteristics in an organism (typically plant or animal) and the deliberate use of that animal in breeding, e.g. the use of a prize bull (or his semen) to inseminate cows in order to pass the specific characteristics on to the offspring. Repeated selection of individuals with the desired characteristics will increase the gene frequency and can, over time, change the population characteristics. Animal cloning is a specific technology that creates genetically identical offspring by the promulgation of body cells of an individual, e.g. striking cuttings from a plant. *(2 marks)*

(ii) The scientist claims that the featherless chickens have been made by selective breeding as the featherless gene was found in a breed of bird (chicken) that has fewer feathers. By breeding this bird with the hens he has manifested the featherless characteristic in his birds and, as such, not modified the genes or tampered with the original DNA artificially. *(3 marks)*

(d) Traditional inheritance studies look at the way a phenotype is transmitted within a population; they do not analyse the specific genes present in an individual's DNA. They can be used to create a pedigree and thus can be used to provide a means of predicting the possibility of the inheritance patterns. As such they have been used over the years to try to identify the possible combinations of specific characteristics being passed on in a family or in animal breeding. They have been used to try to identify the possibility of disease transmission, where diseases have been found to be inheritable; however, they can only be used to determine the likelihood of transmission if the breeding pair is determined to be true breeding for that characteristic.

Mapping the genome in the Human Genome Project has made it possible to physically identify the genes responsible for the cause of specific diseases. By identifying the actual gene, the sequence and the position on the chromosome it has been possible to revolutionise the treatment and prevention of human diseases and provide understanding of evolutionary relationships with other organisms. *(6 marks)*

(e) Some of the technology that is available to assist with the classification of the dingo could include:

- Genetic fingerprinting could be used on each of the Australian dingo, Asian wolf and domestic dog to identify possible relationships or their frequency. Some DNA is extracted from a sample tissue (e.g. blood, skin or hair) and this is then cut using restriction enzymes into fragments that can be separated by gel electrophoresis. This involves the DNA fragments being placed on a gel plate across which a current is applied. The negatively charged DNA moves towards the positive electrode, the fragments separating according to size as the smaller ones move faster. The separated gels can then be analysed using the southern blotting technique for similarities in the band locations or 'fingerprints'.
- Genome mapping of the Australian dingo, Asian wolf and domestic dog could be undertaken. The gene sequences and the possible similarities are compared using a computer, in order to find the specific lineage or the closest match.

- Recombinant DNA mapping could be used to tag and identify specific genes by using Fluorescent in situ Hybridisation (FISH) to see if the genes exist in any of the three possible ancestors.

The relative percentages of matches derived by any of these methods could determine the likelihood of the dingo being more closely related to either the Asian wolf or the domestic dog, depending on which had the higher correlation. *(7 marks)*

Question 35—The Human Story

(a) (i) The oldest layers are C and F, while the youngest layer is A. (2 marks)

(ii) Using the principle of superposition, stratigraphic correlation can be used as a relative dating technique. (1 mark)

(b) Different interpretations of evidence result in alternative views of hominid evolution. These can change over time as new technology makes it possible to re-examine the evidence. Below are two possible alternatives for the evolutionary relationships between hominids of the same species. Many different forms of evidence can be used, e.g. skull size and cranial capacity, dentition, presence of brow prominence, size and shape of jaw, nose and face.

The following diagrams show how differences in evidence from skulls can lead to varying opinions about the evolution of hominids.

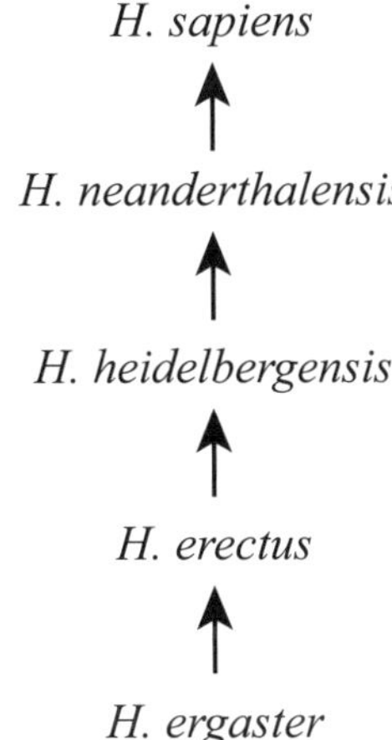

Or an alternative form:

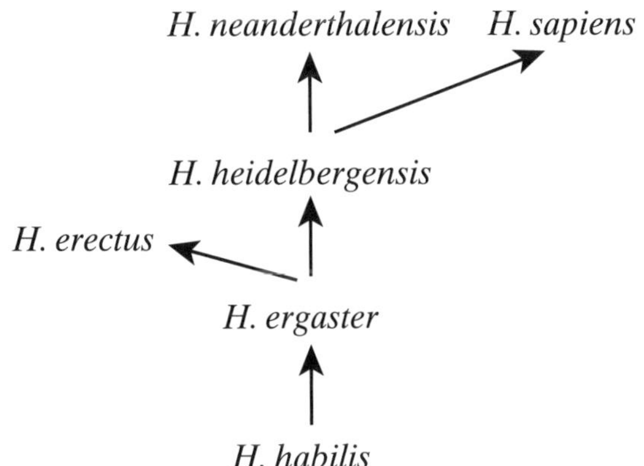

(4 marks)

(c) (i) The discovery of stone tools and cooking hearths suggests cultural development. Stone tools such as cutting and chopping utensils, as well as cooking hearths, suggest a collaborative approach to food preparation. *(2 marks)*

(ii) It is valid to classify *Homo floresiensis* as a species separate from other hominids, including *H. sapiens* and *H. habilis*, because of their markedly different anatomy. For example, the volume of the skull of *H. floresiensis* is 380 cm^2 while the volume of the skull of *H. sapiens* 1305 cm^2. In addition, the arm to leg ratio of *H. floresiensis* is slightly larger than that of *H. sapiens*. Finally, the facial features of *H. floresiensis* are markedly different from those of modern man (*H. sapiens*) but reminiscent of those of earlier hominids. The thick eyebrow ridges, sharply sloping forehead and no chin are features present in both *H. floresiensis* and *H. erectus*. These anatomical differences reflect significant differences between *H. floresiensis* and other hominids, indicating that *H. floresiensis* would be unable to reproduce successfully with other species of hominids. *(3 marks)*

(d) (i) Biological evolution is driven by the process of natural selection. When considering *Homo sapiens* we can observe changes in the species over time. Offspring who varied from their parents and other members of the species and had favourable features survived (i.e. natural selection) and reproduced. Over many generations characteristics such as reduced brow ridges, bowl-shaped pelvis, upright posture, thinner skull bones, and increased cranial capacity and size of the cerebral cortex developed. These enabled the evolution of *H. sapiens* with the capacity to learn, speak and use language and symbols, leading to the development of art, including painting and sculpture, more complex settlements, agriculture and social structures. Such developments allowed for the flourishing of individuals with a range of skills beyond those of earlier hominids, which were out competed. *(4 marks)*

(ii) There is greater genetic diversity between human populations in Africa than between human populations in other continents because of the founders effect. The founders effect results in a loss of genetic variation as a small number of individuals from the original population moves and establishes a new community. The earliest hominids are known to have evolved on the African continent, accumulating genetic variability over many generations. As hominids migrated out of Africa to new continents, they established new and more recent communities with less diversity. *(2 marks)*

(e) When analysing the evolutionary relationships between lemurs and megabats scientists are able to use technologies such as comparative DNA sequencing and DNA hybridisation. These techniques are particularly useful in evolutionary studies in determining how closely related different species are. Specifically, species that have recently diverged from a common ancestor will be expected to show a high degree of similarities in their DNA make up. To test the hypothesis that megabats are more closely related to lemurs than to microbats, scientists will need to take a sample of DNA from the three species and complete the following experiments to confirm the hypothesis.

During DNA hybridisation, DNA samples from each group are heat treated, forming single strands that are subsequently combined. For example, single strands of lemur DNA are combined with that of megabats and then microbats. Complementary DNA strands will combine, with a high degree of pairing occurring between closely related species. If the hypothesis is correct, more DNA strands will combine in samples of DNA from lemurs and megabats than in those from lemurs and microbats. Similarly, in comparative DNA sequencing specific sections of the genome are analysed to determine how closely related different species are. Should the hypothesis be correct, lemurs and megabats will have similar genomes, unlike lemurs and microbats. *(7 marks)*

CHAPTER 11

BOARD OF STUDIES
NEW SOUTH WALES

2012
HIGHER SCHOOL CERTIFICATE
EXAMINATION

Biology

General Instructions

- Reading time – 5 minutes
- Working time – 3 hours
- Write using black or blue pen
  Black pen is preferred
- Draw diagrams using pencil
- Board-approved calculators may be used
- Write your Centre Number and Student Number where required

Total marks – 100

Section I

75 marks

This section has two parts, Part A and Part B

Part A – 20 marks
- Attempt Questions 1–20
- Allow about 35 minutes for this part

Part B – 55 marks
- Attempt Questions 21–30
- Allow about 1 hour and 40 minutes for this part

Section II

25 marks
- Attempt ONE question from Questions 31–35
- Allow about 45 minutes for this section

Section I
75 marks

Part A – 20 marks
Attempt Questions 1–20
Allow about 35 minutes for this part

Use the multiple-choice answer sheet for Questions 1–20.

---

1 What is the name of the process that results in organisms containing DNA from different species?

(A) Transcription

(B) Transgenics

(C) Translation

(D) Translocation

2 The diagram shows different vertebrate embryos at the same stage of development.

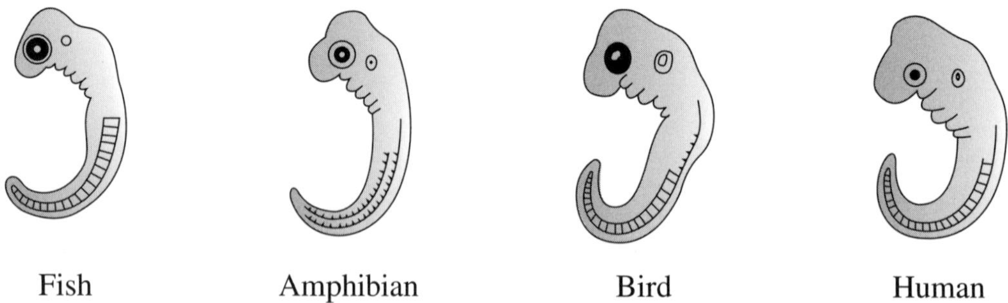

How do the embryos provide evidence for evolution?

(A) The embryos have different shaped eyes.

(B) Different adults evolve from the embryos.

(C) The embryos have structures that look similar.

(D) Divergent evolution results in common characteristics in the embryos.

3 Which of the following are all forms of defence that prevent the entry of pathogens into the body?

(A) Cilia, sweat, saliva

(B) T cells, B cells, antibodies

(C) Inflammation, skin, phagocytosis

(D) Stomach acid, mucus, lymph system

4 The diagram shows a pathogen called Giardia.

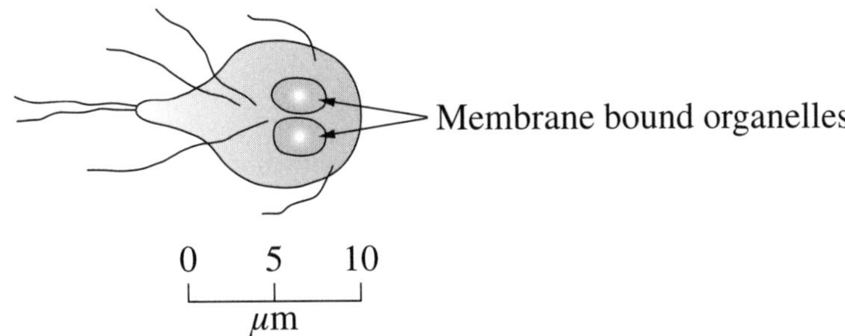

What type of pathogen is Giardia?

(A) Bacterium

(B) Prion

(C) Protozoan

(D) Virus

5

> *Huntington's Disease* is caused by an inherited gene that codes for a toxic protein.
>
> *Kwashiorkor* is a disease caused by a deficiency of proteins in the body.
>
> *Mesothelioma* is a disease caused by a gene mutation in the lungs after exposure to asbestos.

Which row in the table correctly classifies these diseases?

	Huntington's Disease	*Kwashiorkor*	*Mesothelioma*
(A)	Genetic	Nutritional	Environmental
(B)	Nutritional	Environmental	Environmental
(C)	Genetic	Nutritional	Genetic
(D)	Nutritional	Environmental	Genetic

6 How do vaccinations prevent disease?

(A) They increase the inflammation process.

(B) They enable the infected cells to seal off the pathogen.

(C) They increase the number of antibodies against the pathogen.

(D) They decrease the number of antigens that trigger the immune response.

7 Why is it important to continue research into new antibiotics?

(A) New prion diseases have been recently discovered.
(B) Resistant bacteria have evolved from the overuse of antibiotics.
(C) Viral infections require a broad range of antibiotics for eradication.
(D) New diseases are discovered regularly and all require new antibiotics.

8 Identical twins have the same genotype.

Why are there small differences between the phenotypes of identical twins?

(A) Some genes are not co-dominant.
(B) Environment affects the expression of genes.
(C) Both parents are homozygous for those phenotypes.
(D) Chromosomes segregate independently during meiosis.

9 What feature can be used to distinguish mature xylem cells from mature phloem cells?

(A) Phloem cells are located in vascular bundles.
(B) Phloem cells have a cytoplasm.
(C) Xylem cells are located in the leaves.
(D) Xylem cells have cell walls.

10 The diagram shows a model of the movement of ions (represented by X) across a semipermeable membrane.

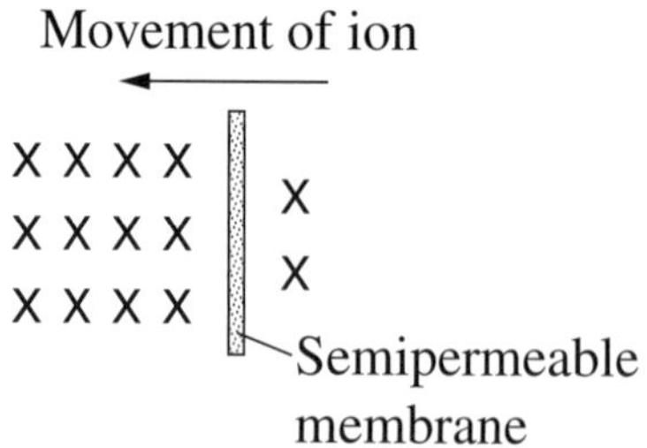

What type of process is modelled in the diagram?

(A) Osmosis
(B) Filtration
(C) Diffusion
(D) Active transport

11 Which of the following is an example of *hybridisation*?

(A) The insertion of a bacterial gene for herbicide resistance into a cotton plant

(B) The culturing of a cell taken from the root of a carrot to form a small plant

(C) Artificial insemination of a domestic cat with wild cat semen to produce a Bengal cat

(D) A cutting taken from one variety of apple tree grafted onto the stem of a different variety of apple tree

12 Nitrogenous waste is at its highest concentration in

(A) plasma in the renal vein.

(B) plasma in the renal artery.

(C) fluid in the collecting ducts of the kidney.

(D) interstitial fluid in the cortex of the kidney.

13 Why do organ transplants trigger an immune response in a recipient?

(A) Antigens in the recipient's body recognise the organ as foreign.

(B) Cell surface markers on the organ attack the recipient's white blood cells.

(C) Antibodies in the recipient stimulate the production of antigens on the organ.

(D) The recipient's white blood cells recognise the antigens on the organ as foreign.

14 What is a role of the kidney in freshwater fish?

(A) To remove water from the fish

(B) To absorb salt from the environment

(C) To excrete concentrated urine from the fish

(D) To decrease nitrogenous waste lost to the environment

15 Which of the following results in an increased absorption of water from the collecting tubule of the nephron?

(A) An increase in the length of the collecting tubule

(B) A decrease in ADH released into the blood by the pituitary

(C) A decrease in glucose moving from the renal tubule into capillaries

(D) An increase in the concentration of dialysate solution in renal dialysis

16 A student carried out an investigation to identify the presence of microbes in water from different sources.

The student's lab notes are shown.

	Control	*Bottled water*	*Tap water*	*Tank water*
Inoculation of an agar plate with water sample	✗	✓	✓	✓
Incubation at 37°C	✓	✓	✓	✓
Appearance of agar plate				

What can be inferred from these results?

(A) The inoculation loop was not sterilised properly.

(B) The water from each of these sources is unsafe to drink.

(C) These water sources are contaminated with the same microbe.

(D) The agar plates were contaminated prior to the beginning of the experiment.

17 Which of the following correctly identifies the relationship between alleles, chromosomes and genes?

(A) Genes contain chromosomes and alleles.

(B) Chromosomes contain genes but not alleles.

(C) Alleles are found in chromosomes but not in genes.

(D) Genes are parts of chromosomes and have different alleles.

18 How does a plant respond in order to keep cool on an extremely hot day?

(A) It grows smaller leaves.

(B) It opens stomata in the leaves.

(C) It grows more hairs on the surface of the leaves.

(D) It decreases the number of stomata on the top of the leaves.

19 Haemoglobin provides an adaptive advantage to an endotherm in a cold environment because it allows

(A) more oxygen to be dissolved in plasma.

(B) the organism to decrease its metabolic rate.

(C) more energy to be available to the organism.

(D) less carbon dioxide to be transported in the blood.

20 A student performed a first-hand investigation in an attempt to model natural selection.

> Each week, a jar was refilled with a fresh packet of cream biscuits containing five different types for people to eat. After a month, there were mostly lemon cream biscuits in the jar.

What is the limitation of this investigation as a model for natural selection?

(A) There is no variation in the 'species'.

(B) Characteristics are not transmitted to successive generations.

(C) Variants of the 'species' do not have the same chance of 'survival'.

(D) Unfavourable characteristics are selected out of the population over time.

2012 HIGHER SCHOOL CERTIFICATE EXAMINATION

Biology

Centre Number

Student Number

Section I (continued)

Part B – 55 marks
Attempt Questions 21–30
Allow about 1 hour and 40 minutes for this part

Answer the questions in the spaces provided. These spaces provide guidance for the expected length of response.

Question 21 (5 marks)

You performed a first-hand investigation to estimate the size of blood cells.

(a) How did you estimate the size of the cells? **2**

....

....

....

....

(b) Draw a scaled diagram that shows the features of both a red blood cell and a white blood cell. **3**

Question 22 (7 marks)

Students in a class conducted a first-hand investigation to test the hypothesis that if CO_2 were continually bubbled in water then the pH would decrease over time.

One student presented the data in the graph, as shown.

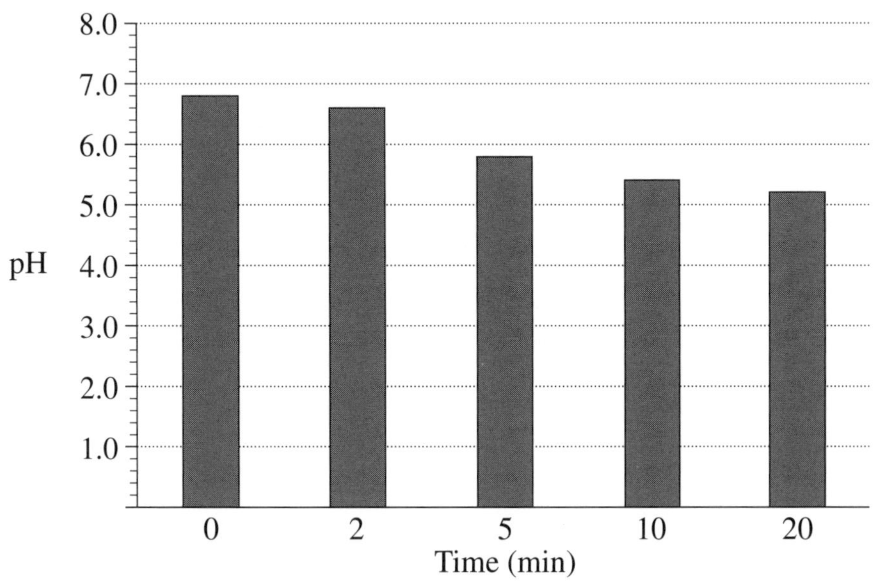

(a) Present these data in a table. **2**

Question 22 continues

Question 22 (continued)

(b) On the grid below, plot the data from the table in part (a) and draw a curve of best fit. **3**

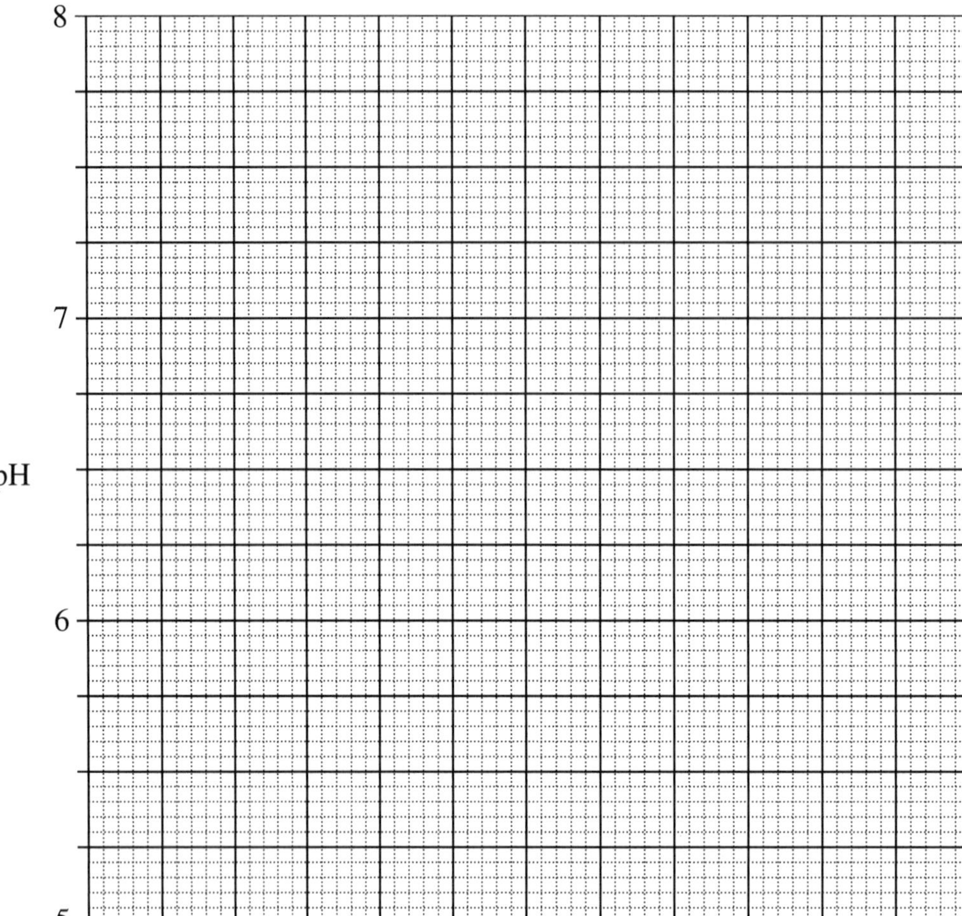

(c) Why is it better to represent these results as a curve of best fit rather than as a column graph? **2**

....

....

....

....

End of Question 22

2012 HIGHER SCHOOL CERTIFICATE EXAMINATION

Biology

Section I – Part B (continued)

Centre Number

Student Number

Question 23 (5 marks)

A non-infectious disease was observed in a mother and her four sons who live with her. She has no daughters. The father of these children does not have the disease and does not live with them. The woman's parents and her two sisters who live overseas do not have the disease.

(a) Her doctor suspects that the disease is NOT inherited. Identify data that could be collected to investigate a non-inheritable cause of the disease. **1**

....

(b) A geneticist suspects that the disease is inherited.

(i) Draw the family pedigree for this disease. **2**

(ii) From the evidence, what indicates that the disease could be the result of a recessive allele and not be sex linked? **2**

....

....

....

....

Question 24 (5 marks)

You conducted first-hand investigations to test the effects of temperature, pH and substrate concentration on enzyme activity.

(a) Complete the following table by identifying the variables for ONE of your investigations. 1

Independent variable	*Dependent variable*	*Kept constant*
....	....	pH, substrate concentration, enzyme concentration

(b) Outline how you measured the activity of an enzyme in your investigation. In your answer, name the enzyme. 2

....

....

....

....

(c) Describe how a condition needed for optimal enzyme activity would be expected to vary between endotherms and ectotherms. 2

....

....

....

....

....

....

2012 HIGHER SCHOOL CERTIFICATE EXAMINATION

Biology

Section I – Part B (continued)

Centre Number

Student Number

Question 25 (6 marks)

Beadle and Tatum's experiment involved the analysis of bread mould growth. Bread mould uses an enzyme to make the amino acid arginine.

The diagram shows bread mould growth after culturing on two different media. In one part of the experiment, the bread mould had been irradiated before culturing.

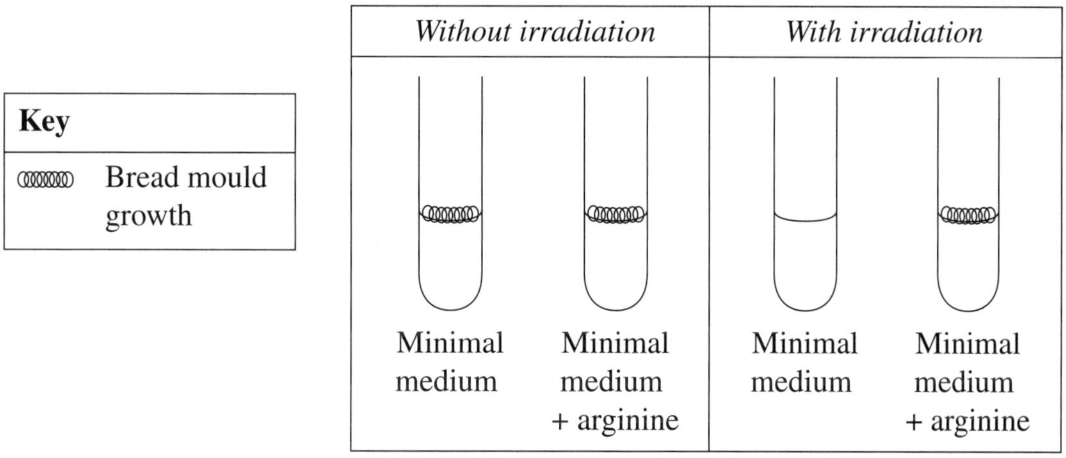

(a) How do these results support the 'one gene – one protein' hypothesis? **4**

....

....

....

....

....

....

(b) Justify the types of secondary sources that would be acceptable to use to research Beadle and Tatum's experiment. **2**

....

....

....

Question 26 (5 marks)

A scientist performed an epidemiological study to investigate the cause and effect relationship of smoking and lung cancer as follows: 5

1. Handed out a scientifically valid questionnaire to all colleagues (n = 144) at work
2. Checked that there were an equal number of male and female respondents
3. Discovered that there were more non-smoking respondents than smoking respondents. Removed some of the non-smokers until both groups had equal numbers
4. Checked that all the respondents had a medical check-up in the past year
5. Analysed data, wrote the paper and published it in a scientific blog.

From the information provided, analyse the methodology used by this scientist.

....

....

....

....

....

....

....

....

....

....

2012 HIGHER SCHOOL CERTIFICATE EXAMINATION

Biology

Section I – Part B (continued)

Centre Number

Student Number

Question 27 (4 marks)

During a major international horse event in Australia, a group of horses, including some from overseas, is discovered to be infected by a deadly virus. This virus is only found in Australia. 4

Give reasons for strategies that could be carried out to control this disease outbreak.

Question 28 (5 marks)

Explain the relationship between replication of DNA and evolution. 5

....

....

....

....

....

....

....

....

....

....

2012 HIGHER SCHOOL CERTIFICATE EXAMINATION

Biology

Centre Number

Section I – Part B (continued)

Student Number

Question 29 (5 marks)

Question 29 (5 marks)

(a) Complete the following diagram to show the process by which gametes are formed. **3**

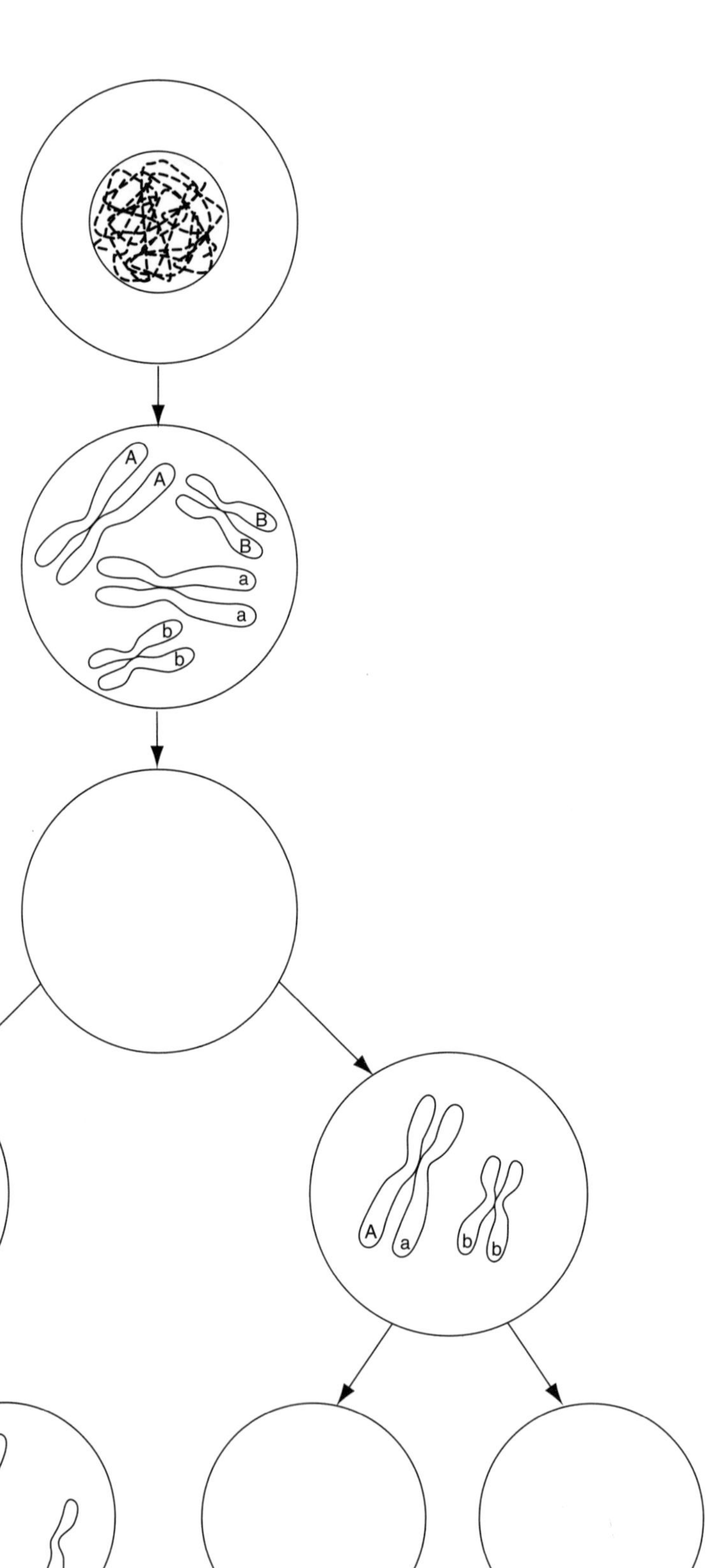

Question 29 continues

Question 29 (continued)

(b) How does the segregation of chromosomes during meiosis lead to a wide variety of gametes being produced? **2**

....

....

....

....

End of Question 29

Question 30 (8 marks)

Explain how the contributions of Louis Pasteur, Robert Koch and MacFarlane Burnet have increased our understanding of the nature and prevention of infectious disease. **8**

....

....

....

....

....

....

....

....

....

....

....

....

....

....

....

....

....

....

....

2012 HIGHER SCHOOL CERTIFICATE EXAMINATION

Biology

Section II

25 marks
Attempt ONE question from Questions 31–35
Allow about 45 minutes for this section

There are four Section II Answer Booklets labelled Part (a), Part (b), Part (c) and Part (d).

Answer each part of the question in the relevant Answer Booklet.

Extra writing booklets are available.

Question 31	Communication
Question 32	Biotechnology
Question 33	Genetics: The Code Broken?
Question 34	The Human Story
Question 35	Biochemistry *(Not included in this reproduction)*

Question 31 — Communication (25 marks)

Answer part (a) in Section II Answer Booklet – Part (a).

(a) Parts of the ear are shown below.

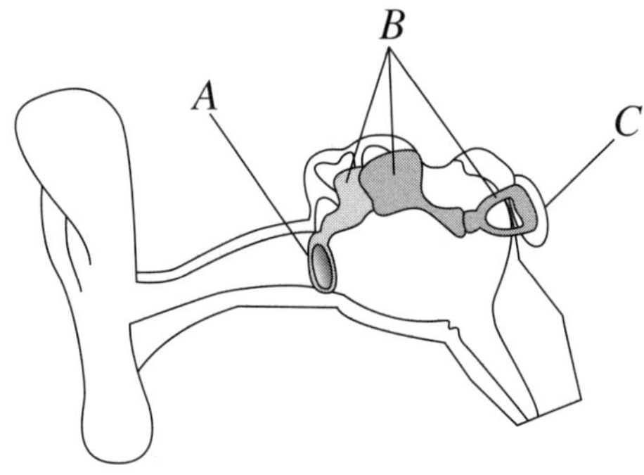

(i) Identify the parts *A*, *B* and *C*. **2**

(ii) Outline the functions of part *B*. **2**

Answer part (b) in Section II Answer Booklet – Part (b).

(b) An experiment was performed to model accommodation in the eye:

- a round bottom flask was filled with a solution
- the flask was placed on a stand on which it was able to rotate
- three lenses (I, J and K) of different thicknesses were attached to the surface of the flask
- a layer of fine paper was stuck around the opposite side of the flask
- a candle was placed near the flask
- different lenses could be brought into the light path by rotating the flask
- an image of the candle was observed on the layer of fine paper.

Question 31 continues

Question 31 (continued)

(i) Match THREE pieces of apparatus used in this experiment with THREE identified parts of the eye. **2**

(ii) Describe the quantitative data that could be collected in the experiment. **2**

(iii) Compare this model to mechanisms of accommodation in the eye. **2**

Answer part (c) in Section II Answer Booklet – Part (c).

(c) (i) Relate specialised features of the eyes of TWO named animals to their environment. **2**

(ii) Describe the detection of vibrations by fish and insects, and the events which lead to a response. **5**

Answer part (d) in Section II Answer Booklet – Part (d).

(d) Demonstrate how technologies help correct specific biological problems associated with human communication. **8**

End of Question 31

Question 32 — Biotechnology (25 marks)

Answer part (a) in Section II Answer Booklet – Part (a).

(a) With reference to TWO biotechnological practices that use yeast, copy and complete the following table in your answer booklet. **4**

Biotechnological practice	*Name of chemical produced by yeast*	*Purpose of chemical produced*

Answer part (b) in Section II Answer Booklet – Part (b).

(b) (i) Explain the purpose of a named transgenic organism. **3**

(ii) Construct a diagram that demonstrates the sequence of events that results in the formation of recombinant DNA in a transgenic organism. **3**

Answer part (c) in Section II Answer Booklet – Part (c).

(c) (i) Why could aquaculture be considered a biotechnology? **2**

(ii) Compare aquaculture with early biotechnologies applied to agriculture. **5**

Answer part (d) in Section II Answer Booklet – Part (d).

(d) Explain how a detailed understanding of cell chemistry has resulted in society making more effective use of biotechnology. Support your answer with examples. **8**

Question 33 — Genetics: The Code Broken? (25 marks)

Answer part (a) in Section II Answer Booklet – Part (a).

(a) With reference to TWO types of cloning, copy and complete the following table in your answer booklet. **4**

Type of cloning	*Process used*	*Example*

Answer part (b) in Section II Answer Booklet – Part (b).

(b) (i) Construct a model of a section of a double stranded DNA molecule containing the sequence GAT using a selection of the symbols below. Include a key in your answer. **3**

(C) (P) (S) (A) (G) (T) (U)

(ii) Outline how genes are expressed. **3**

Answer part (c) in Section II Answer Booklet – Part (c).

(c) (i) What is the Human Genome Project? **2**

(ii) Explain how recombinant DNA technologies can be used to identify the position of a gene on a chromosome. **5**

Answer part (d) in Section II Answer Booklet – Part (d).

(d) Compare and contrast the effects of germ line mutation and transposable genetic elements on whole organisms. **8**

Question 34 — The Human Story (25 marks)

Answer part (a) in Section II Answer Booklet – Part (a).

(a) (i) Using the diagram, name TWO features of the *Homo sapiens* skeleton which make it unique among the primates. **2**

(ii) Outline how ONE named non-skeletal characteristic has assisted primates in their evolution. **2**

Question 34 continues

Question 34 (continued)

Answer part (b) in Section II Answer Booklet – Part (b).

(b) Examine the data collected on *Homo neanderthalensis* fossils found in recent years.

Date	*Data*
2008	• Full sequence of mitochondrial DNA of a range of Neanderthal fossils from within and between different fossil sites • The sequence is nearly identical within one fossil site • The sequence is very different between fossil sites
2010	• Full sequence of Neanderthal nuclear DNA • 1%–4% of genes in European modern humans are specific Neanderthal genes • No identifiable specific Neanderthal genes in modern sub-Saharan African humans • No specific modern human genes in nuclear DNA of Neanderthal fossils

(i) What inferences can be made about Neanderthal populations, based on the data collected in 2008? **3**

(ii) What inferences can be made about migration and breeding, based on the data collected in 2010? **3**

Answer part (c) in Section II Answer Booklet – Part (c).

(c) How have polymorphism and clinal gradation contributed to the diversity of human populations? In your answer, use at least one human phenotype as an example. **7**

Answer part (d) in Section II Answer Booklet – Part (d).

(d) (i) For TWO named hominids, describe fossil evidence that infers the use of technology which led to cultural development. **2**

(ii) Predict and justify future directions of human biological evolution in the context of TWO technologies already developed by modern humans. **6**

End of Question 34

End of paper

2012 HSC Examination Paper

Sample Answers

Section I Part A *(Total 20 marks)*

1 B Transgenic species are made by the insertion of another species' genetic information. This is passed onto the next generation.

2 C Similar structures on the embryos show that they have evolved from a common ancestor.

3 A These are all physical barriers and can prevent the entry of pathogens into the body.

4 C It is a protozoan as it is a single-celled organism with membrane-bound organelles.

5 A Inherited characteristics are genetic. Protein deficiency can be caused by poor diet, so it is nutritional. Mutations caused by exposure to asbestos are environmental.

6 C Vaccinations, through the introduction of harmless/dead strains of the disease, enable the body to build up antibodies so that it can fight the disease should it be exposed to it.

7 B Due to natural selection of bacteria, some antibiotics are ineffective and we have a problem with combating some severe bacterial infections.

8 B While identical twins may be identical genetically, they may express these genes differently (phenotype) due to the environment in which they grow up.

9 B Mature phloem cells are living tissue; mature xylem cells are not.

10 D The ions are moving across the semipermeable membrane against the diffusion gradient.

11 C The use of two varieties to produce unique characteristics of apple tree is an example of hybridisation. Under usual circumstances the act of mixing different species or varieties of animals or plants produces hybrids.

12 C The waste is taken from the collecting ducts to the bladder for removal from the body.

13 D The body produces T lymphocytes which would identify foreign tissue and destroy it.

14 A Freshwater fish release dilute urine.

15 A This is the only possible answer as the other options are incorrect.

16 D The control had the same growth as that of each sample of water.

17 D This is the correct definition for each term.

18 B This promotes transpiration, cooling the plant.

19 A More oxygen means that there can be grater rates of respiration to produce heat.

20 B The lemon cream biscuits are not able to reproduce, passing on their favourable characteristics.

Section I Part B

21 (a) The size of cells can be estimated by using a mini grid slide with known dimensions. The specimen is placed on this when under the microscope. *(2 marks)*

or The size of cells can be estimated by dividing the size of the field of view's diameter by the number of cells you can count across the field of view.

(b)

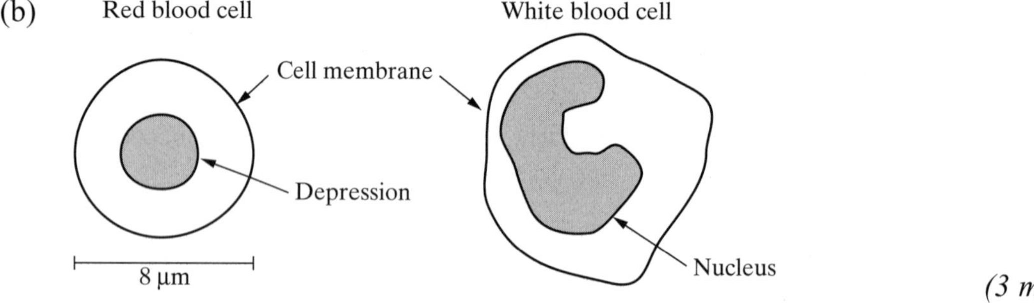

(3 marks)

22 (a)

Time (min)	*pH*
0	6.8
2	6.6
5	5.8
10	5.4
20	5.2

(2 marks)

(b)

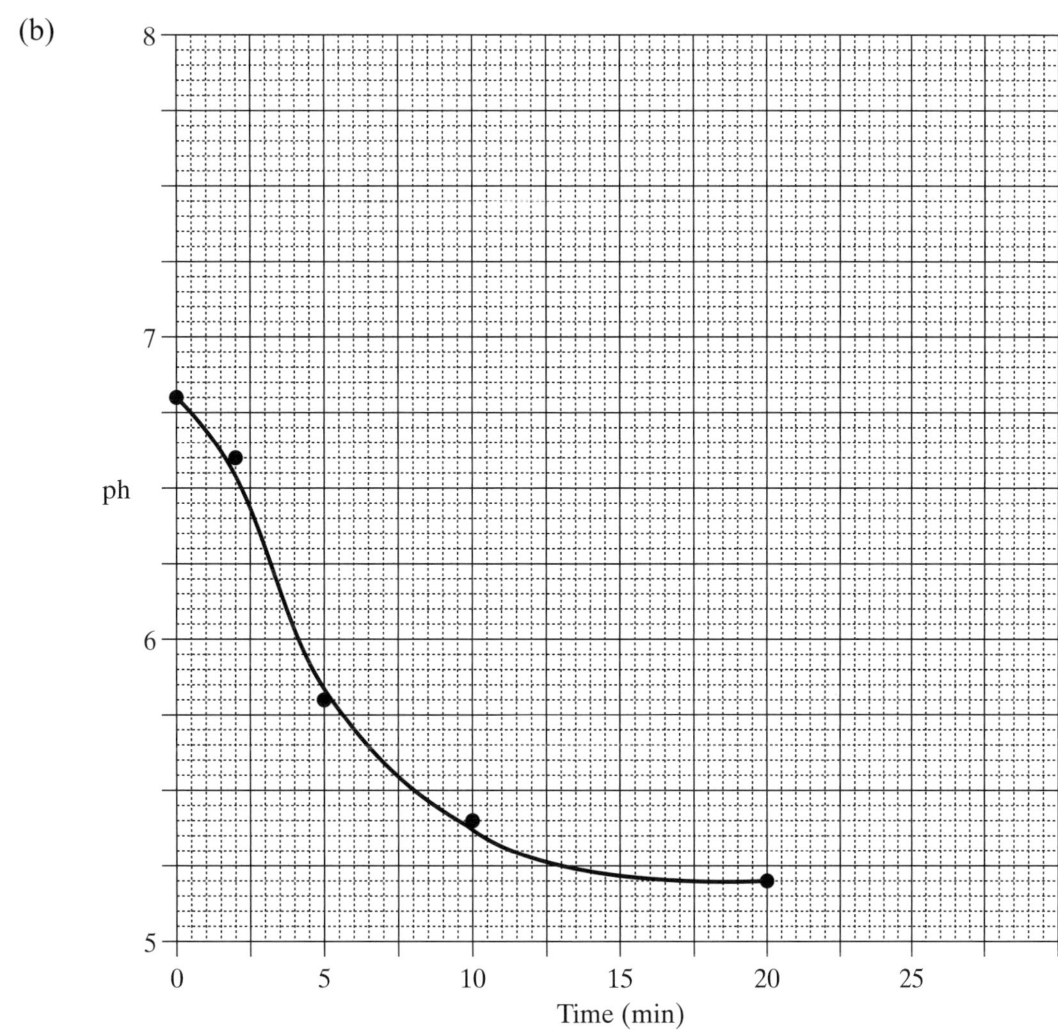

(3 marks)

(c) Presenting the result as a curve of best fit rather than a column graph makes it easier to see the relationships between the two variables, and the trend in the data becomes more obvious. *(2 marks)*

23 (a) Data that would be useful in determining that the disease was not inherited could include checking the grandparents of the mother and the parents or siblings of the father. The disease could be recessive. *(1 mark)*

(b) (i)

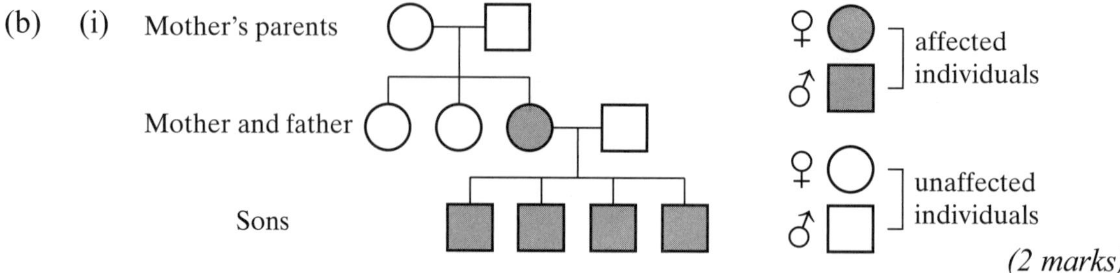

(2 marks)

(ii) The disease could be the result of a recessive allele rather than sex linked, because the father does not have it and each of the sons does. The father passes a Y to each of the sons and they inherit the affected X from their mother. *(2 marks)*

24 (a) Answers will vary, but the following is one possibility.

Independent variable	*Dependent variable*	*Kept constant*
Temperature	Time for substrate to clot	pH, substrate concentration, enzyme concentration

(1 mark)

(b) Answers will vary. The following is one example.

Rennin was used to clot milk at different temperatures. A stopwatch was used to measure how long it takes for the rennin to clot the milk at different temperatures. *(2 marks)*

(c) In endotherms, the temperature that the enzyme requires for optimum function will remain relatively constant. In an ectotherm, the temperature will vary with the ambient temperature and the activity will change relative to the temperature. *(2 marks)*

25 (a) The results resemble the work of Beadle and Tatum because the bread mould did not grow in the sample that was irradiated. This represents a change that would have occurred to an enzyme that prevented growth when it was irradiated. Beadle and Tatum exposed strains of mould to X-rays, causing the creation of strains lacking enzymes that enabled the production of essential nutrients required for normal growth. Each change was on a particular location on the chromosome and hence the specific enzyme. This led to the 'one gene – one enzyme' hypothesis, which was later modified to 'one gene – one polypeptide' to account for the fact that not all proteins are enzymes. *(4 marks)*

(b) The secondary sources that would be acceptable to use to research Beadle and Tatum's experiment would be those that come from government/university institutions or from reputable scientific journals that have been peer reviewed. *(2 marks)*

26 The methodology described will provide general results only and would not be considered valid in terms of the number studied (144). What is more, the scientist removed non-smoking subjects to even out the numbers of smokers and non-smokers, thereby reducing the numbers even more. There was no indication of the basis on which the scientist selected the subjects that were eliminated from the study other than the fact that there were more non-smokers than smokers. In order for the study to be valid, the larger sample size without bias being introduced is essential. Removing some of the non-smokers can skew the results by removing links to other possible sources for the cancers. Checking whether the respondents had medical checks in the past year may only identify that the symptoms were not present and not accurately identify whether the subjects had any changes that were not exhibited yet. You would require specific tests on the lungs themselves to determine this accurately. Other useful information that needs to be included about smoking habits and history of the participants includes history about living/working/socialising with smokers. Without such information, the scientist's conclusions would not be very significant in terms of medical findings. *(5 marks)*

27 Strategies that could be implemented to prevent the spread of the disease would include immediate quarantine of the infected horses to isolate the infected host animals and thereby reduce the chance of contact with non-infected horses. Restriction of movement of all other horses in and out of the country is important to prevent further spread outside of the country via the international horses, possibly even restricting them to local areas to stop spread across the country. *(4 marks)*

28 Replication of DNA is a crucial part of evolution. It is through the replication of DNA that changes arise through mitosis. The mutations that occur to the genetic code result in some changes that can be harmful and others that can turn out to favour the survival of the individual. It is this survival of genetically different organisms that is central to natural selection, where those that are best suited to their surroundings survive to reproduce and manifest the genes, resulting in new species over extended periods of time. Without mutation, there can be no evolution. *(5 marks)*

29 (a) The completed diagram should look like the following.

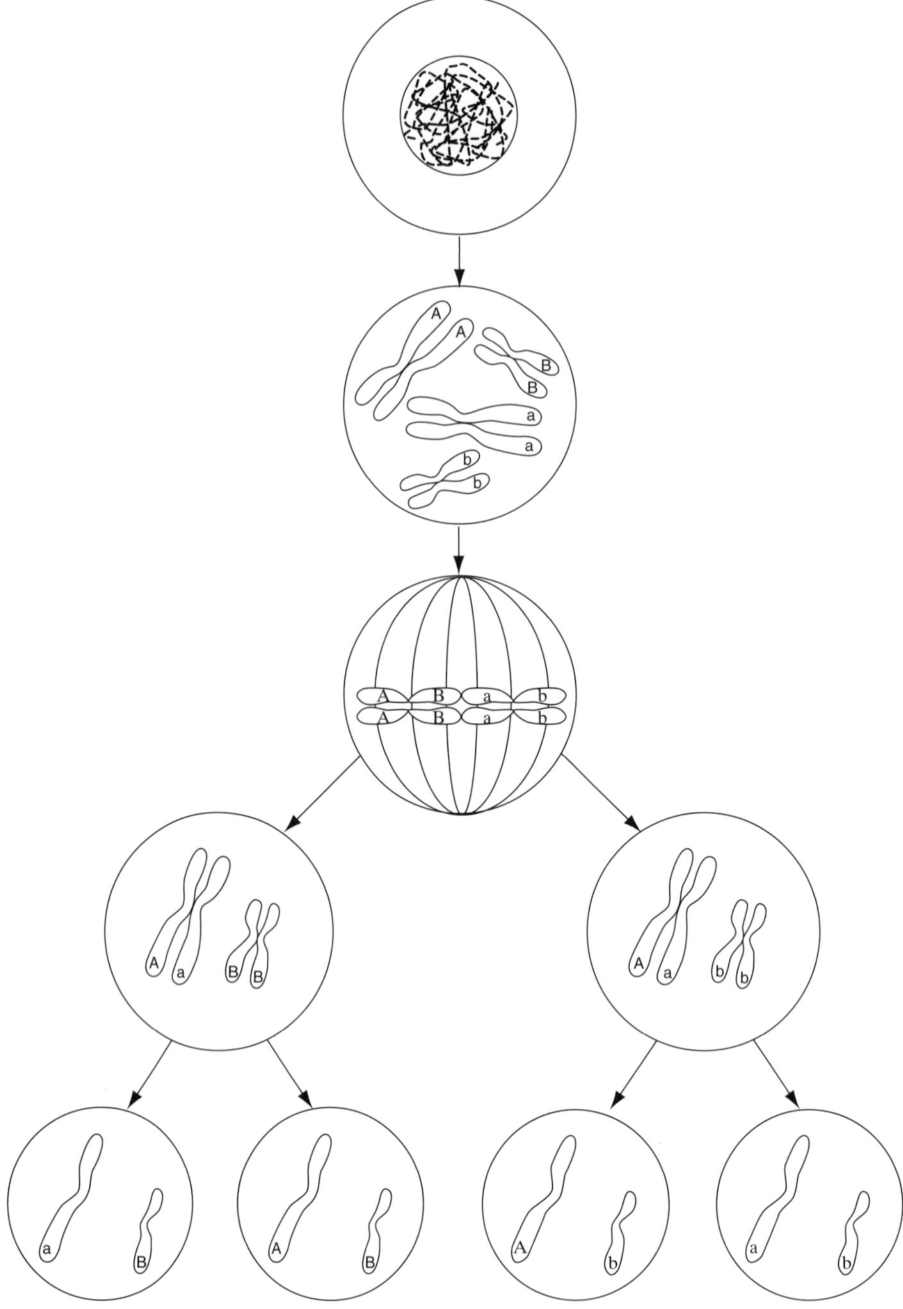

(3 marks)

(b) Segregation in meiosis leads to a wide variety of gametes being produced because there are a number of random ways in which the maternal and paternal chromosomes are combined in the daughter cells. The number of possible combinations in the haploid nuclei is very large. During the segregation, crossing over (where the tips of adjoining chromosomes are exchanged) can also take place, thereby increasing the variation. *(2 marks)*

30 MacFarlane Burnet was one of the key scientists associated with the understanding of immunology. He proposed the idea that the body recognises its own tissue and will not destroy it. This is crucial to the treatment of transplant patients that have new tissue needing to be kept and not destroyed as foreign material (clonal selection theory). This has resulted in the use of immunosuppression drugs in transplant patients to prevent tissue rejection. His work also dealt with the study of viruses and, in particular, influenza A, resulting in the development of a vaccine saving countless lives, preventing the spread of virulent strains of disease and dramatically reducing the incidence of some diseases.

Louis Pasteur has been attributed with demonstrating that spontaneous generation does not occur and most infectious diseases are caused by microscopic pathogens. This is now called the germ theory of disease. He demonstrated this by boiling broth in swan-necked flasks and then braking the top off one of the flasks, exposing the broth to air. The flask that was exposed to air then spoiled and the other remained uncontaminated, thereby showing that germs/pathogens are in the air and must be in contact with the host in order to cause disease. He went on to use methods of treating wine by heating it to 55°C for several minutes to destroy the microbes. This method (known as pasteurisation) is now widely used and has been adapted to treat products including beer and milk. Pasteurisation is commonly used to prevent microbes being consumed by people and causing illness, and has been adapted for sterilisation procedures to prevent the spread of disease.

Robert Koch was a German scientist who studied the cause of disease. In particular, he investigated the possible causes of anthrax, eventually isolating the bacterium that causes it. From his work he established a series of steps that need to be followed when identifying the cause of a particular disease. First, the pathogen must be present in every host with the disease. Second, the microorganism must be isolated from the host and grown in a pure culture. Third, when a new host is inoculated with the microorganism, the new host must develop the same symptoms as the original host. Finally, this microorganism must be able to be isolated and compared to, and be the same as, the original host.

Together these scientists have contributed significantly to the study of microorganisms, enabling the development of many practices that are vital to the nature and prevention of disease such as identification of pathogens, sterilisation and vaccination. *(8 marks)*

Section II—Options

Question 31—Communication

(a) (i) *A* ear drum/tympanic membrane, *B* ossicles, *C* oval window *(2 marks)*

(ii) The ossicles (made up of hammer, anvil and stirrup) are intended for the transmission of vibrations of air molecules under pressure through the middle ear to the oval window. This sets up pressure changes that vibrate the fluid in the inner ear. From here, the vibrations are received by the organ of corti within the cochlea in the inner ear and connected to the auditory nerve. *(2 marks)*

(b) (i) Answers will vary. The following is one example.

The fine paper represents the retina. *I*, *J* and *K* represent the eye's lens adjusted to different thicknesses. The round bottom flask filled with solution represents the eyeball filled with vitreous humour. *(2 marks)*

(ii) The quantitative data in the experiment would be collected by using the ruler to measure the distance (in cm) between the candle and the flask when the image is in focus on the fine paper at the back of the flask, as the different lenses (*I*, *J* and *K*) are placed in the path of the light. *(2 marks)*

(iii) The model is used to demonstrate accommodation in the eye whereby the lens changes thickness in order to produce a focused/sharp image on the retina, depending on how far away from the eye the image is located. The changing lens thickness correlates to the degree of bending or refraction of the light entering the eye that is required to focus the image. *(2 marks)*

(c) (i) Answers will vary. The following is an example.

Compound eyes in insects are made up of large numbers of individual light-detecting units called ommatidia, each with its own cornea and crystalline cone/lens that are connected to light-sensitive receptor cells that send messages to the brain. Having potentially thousands of these ommatidia in the compound eye enables the insect to have a large field of view.

Bees have three types of cones but these are sensitive to blue, green and ultraviolet rather than red wavelengths. This enables them to detect the coloured patterns in different species of plants more accurately. *(2 marks)*

(ii) Fish posses a lateral line that runs down the length of each side of the fish. It is a sensory canal (connected to nerves) that detects the pressure changes/vibrations in the waves/currents that surround the fish. It is thought that this method of detection is a method of the fish perceiving sounds from its surroundings. Fish also posses otoliths found at the back of the head, in a chamber that is lined with hair cells. Auditory nerves detect the differences in vibration between the hairs and the otolith, recorded as a nerve impulse carried to the brain. Insects can detect vibrations in their surroundings using their cuticle and antennae. Crickets have a tympanum (a cavity containing air) on each leg below the knee. This is connected by a release valve on the inside, with nerves detecting the pressure changes directly. Cicadas have a pair of tympana connected to an auditory organ by tendons at the base of the abdomen. *(5 marks)*

(d) There have been many technologies that have assisted in improvements to overcome biological problems encountered with communication between humans.

The verb *demonstrate* means to show by example. The following table summarises some possible examples that could be elaborated in producing an answer to this question.

Problem	*Cause*	*Technology*	*Method of improvement*	*Limitation*
Myopia/ hypermetropia	Ciliary muscles/ suspensory ligaments attaching to the lens do not work correctly	Spectacles/ contact lenses	Artificial lenses of different shapes and thicknesses can ensure that the image is focused onto the retina, worn as spectacles or directly on the eye as contact lenses. More recently, refractive laser surgery can be used to change the curvature of the cornea to alter its refractive power to focus the image on the retina.	These will only overcome visual difficulty if the retina and optic nerve are intact.
Cloudy vision	Cataract	Surgical removal or lens replacement	Protein deposits in the lens can be removed surgically using a laser. Another option is to substitute intraocular lenses for the original lens, which is removed.	These will only overcome visual difficulty if the retina and optic nerve are intact.
Impaired hearing	Damage to outer tympanic membrane or ossicles in the middle ear	Use of hearing aid	A hearing aid can be used. This is a battery-operated device that amplifies the vibrations entering the ear so that they can stimulate the nerve endings in the cochlea.	This requires battery replacement and will only work if the nerve endings in the inner ear are functional. Some frequencies will not be picked up and background noise causes interference.

Continued

Problem	*Cause*	*Technology*	*Method of improvement*	*Limitation*
Profound deafness	Damage to the hair cells in the cochlea within the inner ear	Cochlear implant	This is used when the normal auditory pathway does not function. Electrical stimulation of the nerve endings in the cochlea occurs after the attached implant array receiver detects sound waves from a microphone that has converted them to an electrical signal.	This requires battery replacement and will only work if the nerve endings in the cochlea are functional. It requires programming and surgical implantation.

(8 marks)

Question 32—Biotechnology

(a) Answers will vary. The following table shows two examples.

Biotechnological practice	*Name of chemical produced by yeast*	*Purpose of chemical product*
Beer/wine making	Ethanol	Preserves the beverage and prevents spoilage
Bread manufacture	Carbon dioxide	Assists in making the dough rise

(4 marks)

(b) (i) Transgenic species have been used to create plants and animals for agricultural purposes. Specific genes from one species are inserted into different species to change and/or enhance their characteristics. For example, a gene for producing protein in beans has been introduced into sunflower plants, resulting in a higher quality of protein in these plants. Similarly, genes for the production of human proteins have been inserted into milk-producing animals, which in turn produce biopharmaceuticals in their milk that are later extracted and purified for use. *(3 marks)*

(ii)

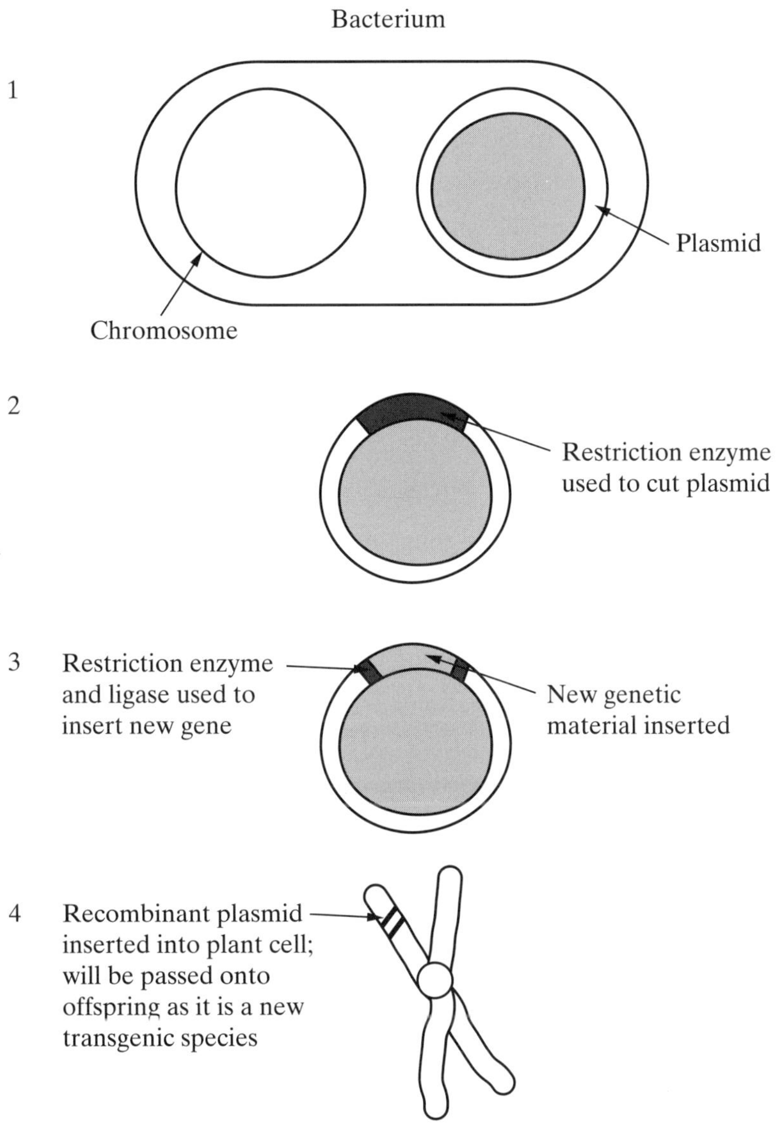

Bacterium

1

Plasmid

Chromosome

2

Restriction enzyme used to cut plasmid

3 Restriction enzyme and ligase used to insert new gene

New genetic material inserted

4 Recombinant plasmid inserted into plant cell; will be passed onto offspring as it is a new transgenic species

(3 marks)

(c) (i) Aquaculture is the cultivation of aquatic organisms, such as various species of fish, crayfish and algae, shellfish such as oysters and abalone, crayfish and algae. This has been practiced by humans for thousands of years. The farming of these species for various purposes such as food, jewellery, building or industrial material, or pharmaceuticals requires an understanding of biological processes. Therefore, it is considered a form of biotechnology. More recently, genetically modified organisms have been used in aquaculture to prevent the release of these organisms into the wild. *(2 marks)*

(ii) Early biotechnologies in agriculture involved the artificial section of characteristics that were favourable such as higher yielding grains and better quality fruit, resulting in more reliable and better yielding crops to ensure quality food supply. Domesticating animals and permitting interbreeding proved to be a reliable source of food and once again specific characteristics could be encouraged rather than leaving breeding to chance. This was seen as the first attempt to manipulate other organisms for human advantage, making it a rudimentary form of biotechnology. In contrast, the aquaculture that is occurring today is far more specialised and advanced, in that artificially modified organisms with manipulated genetic material are being used to cultivate very specific products for consumption, materials production or use. Oysters are seeded to produce pearls or high quality or genetically enhanced growth rates. Genetically modified prawns are being farmed that are resistant to viral disease prevalent in local waters. *(5 marks)*

(d) Greater understanding of cell chemistry has resulted in society making better use of biotechnology. This is evident in the many ways that organisms in the natural environment have been modified to produce more useful products for society's use.

Answers will vary and must outline multiple examples. These may include the following.

- Understanding of breeding mechanisms/patterns. This has enabled the understanding of inheritance patterns resulting in selective breeding and production of hybrids with desirable characteristics, including higher yield in wheat and higher butterfat content in cows.
- Knowledge of DNA structure/genome mapping and genetic screening. This relates to an understanding of the cause and effect of mutation or changes in the DNA sequence.
- Identification of stem cells and how they differentiate. Stem cells have the potential to be used for the creation of desired tissue or organs that can be used to potentially save lives and overcome rejection issues with transplant recipients.
- Understanding of DNA replication mechanism and development of technology such as cloning and recombinant DNA. This results in the ability to produce genetically modified organisms that can, for example, produce disease resistant crops (such as Bt cotton/corn), higher yields resulting from frost-free strawberries that contain a salmon gene, better quality wool, milk with higher protein, and biopharmaceuticals such as insulin. *(8 marks)*

Question 33—Genetics: The Code Broken?

(a) Answers will vary. The following table shows two types.

Type of cloning	*Process used*	*Example*
DNA cloning	Known genes are inserted into DNA of host	Used to produce unlimited amounts of identical copies of genes for study, production of useful protein, hormones, production of transgenic organisms (such as transferring the blue gene from a petunia into a carnation, creating the first light mauve carnation called Moondust) or insertion directly into plants such as tomatoes, cotton, corn of the *Bacillus thuringiensis* (Bt) gene which makes toxins that kill many pests
Whole organism cloning	A single cell is used to make an entire organism through recombinant DNA technology to create genetically identical offspring to the parent	Propagation of plants and production of animals for agriculture, such as fruit trees, sheep and pigs

(4 marks)

(b) (i)

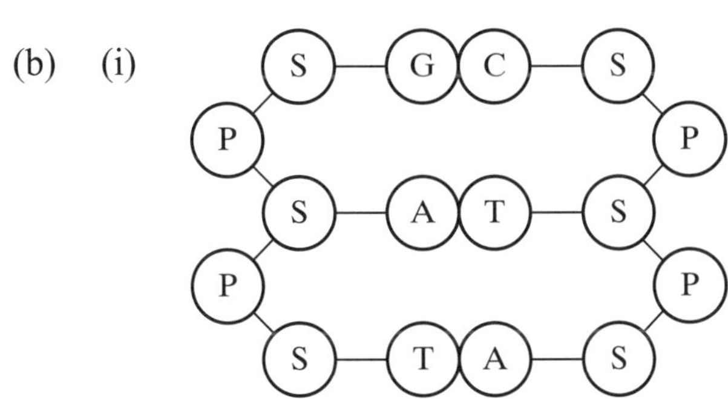

(3 marks)

(ii) Gene expression occurs when information from a gene is transcribed and translated into a polypeptide. Many factors manage the expression, which controls the process through splicing, translation and activation/inactivation of proteins. Gene expression is a regulated process in which information that is encoded on a gene is converted and used in the production of a specific protein or gene product. *(3 marks)*

(c) (i) The Human Genome Project was completed in 2003. It identified the position of genes on chromosomes through sequencing the whole genome, increasing understanding of biological and in particular genetic processes in humans and other animals. It has enabled us to understand more fully the evolutionary relationships between humans and other organisms together with an insight into the treatment and prevention of disease. *(2 marks)*

(ii) Restriction mapping has assisted in the physical mapping of chromosomes by using restriction enzymes to cut DNA into fragments. The fragments are labeled with radioactive phosphorous as a reference point, then cut by using restriction enzymes and the size of the fragment determined by gel electrophoresis. The position on the original fragment can then be deduced. The production of genetic probes has been used to identify genes; these are specific sequences of DNA that are complimentary to a gene or part of it. The probe can be cloned and tagged with fluorescent dye (FISH, fluorescent in situ hybridisation). These are then added to the separated chromosome strands that bond with the complimentary DNA, creating a hybrid. The tag will glow under a fluorescent light, showing its position. This can, for example, be used in genetic screening to identify abnormal chromosomes. *(5 marks)*

(d) The gonads contain germ line cells that divide by meiosis to produce sex cells or gametes that are haploid. Mutation in germ line cells is passed on to offspring, resulting in sources of variation. If advantageous to the organism, these mutations or changes can manifest themselves and become the predominant characteristic through natural selection. Transposable genetic elements were identified about 50 years ago. Although not greatly understood at the time, it is now accepted that genes can change position on a chromosome. The transposable elements of genes are called transposons or 'jumping genes'. These represent segments of DNA coding for enzymes that can move from one part of a chromosome to another. This may sometimes occur after being stable for many generations, or cause an insertion into another gene resulting in a mutation. For example, the phenotype of wild corn can change by the movement of the colour genes resulting in red striping in the kernels, the transposons being the red stripes. It is also thought that these transposons are responsible for the transfer of resistance to antibiotics in bacteria, by transferring resistant genes into the DNA of plasmids, which can then be transferred even among different bacteria and back to the main DNA strand, carrying the resistance with it. Previously, it was thought that mutation alone was causing the build-up of resistance to antibiotics, but now it is believed to be the result of a combination of mutation and transposon movement. Scientists are looking for ways of suppressing jumping genes to limit the exchange of genetic information between pathogens. *(8 marks)*

Question 34—The Human Story

(a) (i) Two features are the wide and bowl-shaped pelvis and the knee bent inwards to bring the foot under the centre of gravity. *(2 marks)*

(ii) Forward-facing eyes that are close together have enabled primates to develop stereoscopic vision. This has enabled the development of depth perception which is advantageous for greater accuracy when hunting and when brachiating/climbing in trees. *(2 marks)*

(b) (i) The identification of a full sequence of mitochondrial DNA being identical in one site only suggests that the individuals were descended from a single matriarchal female because the mitochondrial DNA can only ever be inherited from the female and not the male. The females did not tend to move between sites as the mitochondrial DNA is very different between the sites. *(3 marks)*

(ii) The data collected in 2010 suggests that European modern humans are descended from Neanderthals because 1%–4% of genes are present in the population. As there are no specific Neanderthal genes in the modern sub-Saharan Africans, it is fair to say that the Neanderthals did not migrate outside of Europe. Without the presence of any specific modern human genes in the nuclear DNA of Neanderthal fossils, it could be said that ancestral species of the two did not interbreed but were separate lines. *(3 marks)*

(c) Polymorphism has resulted from a wide geographical distribution across the world. Three major groups of races have formed, but subgroups of these have, in recent history, interbred. Three major races have been identified based on the continent of origin: Europe, Africa and Asia. Gradual changes or clinal gradation have been recognised in different geographical areas, resulting in more diversity and further renaming of groups that has continued to be accentuated by the migration and mixing of races over time. For example, blood groups have varied in native populations in various geographical areas. Three blood group alleles exist: A, B and O. These combine as a pair of genes and control the blood phenotype that prevails in particular races. This is an example of polymorphism in humans. The B allele is the rarest. It occurs most frequently in central Asia and in pockets within Africa, with the lowest frequency being in the New World and Australia. The blood group variations are an indicator of evolution in isolation due to the fact that A and B are dominant and O is recessive.

Skin colour evolved largely due to isolation. The earliest human populations were located in temperate regions further from the equator and were able to absorb more sunlight without burning for this reason. Lighter colour enabled the deeper penetration of sunlight into skin tissue and as such they were able to manufacture more vitamin D. Those located at the equator needed darker skin in order to protect themselves from sunburn, yet still manufacture enough vitamin D. This ability was decreased as darker-skinned people moved away from the equator. Natural selection would have played a significant role in the development of the specific gene pools that arose in different regions. *(7 marks)*

(d) (i) Specific fossil evidence of hominids which infers the use of technology that led to cultural development could include the following.

- *Homo heidelbergensis*: fossil core tools, hunter-gatherer lifestyle, cave dweller, used animal skins for shelter and clothing.
- *Homo ergaster*: fossil stone tools such as hand axes dated 1.8 to 1.6 million years old, evidence of a hunter-gatherer lifestyle with some permanent dwellings. *(2 marks)*

(ii) Human population in 10 000 years has increased to over 6 billion. The hunter-gatherer culture has been replaced by agricultural communities and subsequently the urbanisation of populations has occurred parallel to agriculture, with dense populations building up as people settled together and began to work together. Technology has improved dramatically and the reliance on complex machinery over the last 2000 years has resulted in an exponential increase in population. Populations have become very mobile, resulting in multiculturalism and the mixing of the gene pool across different races. Understanding of hygiene and advances in medical technology have increased, incorporating the use of drugs to treat and prevent disease. This has resulted in better health and extended the general life expectancy of people. Genetic technology and mapping of the human genome (identification of most genes in the human body and a location map of gene sequences) has enabled the identification of genes that cause disease and the ability to manipulate specific genes. The Human Genome Project has enabled us to compare the evolutionary relationships between us and other primates. Together with the ability to clone and advances in stem cell research, the potential to eliminate harmful genes from the gene pool is significant and may result in a great potential to eliminate or at least cure inherited diseases and live longer. *(6 marks)*

CHAPTER 12

BOARD OF STUDIES
NEW SOUTH WALES

2013
HIGHER SCHOOL CERTIFICATE
EXAMINATION

Biology

General Instructions

- Reading time – 5 minutes
- Working time – 3 hours
- Write using black or blue pen
  Black pen is preferred
- Draw diagrams using pencil
- Board-approved calculators may be used

Total marks – 100

Section I

75 marks

This section has two parts, Part A and Part B

Part A – 20 marks
- Attempt Questions 1–20
- Allow about 35 minutes for this part

Part B – 55 marks
- Attempt Questions 21–30
- Allow about 1 hour and 40 minutes for this part

Section II

25 marks

- Attempt ONE question from Questions 31–35
- Allow about 45 minutes for this section

Section I
75 marks

Part A – 20 marks
Attempt Questions 1–20
Allow about 35 minutes for this part

Use the multiple-choice answer sheet for Questions 1–20.

---

1 Infectious diseases can be spread when microbial experiments are conducted.

Which of the following would be an effective barrier to stop this from happening?

(A) Using chlorinated water during the experiment

(B) Washing hands before performing the experiment

(C) Wearing plastic or latex gloves during the experiment

(D) Disposing carefully and safely of waste materials produced during the experiment

2 Antibodies are proteins that

(A) break down pathogens.

(B) bind with a specific antigen.

(C) catalyse biochemical reactions.

(D) are produced by T cells to kill disease-causing viruses.

3 Which of the following can cause an imbalance of microflora in humans?

(A) Overuse of antibiotics

(B) Excessive use of antiviral drugs

(C) Consumption of genetically modified foods

(D) Immunisation against different diseases

4 Which of the following prevents the entry of pathogens into the human body?

(A) Cell death to seal off a pathogen
(B) Mucus lining the respiratory tract
(C) Phagocytosis performed by B cells
(D) Destruction of pathogens by the lymphatic system

5 What is the genotype of pure breeding plants?

(A) Heterozygous
(B) Homologous
(C) Homozygous
(D) Monohybrid

6 What are the main components of the immune response involved in organ rejection?

(A) Antibodies, T cells and B cells
(B) Antibiotics and white blood cells
(C) T cells, B cells and red blood cells
(D) Red blood cells, white blood cells and antigens

7 Vaccination can control the spread of

(A) a genetic disorder.
(B) an infectious disease.
(C) a nutritional deficiency.
(D) an environmental disease.

8 Sutton and Boveri contributed to our understanding of the importance of chromosomes.

What was one of their findings?

(A) Chromosomes carry hereditary factors.
(B) Sex is genetically determined by chromosomes.
(C) Radiation can cause mutations in chromosomes.
(D) The structure of chromosomes is a double helix.

9 What is a feature of an estuarine organism (eg mangroves) that allows it to survive?

(A) High rate of salt excretion

(B) Low rate of osmosis into its cells

(C) High uptake of salt from its environment

(D) Low uptake of oxygen from its environment

10 What does the structure of arteries allow them to do?

(A) Transport oxygen rich blood

(B) Withstand high blood pressure

(C) Release carbon dioxide to the lungs

(D) Remove nitrogenous waste via the kidneys

11 Why are quarantine measures needed when there is an outbreak of an infectious disease in Australian farm animals?

(A) To prevent the spread of the disease to food imported into Australia

(B) To prevent Australian farm animals from becoming immune to the disease

(C) To prevent introduced plants threatening the survival of Australian farm animals

(D) To prevent the disease from spreading to farm animals in different regions of Australia

12 Two experiments were conducted where either cold air or hot air was blown continuously onto a student's legs while the skin temperature on the student's arm was being measured. The graph shows the change in skin temperature on the arm of the student for each experiment.

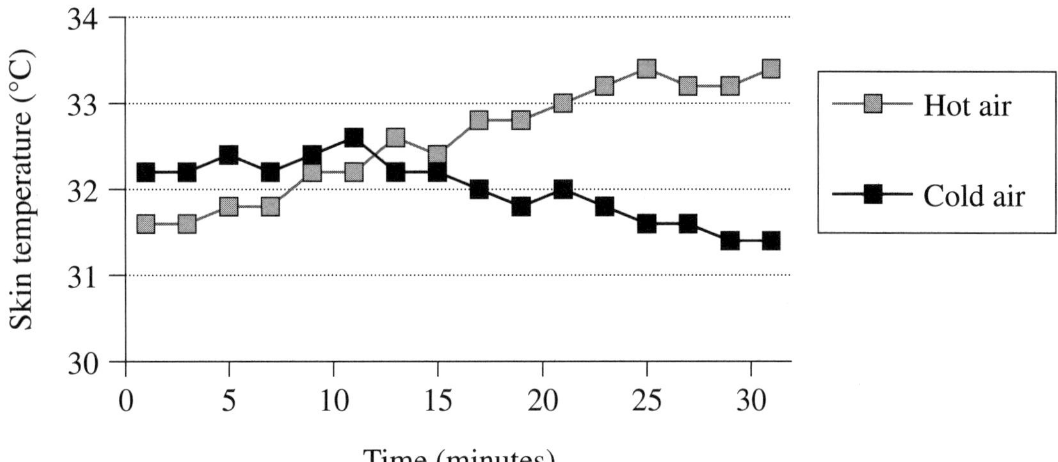

Which process best accounts for the trends shown in the graph?

(A) Diffusion

(B) Enantiostasis

(C) Homeostasis

(D) Inflammation

13 Why is carbon dioxide removed from cells?

(A) To decrease osmosis

(B) To allow oxygen to replace it

(C) To prevent an increase in blood pH

(D) To prevent cells from becoming acidic

14 Hormone replacement therapy was used to treat a patient who had low salt levels in their blood.

Which hormone was used for the treatment?

(A) Aldosterone

(B) Anti-diuretic hormone

(C) Insulin

(D) Oestrogen

15 Which is the correct sequence of a feedback mechanism for the control of blood pressure?

(A) blood pressure → blood vessels → brain → receptors → blood pressure

(B) blood pressure → blood vessels → receptors → brain → blood pressure

(C) blood pressure → receptors → blood vessels → brain → blood pressure

(D) blood pressure → receptors → brain → blood vessels → blood pressure

16 The theory of evolution has been supported by studying the structures of vertebrate forelimbs from the fossil record.

This type of study is best described as

(A) biogeography.

(B) comparative biochemistry.

(C) comparative embryology.

(D) palaeontology.

17 The photo shows insect mounds (nests) in hot northern Australia. The mounds are built with a narrow side facing the midday sun and the wide sides facing towards the morning and late afternoon sun.

The structure of these nests helps the insects in them to

(A) become endothermic.

(B) cool down quickly in the afternoon.

(C) maintain a constant body temperature.

(D) develop enzymes that operate at a high temperature.

18 A family tree is shown.

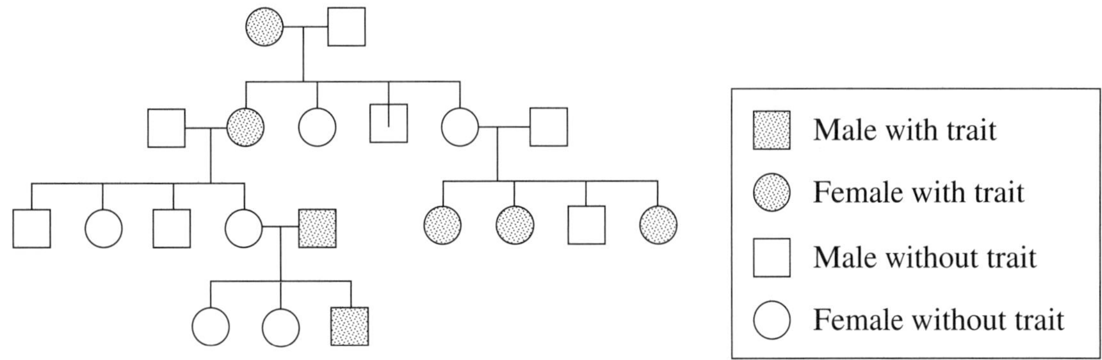

What is represented by this family tree?

(A) Sex-linked inheritance

(B) Co-dominant inheritance

(C) Inheritance of a recessive trait

(D) Inheritance of a dominant trait

19 An enzyme was extracted from a mammal. The graph shows the rate at which bubbles are produced in a reaction at 36ºC using the extracted enzyme.

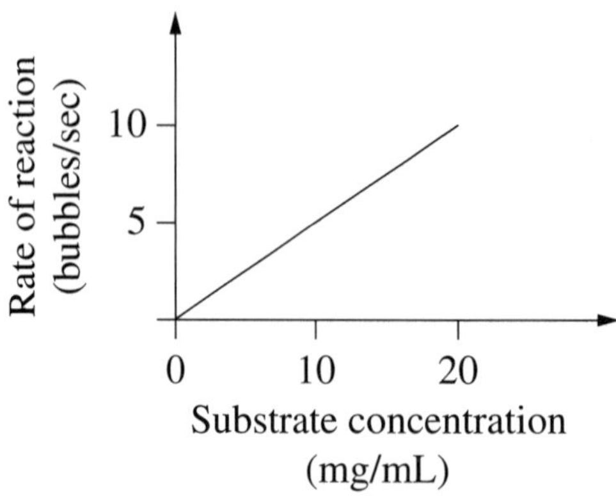

Which of the following graphs would show the results if the enzyme reaction were carried out at 18°C?

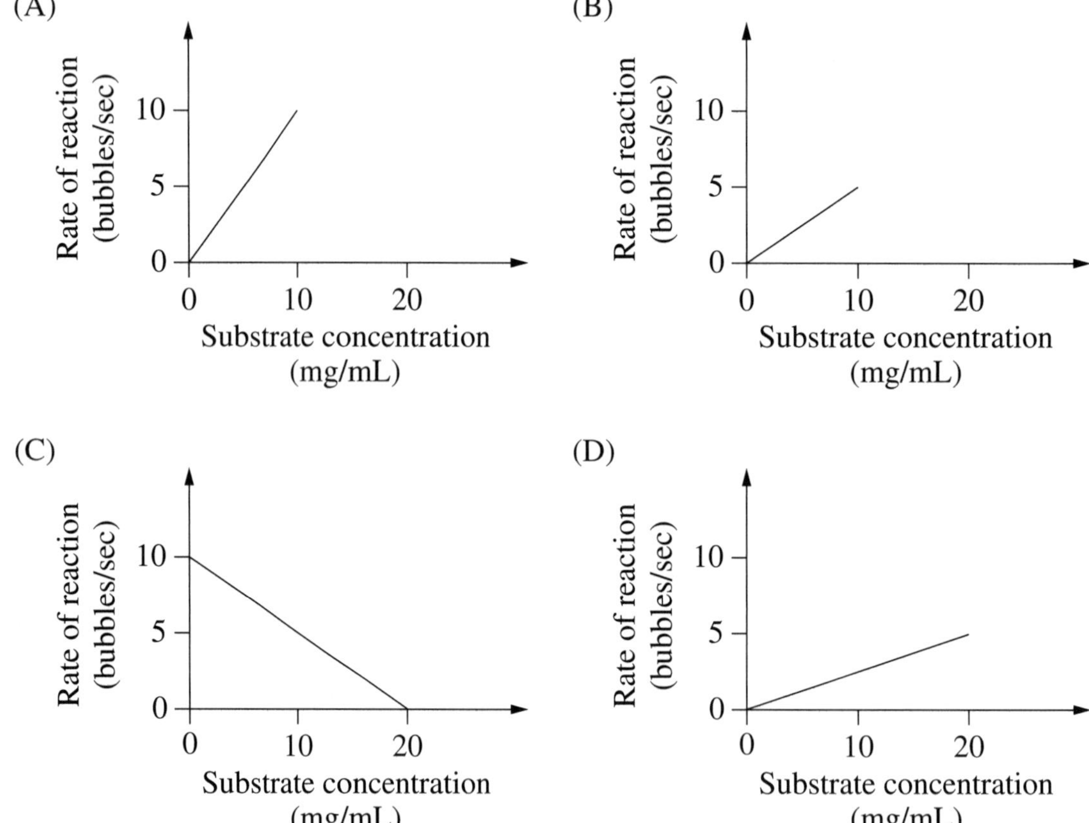

20 Which of the following statements about DNA provides support for Darwin's theory of evolution?

(A) Genes are inherited from both parents.

(B) Mutation of DNA may lead to new alleles.

(C) DNA contains the code for proteins and polypeptides.

(D) The sequence of bases in DNA differs between species.

2013 HIGHER SCHOOL CERTIFICATE EXAMINATION

Biology

Centre Number

Student Number

Section I (continued)

Part B – 55 marks
Attempt Questions 21–30
Allow about 1 hour and 40 minutes for this section

Answer the questions in the spaces provided. These spaces provide guidance for the expected length of response.

Write your Centre Number and Student Number at the top of this page.

Question 21 (4 marks)

Describe the difference between the roles of TWO named types of T cells in the immune response in humans. **4**

....

....

....

....

....

....

....

....

Question 22 (3 marks)

Outline the work done by Morgan that has led to our understanding of sex linkage. **3**

....

....

....

....

....

....

Question 23 (4 marks)

Spinifex is a grass common across central Australia where soils are nutrient deficient. It frequently grows as a circular clump of stems and the diameter of the clump increases slowly each year. Initially the leaves are flat and the roots are shallow. As the plant matures, the leaves curl inwards to form long thin tubes with the stomates on the inside, while the roots grow deep into the soil to obtain nutrients and water. Silicon granules make the stems tough. **4**

Explain how TWO of the adaptations outlined above allow Spinifex to survive in hot, dry conditions.

....

....

....

....

....

....

....

....

Question 24 (7 marks)

(a) Draw a flowchart showing the steps you could take to model Pasteur's experiment to identify the role of microbes in decay. **4**

(b) Explain the importance of using controls in microbial experiments. **3**

Question 25 (7 marks)

The graph below shows the results obtained from testing the activity of a bacterial enzyme.

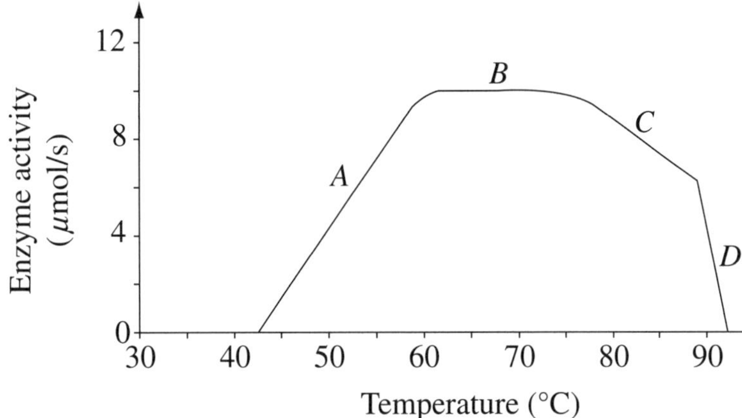

(a) Name ONE variable, other than temperature, that would have been controlled in the experiment. **1**

....

(b) For what temperature range does the enzyme display the maximum rate of change in activity? **1**

....

(c) Account for the activity of the enzyme at the parts of the graph labelled *A*, *B*, *C* and *D*. **4**

....

....

....

....

....

....

....

....

(d) Based on the information in the graph, suggest the type of environment in which these bacteria might survive. **1**

....

....

Question 26 (3 marks)

A virus was used to kill rabbits in Australia. After first release of the virus nearly all rabbits were killed, but over time the numbers recovered. **3**

Outline how Darwin/Wallace's theory of evolution could be used to explain the recovery of rabbit numbers.

....

....

....

....

....

....

....

Question 27 (4 marks)

Explain how ONE advance in technology has provided support for the theory of evolution. **4**

....

....

....

....

....

....

....

....

Question 28 (7 marks)

(a) As part of your Biology course you were required to examine plant shoots and leaves for evidence of pathogens and insect pests. **2**

List TWO observations that indicated the presence of pathogens on plant shoots and/or leaves.

....

....

(b) Describe how TWO named methods have changed the way we manage disease and/or insect pests. **5**

....

....

....

....

....

....

....

....

....

....

....

Question 29 (8 marks)

(a) What chemicals are filtered out of the blood by the kidney? **2**

....

....

(b) What chemicals are reabsorbed into the blood by the kidney? **2**

....

....

(c) Explain the steps involved in the formation of urine. **4**

....

....

....

....

....

....

....

....

....

....

....

....

....

....

Question 30 (8 marks)

Explain how our knowledge of chromosome structure has led to reproductive technologies that have the potential to alter the path of evolution. **8**

2013 HIGHER SCHOOL CERTIFICATE EXAMINATION

Biology

Section II

25 marks
Attempt ONE question from Questions 31–35
Allow about 45 minutes for this section

Answer the question in the Section II Writing Booklet. Extra writing booklets are available.

Question 31	Communication
Question 32	Biotechnology
Question 33	Genetics: The Code Broken?
Question 34	The Human Story
Question 35	Biochemistry *(Not included in this reproduction)*

Question 31 — Communication (25 marks)

Answer parts (a), (b) and (c) of the question in the Section II Writing Booklet. Start each part of the question on a new page.

(a) Outline how the larynx produces sounds of different frequencies. **3**

(b) Outline how sound is transmitted to the brain in humans. **4**

(c) (i) Describe the process of accommodation. You may include diagrams in your answer. **4**

(ii) Outline how ONE technology is used to treat hyperopia. **1**

Answer parts (d) and (e) of the question in the Section II Writing Booklet. Start each part of the question on a new page.

(d) Describe the sequence of events that occurs when photoreceptors in different regions of the retina are stimulated by red light. **6**

(e) (i) Outline ONE essential step that must occur for the photoreceptor signal to be transmitted along the optic nerve to the brain. **1**

(ii) Microelectronic chips containing light sensitive detectors have been implanted in the retinas of blind volunteer patients. These implants do not give normal vision but allow these patients to see only areas of light and dark. **6**

Relate your understanding of the use of hearing aids and cochlear implants to possible advantages and limitations of microelectronic chips to vision.

Question 32 — Biotechnology (25 marks)

Answer parts (a), (b) and (c) of the question in the Section II Writing Booklet. Start each part of the question on a new page.

(a) Outline the function of RNA. **3**

(b) Using a named example, outline how greater knowledge of science has led to a change in traditional methods of fermentation. **4**

(c) (i) Outline the steps in the extraction process of DNA from a named source. **2**

(ii) How could the final product be identified as DNA? **1**

(iii) Describe how a specific gene from extracted DNA could be amplified. **2**

Answer parts (d) and (e) of the question in the Section II Writing Booklet. Start each part of the question on a new page.

(d) Analyse the contribution that ONE of the following applications has made to medicine. **6**

- tissue engineering
- gene delivery by nasal sprays
- production of a synthetic hormone

(e) The domestication of seeds produced the most cultural change in human history, transforming most human societies from hunter gatherers to permanent settlements anchored by agriculture.

(i) Outline how the collection of seeds could be seen as an early example of biotechnology. **1**

(ii) Using your knowledge of the domestication of one plant or animal species, evaluate relevant ethical issues raised by the use of current biotechnology. **6**

Question 33 — Genetics: The Code Broken? (25 marks)

Answer parts (a), (b) and (c) of the question in the Section II Writing Booklet. Start each part of the question on a new page.

(a) Outline *polygenic inheritance*. Include an example in your answer. **3**

(b) The diagram shows part of the process of protein synthesis. **4**

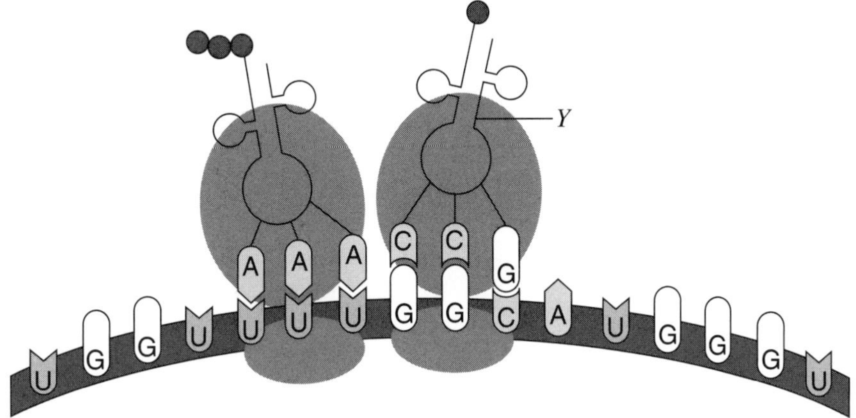

Outline how the structure *Y* enables information from DNA to be translated into a specific polypeptide.

Question 33 continues

Question 33 (continued)

(c) (i) In the early 1900s, Bateson and Punnett studied inheritance in the sweet pea plant. They studied the inheritance of two genes: **2**

- flower colour (*P*, purple, and *p*, red)
- shape of pollen grain (*L*, long, and *l*, round).

They crossed pure lines *PPLL* (purple, long) × *ppll* (red, round). In the F1 generation all offspring had purple flowers and long pollen grains. These offspring were then crossed with the expectation that the Mendelian ratios would occur.

Construct a Punnett square to show the phenotypic ratios that would have been obtained if these genes showed typical Mendelian dihybrid inheritance.

(ii) The table shows the actual results that Bateson and Punnett obtained. **3**

Phenotype	*Number of offspring*
Purple, long	4831
Purple, round	390
Red, long	393
Red, round	1338
Total number of offspring	6952

Explain how the results of cross-breeding experiments could be used to identify the relative positions of genes.

Question 33 continues

Question 33 (continued)

Answer parts (d) and (e) of the question in the Section II Writing Booklet. Start each part of the question on a new page.

(d) Analyse how the understanding of mechanisms of genetic change has influenced the use of genetic technology in society. **6**

(e)

DNA kits can bring unwanted surprises

'For the price of a night out, individuals can learn key elements of their genetic composition and take treatment, or protect their children from hereditary health risks …

To take the test, a client spits into a test tube or swabs the inside of their cheek, then sends the sample for analysis …

Mrs X took the test to determine the cause of some minor health problems … She learnt that she was at risk of breast cancer but also that the man she has called 'Dad' for 50 years was not her father. She tracked down her biological father and a half-sister, who had breast cancer. A biopsy found Mrs X had cancer.'

The Sydney Morning Herald, 7 January 2013 (adapted)

(i) Outline a method used to gain this genetic profile. **1**

(ii) Using your knowledge of the Human Genome Project, discuss the effects that gene therapy may have on society. **6**

Question 34 — The Human Story (25 marks)

Answer parts (a), (b) and (c) of the question in the Section II Writing Booklet. Start each part of the question on a new page.

(a) Outline the features used to identify humans as the species *Homo sapiens*. **3**

(b) ^{14}C can be used for dating fossils. Its half life is approximately 5750 years. **4**

Compare this method for dating with another dating technique.

Question 34 continues

Question 34 (continued)

(c) This table outlines human population growth.

Human Population (billions)	*Year*	*Time Span* (years)
1	1830	50 000
2	1930	100
3	1960	30
4	1975	15
5	1989	14
6	1999	10
7	2011	12
8 (predicted)	2025	14
9 (predicted)	2045	20

(i) What trends in human population growth do the data illustrate? **2**

(ii) Outline specific examples of technology that could explain the growth rate of the human population before and after 1999. **3**

Question 34 continues

Question 34 (continued)

Answer parts (d) and (e) of the question in the Section II Writing Booklet. Start each part of the question on a new page.

(d) Analyse the impact of cultural development on human evolution. **6**

(e) Different hominid skulls are illustrated.

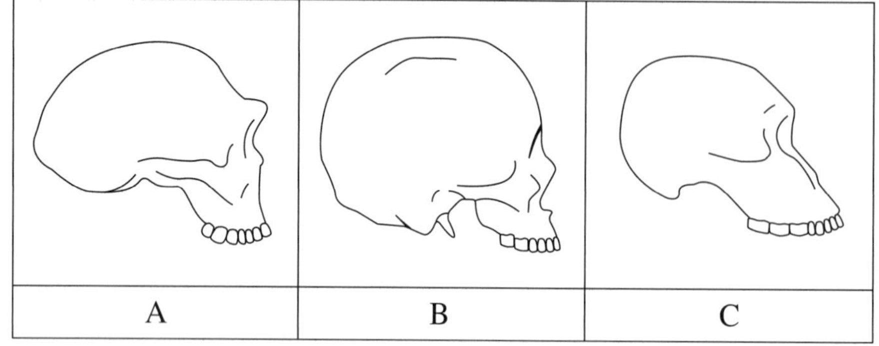

(i) Place the hominid skulls in order from the oldest to the youngest. **1**

(ii) Explain how molecular analysis has changed scientists' opinions about primate classification. In your answer, include specific examples. **6**

End of Question 34

End of paper

2013 HSC Examination Paper

Sample Answers

Section I Part A (*Total 20 marks*)

1 C An effective barrier to prevent the spread of microbes *when experiments are conducted* is to use gloves during the procedure. Chlorinated water may be used as disinfectant but would not be a *barrier*.

2 B An antibody identifies antigens, which are specific markers on pathogens or foreign targets.

3 A Antibiotics destroy bacteria, some of which can reduce fungal growth.

4 B Mucus lining in the respiratory tract is part of the first line of defence and as such *prevents the entry* of pathogens into the body.

5 C Pure breeding plants must be homozygous to ensure that only one type of allele is passed on to the offspring.

6 A This is the only option that offers only components of the immune system.

7 B Vaccinations were developed to prevent infectious disease. All other options are not communicable/infectious diseases.

8 A Sutton and Boveri are responsible for the chromosome theory of inheritance.

9 A Mangroves have a number of adaptations for the highly saline environment they live in. A high rate of salt excretion is the correctly worded option.

10 B Arteries carry blood under pressure from the heart to the body. This is the only answer that relates specifically to the arteries' *structure*.

11 D Quarantine measures are put in place to prevent the spread of disease. Of the options provided, only D takes into account both Australian farm animals and the spread of disease.

12 C The temperature fluctuation on the arm correlates with the temperature of the air being blown onto the legs, reflected by vasoconstriction/dilation in order to maintain body temperature, which is a homeostatic function.

13 D Carbon dioxide reacts in the blood to form carbonic acid, which lowers blood pH levels.

14 A If levels of sodium become too low in the body, the hormone aldosterone is released. Aldosterone increases the amount of sodium held in the body by regulating how much is lost in urine.

15 D A feedback is represented by: stimulus → receptor → CNS → effector → response.

16 D Palaeontology includes the study of similarities in vertebrate structure, such as the pentadactyl limb. Embryology is specific to the study of early developmental stages, biochemistry to cell chemistry, and biogeography to distribution.

17 C Insects are ectothermic (i.e. do not maintain a constant body temperature). The mound shape results in the internal temperature of the mound remaining fairly constant, which in turn helps the insects to maintain a constant body temperature.

18 C The pedigree represents a recessive trait as the trait can skip a generation. It is not sex linked because both sexes display the trait investigated.

19 D This is the only graph that indicates a slower reaction rate due to the cooler temperature.

20 B Mutations are essential to create variation in the gene pool, which facilitates evolution.

Section I Part B

21 T cells mature in the thymus gland and control cell-mediated response by directly attacking antigens. There are a number of types of T cells.

- Cytotoxic T cells destroy infected cells; killer T cells are not as specific. (Some controversy exists about the differentiation of these two and their function.)
- Helper T cells help B cells produce antibodies and assist in the formation of cytotoxic cells.
- Suppressor T cells reduce the output of antibodies and chemicals from cytotoxic cells.
- Memory T cells remain in the lymph node and confer long-term immunity.

(Note: only two types of T cells are required in answer.) *(4 marks)*

22 Thomas Morgan's work on the fruit fly *Drosophila melanogaster* contributed to our understanding of sex linkage. He investigated crosses between white-eyed and red-eyed flies, and found that the results could not be explained using simple Mendelian crosses. Morgan's work demonstrated that some genes were located on the X chromosome and so were only passed on via the X chromosomes, that is, they were sex linked. *(3 marks)*

23 Answers will vary. The following is one example.

Rolling in Spinifex leaves reduces the surface area to volume ratio and protects the stomates from exposure to the direct sunlight and radiant heat, thereby reducing moisture loss through evapotranspiration.

Development of silicon granules in the tissue makes the stems stiff and rigid, providing support for the plant which enables it to withstand the harsh conditions in the hot, arid environment. *(4 marks)*

24 (a)

Take two identical sterile flasks with equal volumes and concentration of yeast extract broth.

↓

Seal one with a stopper and swan neck tube. Leave the other open. Boil the flasks for five minutes each.

↓

Remove the flasks from heat and allow to stand and cool. (This causes condensation to pool in the S-bend of the swan neck tube.)

↓

Observe daily for one week, recording any growth visible in the flasks.

↓

Once complete, destroy the resulting broth and sterilise flasks. (Repeat the experiment or use multiple flasks to start with.)

(4 marks)

(b) Using controls in microbial experiments is important to ensure that you do not introduce additional variables or contamination into the reaction vessel or plate. A control plate or flask is used to enable a comparison between the growth produced/ inoculated plates and also to ensure that the plates were sterile to begin with. Usually the medium plates are sealed to prevent external contamination from pathogens in the air and to prevent the growth from escaping to the surrounding environment. *(3 marks)*

25 (a) Same amount/volume of substrate *(1 mark)*

(b) *D* *(1 mark)*

(c) At *A* temperature increases and reaches the optimum function level of enzyme. At *B* the ideal operating temperature is reached and the enzyme functions optimally. At *C* the temperature increases beyond optimum. Enzyme activity drops because it is getting too hot. At *D* the temperature is too hot for enzyme activity to continue and the enzyme denatures. *(4 marks)*

(d) Based on the data in the graph, the bacteria depicted might thrive in very warm environments such as warm springs or hydrothermal vents. *(1 mark)*

26 Darwin/Wallace's theory would state that after the release of the virus, the rabbits that had a gene combination that gave them immunity to the virus survived, while the others were killed. When the surviving rabbits reproduced they passed on their genetic combination to their offspring. The number of rabbits with this genetic combination increased over time, creating a rabbit population that was resistant to that particular virus. *(3 marks)*

27 Answers could include advances such as biochemical analysis, sophisticated use of X-ray/scanners (CT/MIR), microscopes or radiometric dating. The following is an example.

One technological advance that has provided support for the theory of evolution is the development of the microscope. The microscope has developed from a simple light microscope to a sophisticated electron microscope that has magnification capabilities of about 300 000 times and resolves about 0.0002 microns. Its development has assisted us to compare structures, understand the development of cell organelles and understand the assimilation of single cells, such as chlorophyll-containing plastids (chloroplasts) into plant cells, or those capable of respiration (mitochondria) into cells. It also enabled more detailed comparisons of living things and allowed us to examine the progression of simple structures through to more complex ones. The development of the microscope was important in the discovery of DNA. This discovery has enabled biochemical similarities and relationships between species to be made. The finding that humans and apes share 99% of their DNA supported the theory that humans evolved from apes. *(4 marks)*

28 (a) Your answer may include any two of the following observations that indicate the presence of pathogens on plants: spots, rust, bite marks, galls, discolouration (yellow), scale, distortion and curling, limp foliage, webbing present, small white cotton blobs attached to the stem. *(2 marks)*

(b) Answers will vary. The following is an example.

Transgenic species have been developed to create disease-resistant plants such as Bt cotton, corn and potato. The genes of these plants have been modified to contain material from *Bacillus thuringiensis*, a bacteria found in soil, to protect them from pests such as bollworm. This reduces the need for intensive use of pesticides.

Quarantine laws in Australia restrict the movement of plant and animal matter/organisms across internal and external borders, preventing the introduction of pests and the entry of foreign diseases, such as foot-and-mouth disease. This removes the need for treatment of diseases or use of pesticides and herbicides. International immunisation programs have eradicated diseases such as smallpox. *(5 marks)*

29 (a) The kidney filters water, urea and ions such as Cl^-, Na^+, K^+, Ca^{2+} and HCO_3^- from the blood. *(2 marks)*

(b) Amino acids, glucose, water, ions (Na^+, K^+, Cl^-, Ca^{2+} and HCO_3^-) and some vitamins are reabsorbed into the blood by the kidney. *(2 marks)*

(c) Urine formation begins with the process of *filtration*, when blood passes into a nephron through the glomerulus. Fluid containing both useful chemicals and dissolved waste materials moves out of the blood and filters through the membranes of the glomerulus (by osmosis and diffusion) into the Bowman's capsule. The filtrate passes through the proximal tubule, the descending and ascending loop of Henle and the distal tubule in turn. As the filtrate moves along this pathway, substances that the body requires (e.g. water, glucose, Na^+) are *reabsorbed* back into the capillaries surrounding the renal tubules. Other substances (e.g. H^+, K^+, NH_3) move into the distal and collecting tubules via the process of *secretion*. Here they mix with water and other wastes to produce urine, which is eliminated out of the body. Water can be reabsorbed in the collecting tubule if required by the body. Hormones such as antidiuretic hormone (ADH) and aldosterone are used to control the composition of urine as it passes through the tubules. (4 marks)

30 Our knowledge of chromosome structure has increased significantly over time, and has changed our understanding of chromosome structure and composition. Chromosomes are made of DNA. DNA is a double helix shaped nucleic acid. The strands of the helix consist of four different nucleotides, each made up of deoxyribose sugar, a phosphate molecule and a nitrogen base. Genes, which are identified with specific traits/characteristics, are coded within the DNA on the chromosomes. These genes have the potential to be isolated. Reproductive technologies have moved forward from simple selective breeding for specific desirable characteristics to more complex cloning and transgenics. Through the physical manipulation of an organism's genome it is possible to change the individual gene sequence and to create an organism with desirable foreign genetic material. In these transgenic organisms, that material can be passed on from one generation to the next. Transgenic wheat has been developed to enhance crop yields which can provide extra food in places where it is in short supply. Strawberries have been developed that are frost-resistant, so they can now be grown outside their normal growing season. This was made possible by inserting a gene from a salmon into the strawberries. The cloning of agricultural species, such as fruit trees and livestock, reduces the natural variation within the gene pool of these organisms. Evolution relies on random variation and a struggle for existence selecting the most favourable characteristics. New technologies such as cloning and transgenics mean that natural selection no longer drives evolution. They alter the path that nature would otherwise have taken. The technological advances that are being used to benefit society are possibly reducing the natural biodiversity that would have otherwise been created through evolution. *(8 marks)*

Section II—Options

Question 31—Communication

(a) The larynx is located below the tongue and soft palate. Inside are the vocal cords. The larynx produces sounds of different frequencies by adjusting the position and tension of the vocal cords. Longer cords produce deeper sounds. The faster the cords vibrate, the higher the pitch.

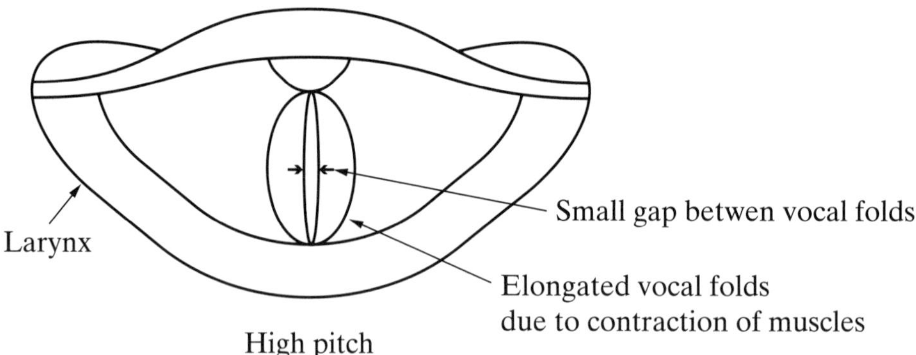

High pitch

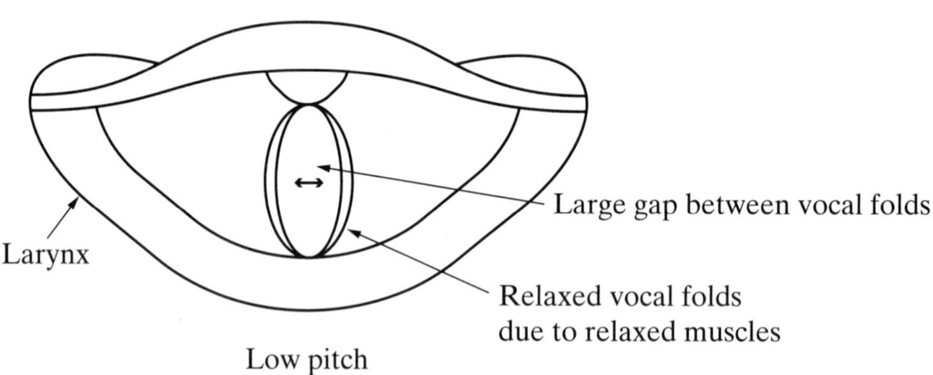

Low pitch

(Please note: diagrams may be used in the answer to this question, but are not essential for full marks.) *(3 marks)*

(b) Sound is transmitted from the environment to the brain along the following pathway. The outer ear (pinna and tympanic membrane) collects and transmits mechanical vibrations into the middle ear (containing the ear ossicles: hammer, anvil and stirrup). From the middle ear the vibration is passed through the oval window into the inner ear (round window, cochlea, organ of Corti). Here the vibrations are changed into an electrical signal that is sent along the auditory nerve to the brain for interpretation. *(4 marks)*

(c) (i) Accommodation is the process by which the lens of the eye changes shape to focus an image onto the retina, irrespective of the distance of the object away from the eye.

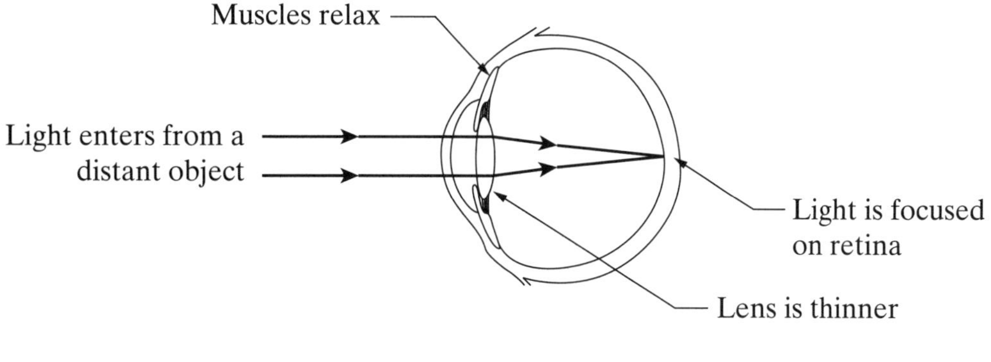

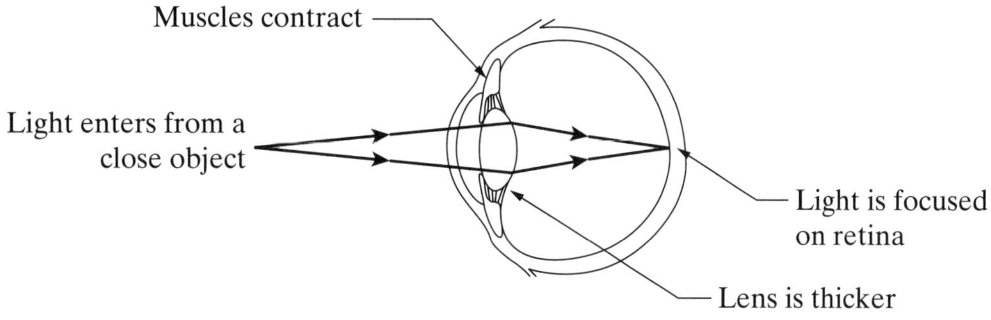

(4 marks)

(ii) Hyperopia, or long-sighted vision, occurs when the eyeball is too short or the lens cannot become round enough to focus the object onto the retina. This makes it difficult to focus on close objects, but can be rectified using a convex lens.

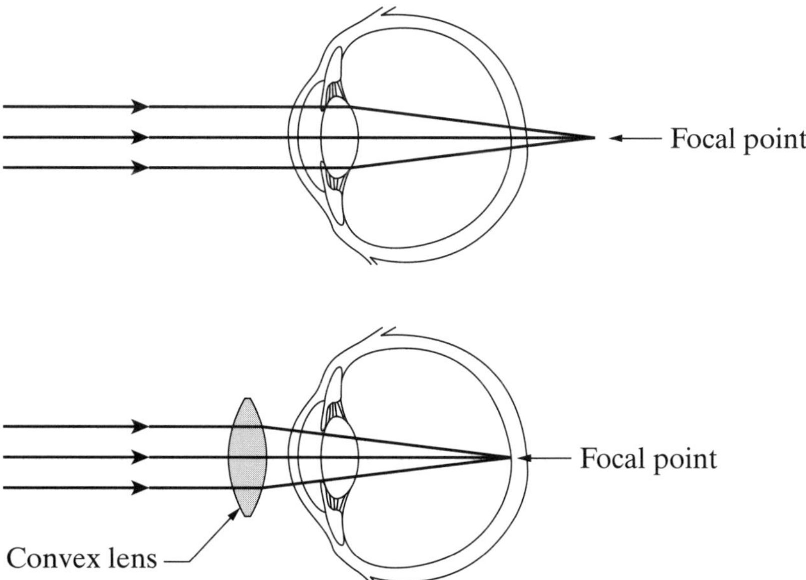

(Please note: diagrams may be used in the answer to this question, but are not essential for full marks.) *(1 mark)*

(d) Photoreceptor cells in the retina contain light-sensitive pigments (e.g. rhodopsin) that are activated by light energy and change the light into electrochemical impulses. The impulses are caused by sodium ions moving out of the neuron and potassium ions moving in. These impulses travel along the optic nerve to the brain where they are interpreted. The retina has two different groups of photoreceptor cells: rods and cones. Rods are long, rod-shaped cells that are linked in groups to single neurons. They tend to be distributed around the periphery of the retina. Cones are conical cells that contain pigment that is sensitive to high light intensity. There are three forms of cones: those sensitive to blue light frequencies; those sensitive to green light frequencies; and those sensitive to red light frequencies. The number of cones tends to increase toward the central part of the retina and is concentrated in the fovea.

As red light enters the eye, it is detected by the red cones. These cones possess photopsins that are sensitive to frequencies of about 584 nm. The photopsins react with the red light and stimulate the nerve endings of the optic nerve. This sends a message to the brain that red wavelengths are present in an image. The brain then recreates an image based on the number of impulses received. If the light enters a rod cell, it splits rhodopsin molecules into two components (opsin and retinal), resulting in an impulse in the neuron attached to the rod. However, rods are only sensitive to blue/green light so no message will be sent to the brain. *(6 marks)*

(e) (i) In order for the photoreceptor signal to be transmitted to the brain, neuron stimulation must reach a threshold. If the threshold is reached, the signal is transmitted. If it is not reached, there is no signal. This is known as an all-or-nothing response. *(1 mark)*

(ii) Hearing aids, and more recently cochlear implants, are used to augment hearing. A hearing aid is an external device that sits on or just inside the outer ear. It amplifies mechanical vibrations so that they will pass into the middle ear. The wearer will only be able to hear if the cochlea still has some sensitivity and the auditory nerve still functions. Hearing aids do not restore normal hearing and cannot eliminate background noise, so the wearer needs to adjust to interpreting sounds under different conditions.

A cochlear implant is an electronic device that provides a sense of sound to profoundly deaf patients. It works by bypassing the damaged parts of the inner ear and directly stimulating the auditory nerve. Each implant consists of two parts. One part, containing electrode wires that are inserted into the cochlea, is surgically implanted in the skull. This is connected to a second, external component consisting of a microphone with a speech processor that converts sound into electrical impulses. Cochlear implants do not restore normal hearing, and recipients must learn to interpret the signals the implant creates. This has to be combined with techniques such as lip reading to truly understand conversation. However, this is better than no hearing at all.

Microelectronic chips containing light-sensitive detectors would be similar to cochlear implants in that they would not restore complete vision, but would enable the recipient to detect changes in light. This would be a significant improvement if the recipient was completely blind. The ability to detect shadows or moving objects, together with the use of a cane, guide dog and braille, would make it easier for the recipient to get around. A limitation of microelectronic chips is that the signals received may appear as shadows, possibly pixelated on a white to black spectrum. As is the case with cochlear implant recipients, interpretation of these signals/images would take some getting used to. *(6 marks)*

Question 32—Biotechnology

(a) RNA plays an important role in the transcription and translation of DNA into proteins. DNA is a large molecule. Because of its size it cannot move out of the cell nucleus to the ribosome that will make the protein. Instead, in a process called transcription, messenger RNA (mRNA) copies a complementary code from the DNA molecule in the nucleus and carries the code out to the ribosome. The code consists of codons, areas of three nitrogenous bases that code for a particular amino acid. Transfer RNA (tRNA) molecules, which have amino acids attached to them, have corresponding anticodons. On the ribosome, the tRNA molecules match up with the mRNA codons. This process is called translation, and is how the order of bases on DNA (and in turn, mRNA) determines the order of amino acids in a protein. *(3 marks)*

(b) Fermentation is the breakdown of glucose and other simple sugar molecules into carbon dioxide and alcohol. Today fermentation is a widely applied process used for the production of many compounds, including antibiotics, enzymes and lactic acid. One example of how greater knowledge of science has led to changes in traditional practices in fermentation is bread making. Yeast is an essential ingredient in the making of bread and has been used for centuries. The carbon dioxide produced from fermentation causes the bread dough to rise. Brewer's yeast was used for this purpose but gave the bread a bitter taste. Toward the end of the 18th century, baker's yeast was isolated from specific strains of *Saccharomyces cerevisiae* grown on molasses then separated using a centrifuge. However, the quantities of yeast produced were small. Later Pasteur discovered that some microbes can grow both anaerobically and aerobically, behaving differently under different conditions. The Pasteur effect, as it was later known, results in less alcohol being produced when air is permitted to enter the fermentation vessel. With this understanding, modern baker's yeast is produced in large bioreactor vats. It is continuously aerated and nutrients are added along the way as required, resulting in faster-growing yeast cells. This yeast is then combined with the bread ingredients. It ferments the carbohydrate, producing carbon dioxide. The alcohol vaporises during baking, resulting in bread with a better flavour. *(4 marks)*

(c) (i) A number of DNA sources could be used. In this example, the steps are given for extracting DNA from onion.

- Place onion in a food processor (to break up cells).
- Add detergent and a salt solution.
- Mix together.
- Filter the mixture.
- Place the filtrate into a test tube and carefully add methylated spirits.

The DNA forms at the intersection of the onion filtrate and the alcohol. *(2 marks)*

(ii) An indicator called diphenylamine (DPA) can be used to confirm the presence of DNA. A blue colour change will indicate a positive result. *(1 mark)*

(iii) A polymerase chain reaction is used to amplify, or make multiple copies of, DNA sequences. The DNA is isolated and fragmented with restriction enzymes, then separated by gel electrophoresis. The DNA is then denatured by heating it to about 95°C, which fragments it into single strands. These strands are exposed to a radioactive DNA probe, with a sequence that has base pairs of the required DNA. The probe will bind to the corresponding sequence of the specific DNA selected under appropriate temperature (50 to 65°C), salt and pH conditions. *(2 marks)*

(d) Many answers are possible. This example is for production of a synthetic hormone, insulin.

The bacterium *Escherichia coli* was first used to host a synthetically derived version of human insulin in 1978 at the University of California. Since 1982 *E. coli* has been used by Genentech to produce recombinant human insulin, the first product of modern biotechnology. This was possible because insulin is a protein, made of amino acids which can be isolated or chemically synthesised. Once introduced into the bacterial cells by insertion of a plasmid (a small circle of DNA that serves as a gene carrier), the human gene can be interpreted by the bacterial cell's protein manufacturing mechanism which then makes the human protein. Originally this was a major technological breakthrough. Today it is more common, and the process has been adapted to be able to produce longer lasting insulin. This revolutionary technology has had significant success and is growing in popularity. It has been adapted for other uses, such as the production of growth hormones. *(6 marks)*

(e) (i) The collection of seeds could be seen as an early example of biotechnology because these early societies were taking control of the genetic make-up of the successive generations of plants. Early societies gathered seeds from specific plant species that had characteristics that they desired. They realised that by sowing these seeds they could be guaranteed characteristics that were desirable and useful to them, such as ears of wheat that mature at the same time or would not be blown away easily. *(1 mark)*

(ii) Many answers are possible. This example considers the domestication of wheat.

The domestication of wheat is a product of both human intervention and natural hybridisation between related species. In the Middle East, *Triticum boeticum* and *Triticum dicoccoides* were ancestral species of wheat. When these species were hybridised, the ears grew thick enough to be worth harvesting. Seeing this, people began to selectively grow and harvest the new variety. This selective breeding, choosing seed that did not blow away easily but stuck to the plant, altered the genetic make-up of the wheat. Because natural dispersion did not occur, it also reduced the natural diversity of the gene pool. As a consequence, the resulting domesticated wheat could not germinate easily without human intervention. Today the wheat has been further manipulated through selective breeding, as well as by genetic engineering, to be husk free and high yielding. Trials of varieties that are disease/fungus resistant or salt/drought resistant continue. Trials for other varieties that may have benefits for society have been approved. However, many people have reservations about the possible impact that biotechnology will have on biodiversity and on the long-term health of consumers. Such reservations have resulted in trials being delayed or protracted, as the stakeholders continue to debate the merits of creating genetically modified organisms that are going to enter the human food chain, either by direct consumption or by being consumed by animals that humans eat. *(6 marks)*

Question 33—Genetics: The Code Broken?

(a) Polygenetic inheritance describes how a phenotype that is influenced by more than one gene is passed on. Polygenetic traits include skin colour and height, the inheritance of which does not demonstrate the typical Mendelian ratios, even though each contributing gene is inherited according to Mendel's descriptions. Characteristics that exhibit polygenic inheritance usually have a normal distribution in the population. These characteristics can also be influenced by environmental factors including diet, disease and exercise, and as such are described as multi-factorial. The increased number of genes controlling a characteristic gives rise to a larger number of possible gene combinations existing, and therefore more phenotypes. *(3 marks)*

(b) Structure *Y* represents a molecule of transfer RNA (tRNA). One part of the tRNA molecule carries a three-base anticodon which will match the three-base codon on the messenger RNA (mRNA). The other side of the molecule holds an amino acid (which is determined by the anticodon). The tRNA is floating in the cellular fluid. The tRNA anticodon binds to the mRNA codon. As the mRNA gets its sequence from the DNA (through transcription) the ultimate order of bases, and thus amino acids, is determined by the DNA. The sequence of amino acids forms polypeptide bonds, forming a specific polypeptide. *(4 marks)*

(c) (i) Crossing the pure lines

Parents: *PPLL* × *ppll*
Possible gametes: *PL*, *PL*, *PL*, *PL* and *pl*, *pl*, *pl*, *pl*
F1 genotypes: all *PpLl*
F1 phenotypes: all purple, long

Crossing two F1 individuals together

Parents: *PpLl* × *PpLl*
Possible gametes: *PL*, *Pl*, *pL*, *pl* and *PL*, *Pl*, *pL*, *pl*

F2 genotypes:

Gametes	***PL***	***Pl***	***pL***	***pl***
PL	*PPLL*	*PPLl*	*PpLL*	*PpLl*
Pl	*PPLl*	*PPll*	*PpLl*	*Ppll*
pL	*PpLL*	*PpLl*	*ppLL*	*ppLl*
pl	*PpLl*	*Ppll*	*ppLl*	*ppll*

F2 phenotypes:

Genotypes	Number	Phenotypes
PPLL	1	purple colour, long pollen
PPLl	2	purple colour, long pollen
PpLL	2	purple colour, long pollen
PpLl	4	purple colour, long pollen
PPll	1	purple colour, round pollen
Ppll	2	purple colour, round pollen
ppLL	1	red colour, long pollen
ppLl	2	red colour, long pollen
ppll	1	red colour, round pollen

The phenotypic ratio of F2 individuals is 9 purple long : 3 purple round : 3 red long : 1 red round. *(2 marks)*

(ii) Cross-breeding experiments have been used to study the relative positions of linked genes on homologous chromosomes. Morgan's student Sturtevant proposed that when the distance between genes on chromosomes was large, there was a greater chance of crossing over occurring between these genes. This was the beginning of chromosome mapping, which indicates the relative positions of genes on the chromosomes. On homologous chromosomes, the genes (and their alleles) are found on corresponding loci. As a result of 'swapping over', the alleles on homologous chromosomes are interchanged. The greater the percentage of recombination, the larger the distance between the genes. For example, if recombination occurs 10% of the time, the genes are thought to be 10 units apart on the chromosome. If recombination occurs 50% of the time, the genes are thought to be 50 units apart. By comparing recombination rates one can develop a map of the chromosome showing the relative positions of genes on that chromosome. *(3 marks)*

(d) *Mutation* refers to any permanent change in the base sequence of DNA in a cell. Genetic changes can occur spontaneously during DNA replication or as a result of exposure to a mutagen, which could be radiation (e.g. gamma, X–ray, ultraviolet) or chemicals (e.g. various hydrocarbons, Agent Orange, DDT). Changes in the base sequences can occur due to rearrangements, changes in chromosome number (including trisomy and polyploidy) and mutations of genes, including base substitution and frameshift mutation.

Rearrangements result from copying mistakes, causing deletion, duplication, inversion or translocation of bases along a sequence of DNA. If these sequences are altered, so too is the code for the specific protein the DNA codes for, and as such also the function that the sequence originally coded for. *Chromosomal mutations* occur during meiosis, resulting in non-disjunction. This creates some gametes with extra chromosomes and others with fewer than the normal number (e.g. trisomy, such as Down syndrome where an extra chromosome number 21 is present). A frameshift mutation occurs when the insertion or deletion of nucleotides shifts the sequence in which the DNA molecule is read.

Understanding the types of mutations that can occur, and what the resulting phenotypes are, has resulted in the use of genetic technology to identify, manipulate, remove and/or replace some of these changes in DNA sequences. For example, now that it is possible to map the chromosomes and the genes on them, possible problems in IVF embryos can be identified. This means that only embryos that are healthy are transferred back in to the uterus, eliminating the risk of the baby having significant or life-threatening health issues. With this knowledge and a human genome map, it may be possible over time to completely eliminate some diseases, such as cystic fibrosis, bone marrow disorders and Huntington's disease, completely from the gene pool. *(6 marks)*

(e) (i) DNA fingerprinting is a method that can be used to obtain a genetic profile. In this process extracts of DNA from a tissue sample (e.g. skin, blood or hair) can be cut into fragments using a restriction enzyme. This is then separated by gel electrophoresis. The fragments of DNA separate out according to size and a radioactive probe is added. The probe bonds with the fragments, allowing an image to form on X-ray film. The fragments can then be observed. *(1 mark)*

(ii) The Human Genome Project (HGP) was a collaborative project undertaken across a number of countries to identify the genes in human DNA and determine the base pair sequence. The project took 13 years to complete, finishing in 2003. The information from this project is stored on database records and will continue to be analysed for years to come. The major benefits for society that have resulted from this work so far include identification of predisposition to genetic disease and diagnosis of hereditable diseases. The information is also being used for manufacturing drugs that are better able to target disease, and for gene therapy that is better able to isolate defective genes and treat and/or remove and replace them. The positive outcomes from the HGP go beyond using information from the project to eliminate disease and disability, and include the application of the technological advances made in the project that enable us to manipulate genes. Already we have seen advances in the experimental treatment of cystic fibrosis, bone marrow disorders and Huntington's disease, in which defective genes in patients are being removed and replaced with success. Such revolutionary treatment has many benefits as not only does it give these individuals a better quality of life, but reduces their dependence on drugs and constant medical care. By completely eliminating certain diseases, or offering effective treatment, the economic burden of health care on society is greatly reduced. There are some drawbacks, however: there may not be equal access to these new treatments; there may be issues with confidentiality of the information and the individual's right to privacy; commercialisation of the drugs and distribution throughout the market place may further restrict access to treatments; and there are ethical considerations regarding the purpose of the information and the integrity of those empowered to govern its use and associated genetic technologies. *(6 marks)*

Question 34—The Human Story

(a) The features used to classify *Homo sapiens* as a species include reduced body hair, large brain, protruding chin, high forehead, capacity for speech and abstract thought, and a complex social organisation system. *(3 marks)*

(b) There are a number of possible dating techniques that could be compared with ^{14}C dating. The following is one example.

^{14}C has a relatively short half life compared to other isotopes, and as such is really only useful for dating organic material up to about 60 000 years old. It is a form of absolute dating and is typically used for dating fossils or sediment that has carbon present. The rate of decay can provide an estimate of the age of the material. The principle of superposition is another way of dating material. It is a relative dating technique that uses the fact that sedimentary layers are youngest at the top and increase in age as you move away from the surface. The fossils found within the layers can be correlated to others found in surrounding layers above or below, giving a relative age. This technique can still be accurate when rocks have been folded or eroded because it is still possible to establish a sequence. Observation over time indicates that similar rocks tend to have similar fossils irrespective of their location, inferring that they were produced at the same time and hence are similar in age. *(4 marks)*

(c) (i) The data in the table indicate that the rate of human population growth was very slow for the first 50 000 years (0.00005/yr), then doubled within 100 years. Population then increased by 50% in only 30 years to 3 billion by 1960. The rate continued to increase until 1999 after which it slowed. Population growth is predicted to continue to slow down, reaching 8 billion in 14 years then 9 billion in another 20 years. *(2 marks)*

(ii) Early human populations lived for thousands of years in small groups. Their numbers did not grow because they were vulnerable to predators and disease, and were limited by the availability of food. Over time, death rates have been reduced and birth rates increased due to expanding knowledge and use of technology, including domestication of animals, extensive agricultural practices, use of machinery, sanitation, water supplies, communication technology, public health education, and medical and reproductive technologies. This has enabled societies to make better use of their resources to house, feed and educate their population. The environment no longer acts as an agent of selection, as we have the ability to control many aspects of the environment around us, changing it to suit our needs. Since 1999 the rate of population growth has decreased. This may be due to more effective methods of contraception and knowledge of birth control. The incidence of smaller family units due to the knowledge that most children will now survive until adulthood may also be a factor in this reduction. *(3 marks)*

(d) The impact of cultural development on human evolution has been significant. Major cultural changes that have taken place over the past 40 000 years include: the development of social structure or hierarchy; changes in diet, ranging from simple hunter-gatherer subsistence (seed eaters) to a varied diet that includes high protein levels obtained through cooperative hunting; sharing of food; increased learning and levels of education; development of more complex maternal behaviours; the ability to move long distances; permanent settlement; specialisation, such as agriculture; effective communication through the use of verbal and written language; and increased complexity of tools, pottery, metals and technology, including machinery and electronics. Without the rise of culture, and the changes associated with it (resulting in the improvements/developments listed above), human evolution would not have progressed well. Today, these changes are moving forward very quickly due to rapid communication and sophisticated technology. Human evolution is no longer dependent upon how good we are as hunters, but on our access to health care. We can now survive with poor eyesight because we have the technology to make glasses to correct these deficiencies. Our survival in plagues is no longer based on a robust immune system, but our access to medical care. These changes may well affect our species' evolution. *(6 marks)*

(e) (i) *C* (oldest skull) → *A* → *B* (youngest skull) *(1 mark)*

(ii) Answers will vary. The following is an example.

Molecular analysis has assisted in changing scientists' opinions about primate classification over time, as various technologies have given rise to information that was previously unavailable. For example, DNA hybridisation can be used to compare how closely related two species are. This is achieved by heating extracted DNA from the two species, separating the strands and using restriction enzymes to cut them into smaller sections. When these are recombined and cooled, they form a realigned double strand where regions of similar code will form strong bonds. The DNA is heated again. The temperature required to cause separation again indicates how closely related they are: a higher temperature indicates more hydrogen bonds, and the more hydrogen bonds, the more genetically similar the species. For example, this technique has shown that humans are more closely related to chimpanzees than to gibbons.

Comparative biochemistry, such as analysis of cytochrome C or haemoglobin, gives an indication of degrees of similarity and common ancestral linage. In order to test for haemoglobin similarity, human blood is injected into rabbits. The rabbits form antibodies that are then used to test for precipitation reactions in other species. The rate at which the precipitate forms is used to calculate the degree of similarity to humans: the faster the rate, the more similar. This has been used to establish links with the primates and exclude close links with other species, such as dogs.

Comparing the DNA sequences from the genomes of different primates has been very useful in recent times. The degree to which mitochondrial DNA differs between primates can be used to establish the length of time since the individuals shared a common ancestor.

As a result of these techniques, along with anatomical, archaeological and genetic evidence, the current view is that all modern human ancestors were evolved from a single African origin and did not share genetic material with the Neanderthals that migrated to Europe. *(6 marks)*

CHAPTER 13

2014 HIGHER SCHOOL CERTIFICATE EXAMINATION

Biology

General Instructions

- Reading time – 5 minutes
- Working time – 3 hours
- Write using black or blue pen
  Black pen is preferred
- Draw diagrams using pencil
- Board-approved calculators may be used

Total marks – 100

Section I

75 marks

This section has two parts, Part A and Part B

Part A – 20 marks
- Attempt Questions 1–20
- Allow about 35 minutes for this part

Part B – 55 marks
- Attempt Questions 21–30
- Allow about 1 hour and 40 minutes for this part

Section II

25 marks
- Attempt ONE question from Questions 31–35
- Allow about 45 minutes for this section

Section I
75 marks

Part A – 20 marks
Attempt Questions 1–20
Allow about 35 minutes for this part

Use the multiple-choice answer sheet for Questions 1–20.

---

1 Exposure to radiation such as X-rays may change the sequence of bases in DNA.

What is this called?

(A) Mutation
(B) Translation
(C) Replication
(D) Transcription

2 The diagram models the movement of particles in a particular process.

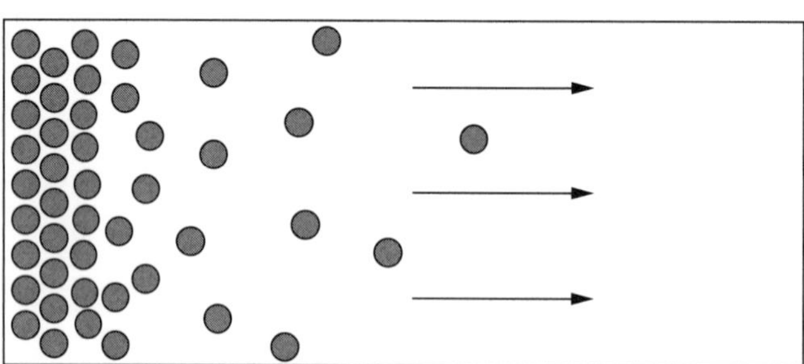

What process is being modelled?

(A) Osmosis
(B) Filtration
(C) Diffusion
(D) Active transport

3 Which of the following is a plant adaptation that minimises water loss?

(A) Small leaves
(B) Large flowers
(C) Large fleshy fruit
(D) Deep root systems

4 Which of the following correctly identifies the nitrogenous wastes of mammals, marine fish and insects?

	Mammals	*Marine fish*	*Insects*
(A)	Uric acid	Urea	Ammonia
(B)	Uric acid	Ammonia	Urea
(C)	Urea	Uric acid	Ammonia
(D)	Urea	Ammonia	Uric acid

5 In which of the following cases would bacteria be described as pathogens?

(A) Bacteria reducing lung function in mammals

(B) Bacteria in the human gut assisting digestion

(C) Bacteria producing insulin as a result of genetic engineering

(D) Bacteria in the mouth preventing growth of harmful bacteria

6 A body organ is shown.

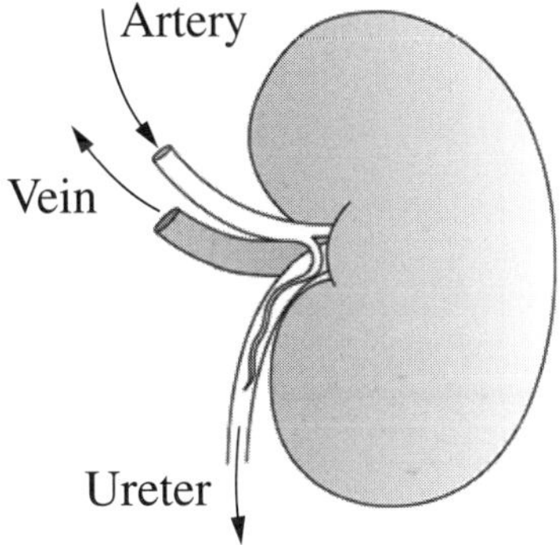

One function of this organ is to

(A) store the products of digestion.

(B) regulate salt levels in body fluids.

(C) remove carbon dioxide from the blood.

(D) release ADH when water levels in the blood are high.

7 Which of the following is essential in any model of natural selection?

(A) High reproductive rates

(B) Random selection of prey

(C) A population of predators

(D) Differences in the population

Refer to the diagram of plant tissue shown to answer Questions 8 and 9.

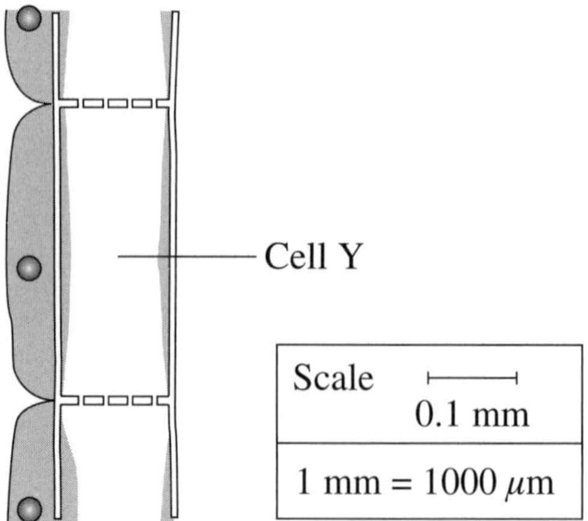

8 What is the most suitable title for this diagram?

(A) A section of xylem tissue

(B) A section of phloem tissue

(C) A transverse section of phloem tissue

(D) A longitudinal section of xylem and phloem tissue

9 What is the width of Cell Y?

(A) 3 mm

(B) 13 mm

(C) 130 μm

(D) 180 μm

10 Why do co-dominant alleles NOT produce simple Mendelian ratios?

(A) Both alleles are expressed in the phenotype.

(B) Neither allele is expressed in the phenotype.

(C) The recessive allele is only expressed in the homozygous genotype.

(D) The expression of the dominant allele is affected by the recessive allele.

11 What is a likely adverse impact of the evolution of bacteria on the management of infectious disease?

(A) Viruses become resistant to antibiotics.

(B) Some treatments lose their effectiveness.

(C) Inherited diseases become more common.

(D) Quarantine of infected individuals no longer works.

Refer to the following information to answer Questions 12 and 13.

A student conducted a first-hand investigation using nutrient broth, beakers and an S-shaped delivery tube in an attempt to model Pasteur's experiment.

The equipment and data collected are shown.

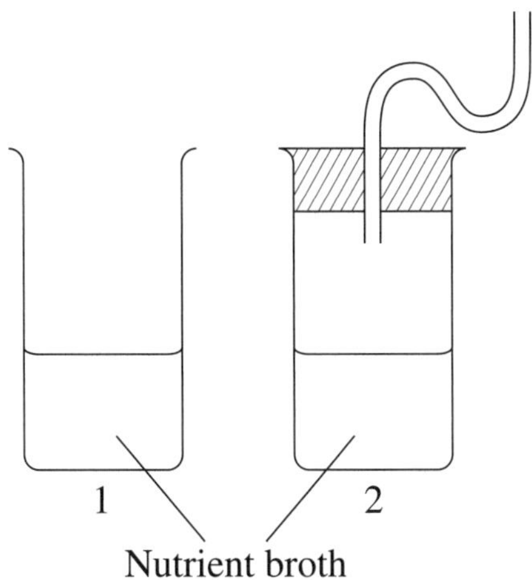

Test tube	*Observation of nutrient broth* Day 1	*Observation of nutrient broth* Day 14
1	Clear	Cloudy
2	Clear	Cloudy

12 The data collected by the student are

(A) quantitative because data were collected for fourteen days.

(B) qualitative because the appearance of the broth is described.

(C) qualitative because this is a model of a past scientific experiment.

(D) quantitative because the results were recorded for two different beakers.

13 The student's results were different from Pasteur's results.

Which of the following provides the best explanation for the difference?

(A) The nutrient broth was different from Pasteur's.

(B) The nutrient broth always goes cloudy as it ages.

(C) The nutrient broth was not boiled thoroughly on Day 1.

(D) The nutrient broths were both exposed to oxygen from the outside air.

14 The current theory to explain the movement of materials within the phloem of a living plant involves the following steps:

1. Osmosis
2. Active transport of sugars into non-photosynthetic cells
3. Active transport of sugars from photosynthetic cells
4. Flow of sugar solution up and down

Which of the following is the correct order of these steps?

(A) 3, 4, 1, 2

(B) 2, 4, 1, 3

(C) 1, 3, 2, 4

(D) 3, 1, 4, 2

15 Goltz Syndrome is a condition in humans that adversely affects the skin. It is inherited as a dominant gene carried on the X chromosome.

A man with Goltz Syndrome and a woman who does NOT have the trait have two children, a boy and a girl.

Which of the following is correct about the inheritance of Goltz Syndrome in these children?

(A) Both children have the syndrome.

(B) The girl has the syndrome and the boy does not.

(C) The girl has the syndrome and the boy is a carrier.

(D) The girl has a 50% chance of having the syndrome and the boy has a 0% chance.

16 What is the best explanation for the successful development of transgenic species?

(A) Artificial pollination works across the plant kingdom.

(B) Nuclear transplantation from cell to cell is easily achieved.

(C) DNA in the biosphere is composed of the same chemical components.

(D) Genes from different animals within the one species are easily combined.

17 When B cells are activated they divide to form a large number of antibody-secreting cells.

How is this best explained?

(A) Mitosis and gene expression produce cytotoxic cells.

(B) Mitosis and differentiation produce B cells that repair tissues.

(C) Mitosis and cell specialisation produce cells that maintain health.

(D) Mitosis and differentiation produce cells for growth and development.

18 Two people were exposed to pathogen *P* on the same day. The graph shows the blood antibody levels for that pathogen over the following 28 days for each person.

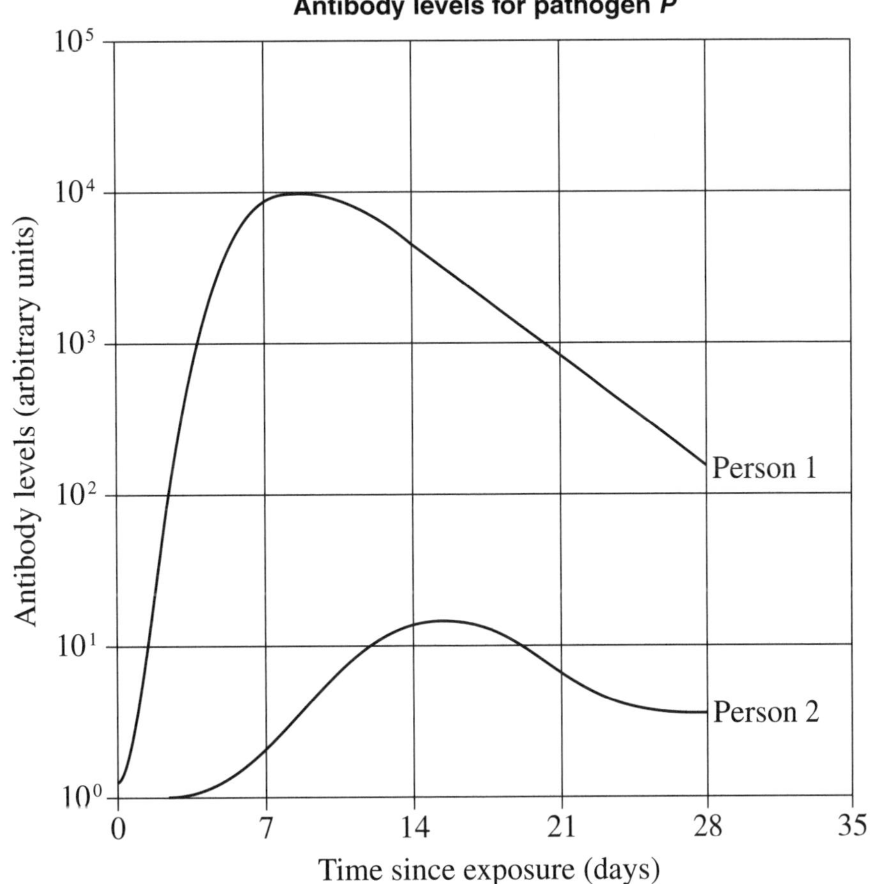

Which of the following best explains these results?

(A) Person 1 had not been previously exposed to pathogen *P* but had a recent organ transplant.

(B) Person 1 had been previously exposed to pathogen *P*.

(C) Person 2 had already been vaccinated against pathogen *P*.

(D) Person 2 had recent contact with a person infected with a similar pathogen.

Refer to the information to answer Questions 19 and 20.

A case study of a fish species

A fish species can survive in environments with a wide range of temperature variation as long as the temperature change is gradual. The fish die if the temperature change is rapid because some metabolic reactions cannot take place.

Scientists have discovered that all the fish in this species have genes to produce four different enzymes (*W*, *X*, *Y* and *Z*) that catalyse the same reaction at different temperatures.

The graph shows the effect of increasing temperature on enzyme activity.

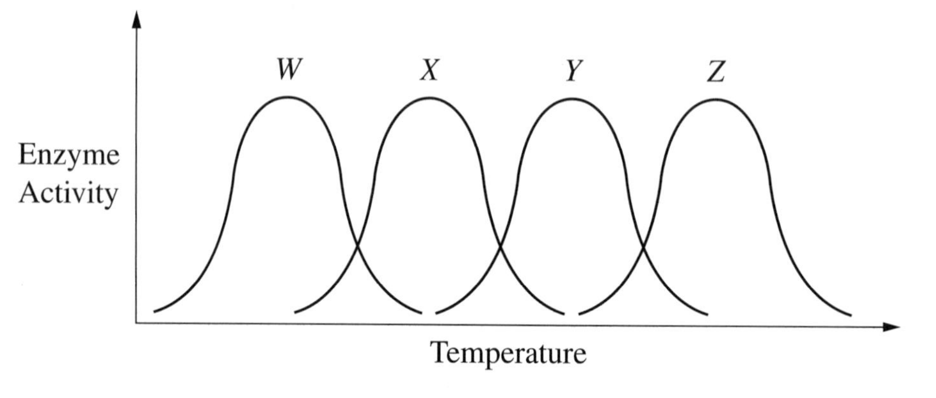

19 What is the best explanation for the fish surviving gradual temperature change but not a rapid temperature change?

(A) It takes time for each gene to be expressed.

(B) Enzyme activity is decreased at low temperatures.

(C) The fish produce different enzymes at different temperatures.

(D) Some enzymes will not denature if the temperature change is gradual.

20 What does this case study demonstrate?

(A) The specificity of substrates to enzymes

(B) The effect of the environment on phenotype

(C) The narrow temperature limitations for individual species

(D) The maintenance of a constant body temperature for metabolic efficiency

2014 HIGHER SCHOOL CERTIFICATE EXAMINATION

Biology

Centre Number

Student Number

Section I (continued)

Part B – 55 marks
Attempt Questions 21–30
Allow about 1 hour and 40 minutes for this part

Answer the questions in the spaces provided. These spaces provide guidance for the expected length of response.

Write your Centre Number and Student Number at the top of this page.

Question 21 (3 marks)

(a) The diagram shows a process that is a part of the immune response. **1**

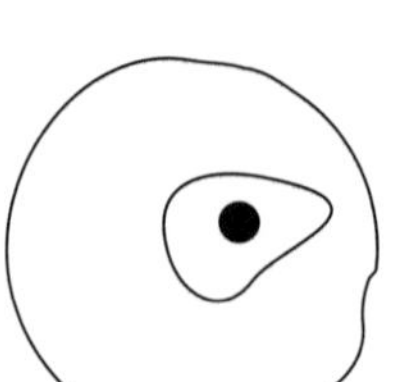

What is the name of the process?

(b) Outline how inflammation contributes to the immune response. **2**

Question 22 (6 marks)

(a) Explain how TWO specific personal hygiene practices reduce the risk of infection. **4**

....

....

....

....

....

....

....

....

(b) Drinking water contaminated with dissolved lead (a heavy metal) can cause a serious disease. **2**

Classify this disease as either infectious or non-infectious. Justify your answer.

....

....

....

....

Question 23 (5 marks)

Fungicides are chemicals that are used to treat fungal diseases such as rust in plants. Rust symptoms include orange-brown patches on the underside of leaves.

A new brand of fungicide claims to successfully treat rust disease in plants.

You designed an experiment to test this claim. Other students also performed your experiment and gave you their results.

(a) Write a valid procedure for your experiment. **4**

....

....

....

....

....

....

....

....

....

....

....

....

(b) How would you assess the reliability of your experimental design? **1**

....

....

Question 24 (4 marks)

(a) Use labelled diagrams to distinguish between the structure of an artery and that of a capillary. **2**

(b) Relate one structure of a capillary to its function. **2**

Question 25 (6 marks)

(a) Name a development in the history of our understanding of malaria and identify a prevention strategy that resulted from that development. **2**

....

....

....

....

(b) Scientists are currently developing a vaccine for malaria. **4**

Outline how a vaccine against malaria would prevent the occurrence and spread of the disease in a population.

....

....

....

....

....

....

....

....

....

....

....

....

Question 26 (5 marks)

Explain how Darwin/Wallace's theory of evolution by natural selection and isolation accounts for convergent evolution. Use an example to support your answer. **5**

....

....

....

....

....

....

....

....

....

....

....

....

....

....

....

Question 27 (7 marks)

A student set up an experiment as shown.

Set-up 1: Rat

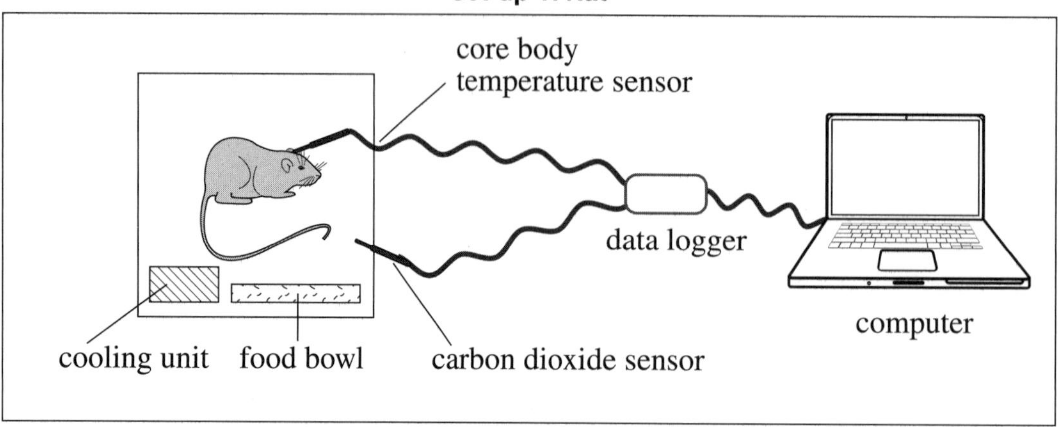

Set-up 2: Snake

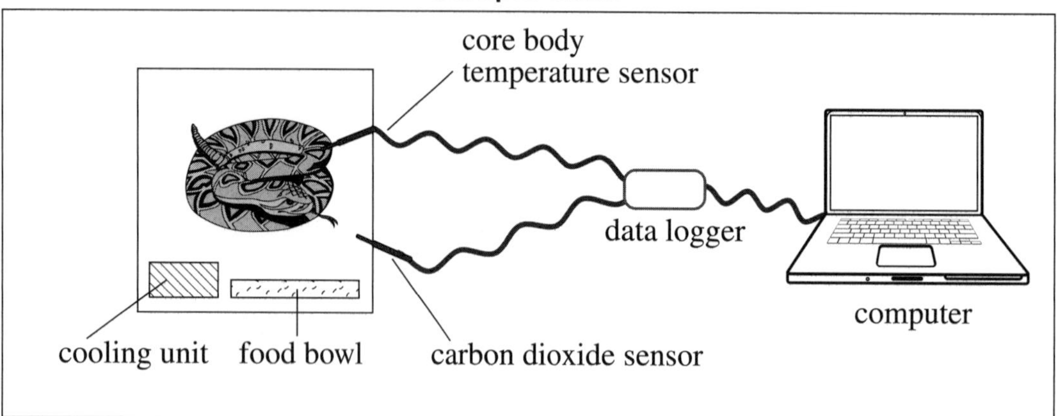

The animals were each placed in 1 m^3 incubators which had glass front doors. The incubators were set at 25°C and the temperature was gradually decreased to 10°C. The initial core temperature of the rat was 37°C and that of the snake was 25°C. The experiment ran for two hours.

(a) The carbon dioxide concentrations rose in each of the incubators. **1**

Explain why.

....

....

Question 27 continues

Question 27 (continued)

(b) Sketch line graphs on the axes below to predict body temperature data that would be collected in each set-up. **3**

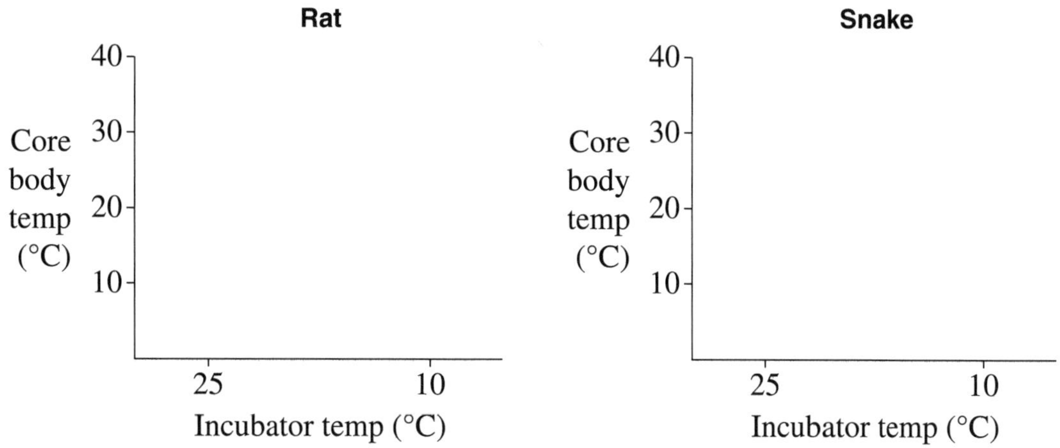

(c) Complete the table to explain TWO predicted observations, other than body temperature, that could be made of Set-up 1: Rat during the experiment. **3**

Predicted observation	*Explanation*

End of Question 27

Question 28 (6 marks)

Rennin is an enzyme found in the stomach of young mammals. Rennin curdles the milk drunk by the mammal and allows the milk solids to stay longer in the stomach to be further digested.

Students conducted an investigation into rennin activity. They bubbled different volumes of carbon dioxide gas into milk samples. Each sample was 50 mL and was kept at a constant temperature. The students then added rennin to each milk sample and recorded the time taken for the milk to curdle.

Volume of CO_2 (number of bubbles)	*Time taken for rennin to curdle milk samples* (seconds)				
	Trial 1	*Trial 2*	*Trial 3*	*Trial 4*	*Average*
100	253	257	250	260	255
150	238	232	241	229	235
200	216	214	219	211	215
250	208	202	212	198	205
300	210	200	199	311	203

(a) Account for the students' calculated average time for 300 bubbles of CO_2. **2**

....

....

....

....

(b) Explain the results of this experiment. **4**

....

....

....

....

....

....

....

....

....

....

....

....

Question 29 (5 marks)

Scientists have tried to achieve a viable embryo by fusing two ova (eggs) from the same female. 5

Explain whether the offspring produced using this process would be a clone of the female whose two ova were used. Use your knowledge of gamete formation and sexual reproduction to support your answer.

....

....

....

....

....

....

....

....

....

....

....

....

....

....

Question 30 (8 marks)

Our knowledge of biology is increased by scientists exploring and testing ideas using available technologies. The explanations of scientists are then verified or modified by the work of later scientists using newer technologies. **8**

Justify this statement using the work of THREE named scientists who have contributed to the development of ideas on inheritance.

2014 HIGHER SCHOOL CERTIFICATE EXAMINATION

Biology

Section II

25 marks
Attempt ONE question from Questions 31–35
Allow about 45 minutes for this section

Answer parts (a)–(e) of ONE question in the Section II Writing Booklet. Extra writing booklets are available.

Question 31	Communication
Question 32	Biotechnology
Question 33	Genetics: The Code Broken?
Question 34	The Human Story
Question 35	Biochemistry *(Not included in this reproduction)*

Question 31 — Communication (25 marks)

Answer parts (a), (b) and (c) of the question in the Section II Writing Booklet. Start each part of the question on a new page.

(a) The biological structure shown is part of one of the systems in the body.

(i) Name the biological structure. **1**

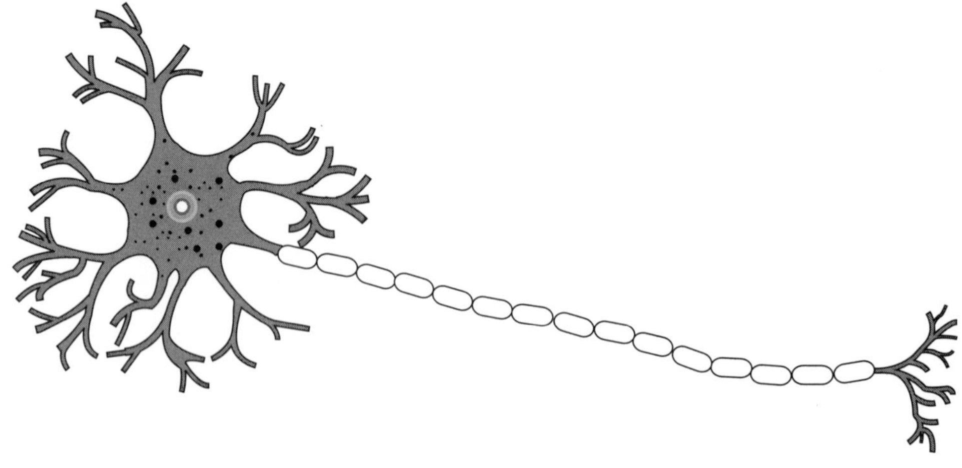

(ii) Outline the steps involved in this system's response to a stimulus. **3**

(b) Construct a flow chart to show the energy pathway of sound waves from the outer ear to the auditory nerve. On your flow chart indicate the locations and types of energy transformation that occur. **4**

(c) Explain how the differences between photoreceptor cells in the eye contribute to human vision. **4**

Question 31 continues

Question 31 (continued)

Answer parts (d) and (e) of the question in the Section II Writing Booklet. Start each part of the question on a new page.

(d) The diagram shows the visual fields of three animals. The visual field is the proportion of the 360° view of the environment which can be seen by the animal when looking straight ahead.

Visual fields can be mapped according to which eye provides the vision in each part of the visual field.

The diagram also shows how the position of the eyes on the skull affects the visual field map.

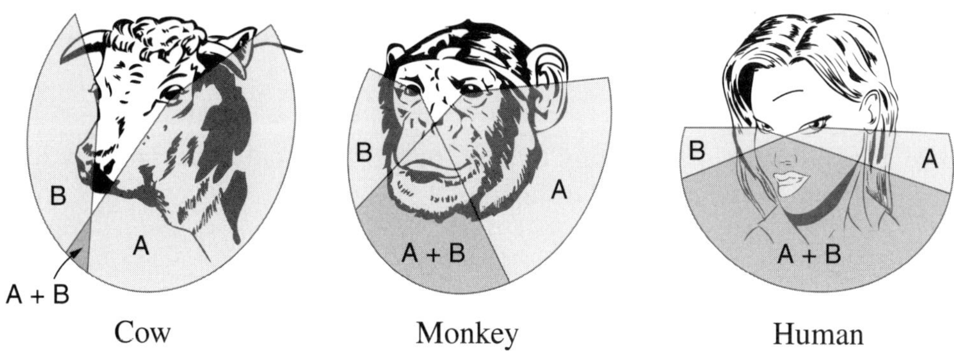

Key to map of visual fields	
A	Field seen by left eye only
B	Field seen by right eye only
A + B	Field seen by left and right eyes combined

(i) Use the data to describe the relationship between the position of the eyes in the head and the visual fields of the animal. **2**

(ii) Explain the effects of the loss of one eye on a monkey's vision and its chance of survival. **4**

(e) **7**

> Sound is a versatile form of communication. It relies on the successful transfer of signals between individuals of the same species, as well as of different species, for a purpose.

Justify this statement with reference to specific animals other than humans.

End of Question 31

Question 32 — Biotechnology (25 marks)

Answer parts (a), (b) and (c) of the question in the Section II Writing Booklet. Start each part of the question on a new page.

(a) (i) Name the process for the synthesis of a polypeptide chain from a messenger–RNA base sequence. **1**

(ii) Outline the steps in the formation of a functional enzyme from polypeptide chains. **3**

(b) Construct a flow chart to summarise the process of the polymerase chain reaction to amplify DNA. **4**

(c) Compare a traditional biotechnology to a biotransformational biotechnology in terms of ONE similarity and ONE difference. **4**

Question 32 continues

Question 32 (continued)

Answer parts (d) and (e) of the question in the Section II Writing Booklet. Start each part of the question on a new page.

(d) Bacteria can be used to produce human blood protein that functions at pH 7.4. Two bacterial species are described in the table.

Species	pH growth conditions			Efficiency of plasmid uptake
	Minimum pH	Optimum pH	Maximum pH	
Sulfolobus acidocaldarius	1.0	2.5	5.0	Low
Thiobacillus novellus	5.7	7.0	9.0	High

(i) Which of these species would a modern scientist choose in order to make a new bacterial species capable of producing a human blood protein? Give reasons for your choice with reference to these data. **2**

(ii) These two species of bacteria were mixed. Describe how a biotechnologist in the 1940s would have isolated them from each other using an experimental procedure based on these data and standard techniques. **4**

(e) The text summarises an ethical framework for decision-making related to the use of biotechnologies. **7**

> *Utilitarian Ethics*
> An ethical activity is one that provides the greatest balance of good over harm for society and the environment.
>
> *The Precautionary Principle*
> If there is doubt about the harm that may be caused by an activity, the proposers of the activity must prove the harm is not significant for society and the environment.

Evaluate the ethics of the use of ONE specific biotechnology in relation to this framework.

End of Question 32

Question 33 — Genetics: The Code Broken? (25 marks)

Answer parts (a), (b) and (c) of the question in the Section II Writing Booklet. Start each part of the question on a new page.

(a) (i) Name the segment of DNA shown in the diagram. **1**

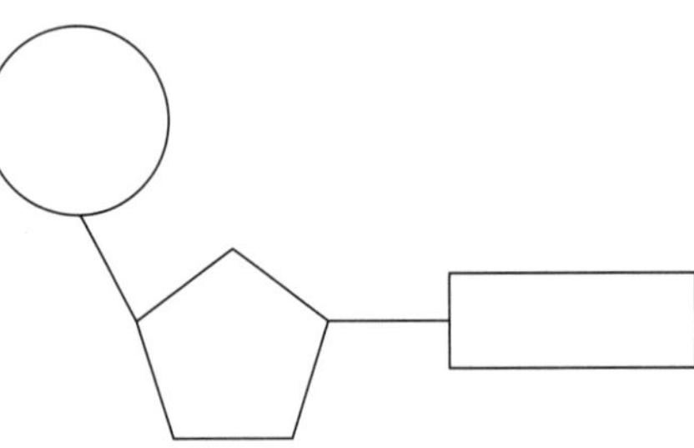

(ii) Outline the ability of DNA to repair itself. **3**

(b) Construct a flow chart to summarise the process of 'gene cloning' in a named example. **4**

(c) Assess the use of blood groups and highly variable genes for paternity testing. **4**

Question 33 continues

Question 33 (continued)

Answer parts (d) and (e) of the question in the Section II Writing Booklet. Start each part of the question on a new page.

(d) The data compare a segment of the eye control gene in mice to the equivalent gene segment in a range of different species. The expression of these genes is necessary for eye development to begin.

Mouse
eye control gene:
GTATCCAACGGTTGTGTGAGTAAAATTCTGGGCAGGTATTACGAGACTGGCTCCATCAGA

Species 1
eye control gene:
GTATCAAATGGATGTGTGAGCAAAATTCTCGGGAGGTATTATGAAACAGGAAGCATACGA

Genetic similarity to mouse: 76.66%
Protein similarity to mouse: 100%

Species 2
eye control gene:
GTGTCCAACGGTTGTGTCAGTAAAATCCTGGGCAGATACTATGAAACAGGATCCATCAGA

Genetic similarity to mouse: 85%
Protein similarity to mouse: 100%

Species 3
eye control gene:
GTCTCCAACGGCTGCGTTAGCAAGATTCTCGGACGGTACTATGAGACGGGCTCCATAAGA

Genetic similarity to mouse: 78.33%
Protein similarity to mouse: 100%

Species 4
eye control gene:
GTGTCTAATGGTTGTGTTAGTAAAATACTTTGCCGATATTATGGAACAGGTTCTATTAAA

Genetic similarity to mouse: 71.66%
Protein similarity to mouse: 100%

Note: grey highlighted bases are the same as those in the mouse gene.

(i) With reference to the data in the table, explain why it is possible for different gene sequences to produce the same protein. **2**

(ii) Discuss ONE strength and ONE limitation of using the data shown to determine the evolutionary relationships between these species. **4**

Question 33 continues

Question 33 (continued)

(e) The text below summarises some recent scientific experiments. **7**

> Scientists, studying the development of human female embryos, recently discovered a gene called XIST. This gene silences one of the two X chromosomes so that they do not over-function in normal human females.
>
> The scientists were then able to insert the XIST gene into human cells grown in tissue culture to successfully silence other chromosomes.
>
> Scientists are now attempting to insert the XIST gene into the extra chromosome of mice that have trisomy.

With reference to genetics and gene technologies, explain these experiments and their implications.

End of Question 33

Question 34 — The Human Story (25 marks)

Answer parts (a), (b) and (c) of the question in the Section II Writing Booklet. Start each part of the question on a new page.

(a) (i) State a feature that is used to classify humans as *Homo*. **1**

(ii) Outline the difficulties of interpreting data from fossils. **3**

(b) Construct a key to identify old world monkeys, apes, new world monkeys and prosimians using observable characteristics. **4**

(c) Demonstrate how applications of the Human Genome Project could affect future trends in human biological evolution. **4**

Question 34 continues

Question 34 (continued)

Answer parts (d) and (e) of the question in the Section II Writing Booklet. Start each part of the question on a new page.

(d) (i) The graph shows the relationship between the average skin pigmentation of humans and the distance they live from the equator. **2**

Identify the TWO general trends shown in the graph.

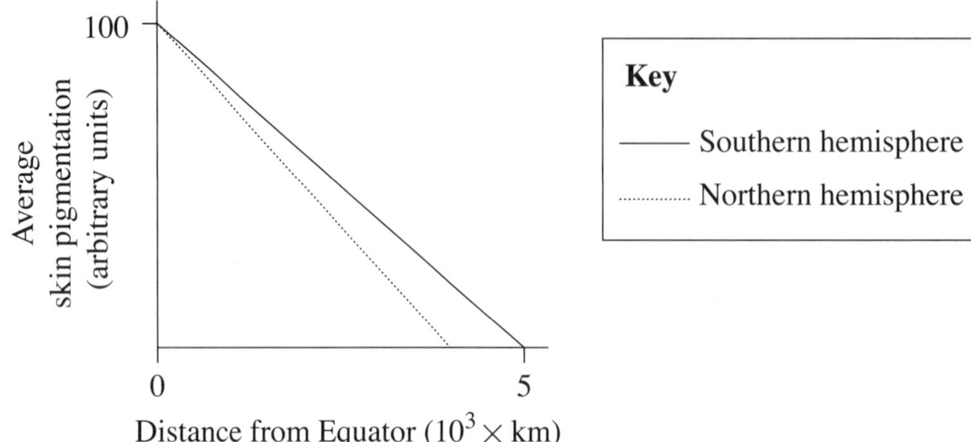

(ii) Explain the evolutionary significance of a named example of polymorphism (other than skin colour) in humans. **4**

(e) This text summarises new ideas on recently discovered fossil hominins called Denisovans. **7**

> Denisovans were identified from a number of separate intact fossil bones discovered in a cave in western Russia. One of these fossils was dated to 41 000 years ago. Very recently another cave site, this time in Spain, revealed even older intact Denisovan fossil bones.
>
> These fossils have been analysed to reveal a story of migration of Denisovans out of Africa 600 000 years ago. They were a large and genetically diverse group of hominins.
>
> Denisovans interbred with Neanderthals and *Homo sapiens*. A group of Denisovan males even migrated to Papua New Guinea and interbred with *Homo sapiens* there.

Using this information, analyse how scientists use technologies and the results they generate, to deduce ideas on hominin history which could modify existing theories.

End of paper

2014 HSC Examination Paper

Sample Answers

Section I Part A

(Total 20 marks)

1 A X-rays are known mutagens.

2 C Diffusion is the only option listed that does not involve a membrane.

3 A Small leaves reduce the surface area through which water loss can occur.

4 D This is the only option that correctly identifies that insects secrete uric acid, which they do to conserve water.

5 A This is the only scenario where the bacteria are detrimental to an organism.

6 B The organ depicted is part of the renal system, and so it is involved in filtering nitrogenous waste from the body and controlling the salt balance.

7 D Variation in the gene pool is essential to evolution; without it, natural selection could not take place.

8 B The diagram is a longitudinal section showing a phloem with sieve plates and a companion cell.

9 C Correct use of scale and conversion is 0.13 mm $\times$ 1000.

10 A Co-dominance is a specific example of inheritance where neither allele masks the other in a heterozygous combination.

11 B The evolution of bacteria renders antibiotics ineffective as resistance develops.

12 B Only observations of appearance have been recorded. No numerical data has been included.

13 C The nutrient broth was contaminated from the start. Oxygen does not cause growth/spoilage on its own.

14 D This sequence describes the source to sink theory: loading of sugar at the source against the gradient; osmosis; movement of sugar through plant; unloading by active transport at the sink.

15 B The girl will inherit the dominant gene on the X chromosome from the father. The boy's X chromosome is from the mother who does not have the gene.

16 C The same chemical composition of DNA enables easier interchange of genetic material between species.

17 C Somatic cells reproduce by mitosis and different cells arise through specialisation.

18 B Person 1 had previously been exposed so was able to quickly produce larger quantities of antibodies.

19 C The fish produce different enzymes at different temperatures but require time to produce them, hence gradual change will enable them to survive.

20 B The expression of the genes depends on the environmental conditions. Different phenotypes occur in different conditions.

Section I Part B

21 (a) Phagocytosis *(1 mark)*

(b) Inflammation is a non-specific immune response. The body brings blood and fluid to the location of damaged cells or infection, causing it to redden, swell and warm. The extra blood carries phagocytes to destroy the pathogen, enables clotting to occur and prevents pathogens spreading. *(2 marks)*

22 (a) Answers will vary. The following are two examples.

Washing hands with soap kills pathogens, reducing their spread from the hands to the mouth. This also reduces the risk of further contamination of food or other surfaces.

Covering the nose and mouth when sneezing prevents the possible spread of pathogens via aerosol droplets to others and onto surfaces that others come in contact with. *(4 marks)*

(b) Heavy metal poisoning is a non-infectious disease because it is not caused by a pathogen, but an environmental factor. *(2 marks)*

23 (a) Validity is achieved when only the independent variable is changed (in this case, the fungicide used). All other variables must be kept constant. The following steps outline a valid experiment to test the effectiveness of the fungicide.

1. Obtain a **large sample** (e.g. 60) of infected plants and an **identical control group** of plants that are not infected.
2. Plant in **identical trays** containing the **same volume and type of soil**. Water with the **same volume of water at regular intervals** of two to three days, using a measuring cylinder. (The trays, soil and water are controlled variables.)
3. Leave the plants in the **same position**, one that is appropriate to promote growth for that type of plant.
4. Spray half of each group of plants with the fungicide from a spray bottle, ensuring the plants are **sprayed identically**. Monitor the plants daily to see if the fungicide makes any difference to the growth of the infected and control groups.
5. Repeat these steps multiple times with the same fungicide to ensure reliability of results. *(4 marks)*

(b) The reliability of the experiment could be assessed by looking at how many times the experiment was repeated by different individuals, the sample size and the average of the results. *(1 mark)*

24 (a)

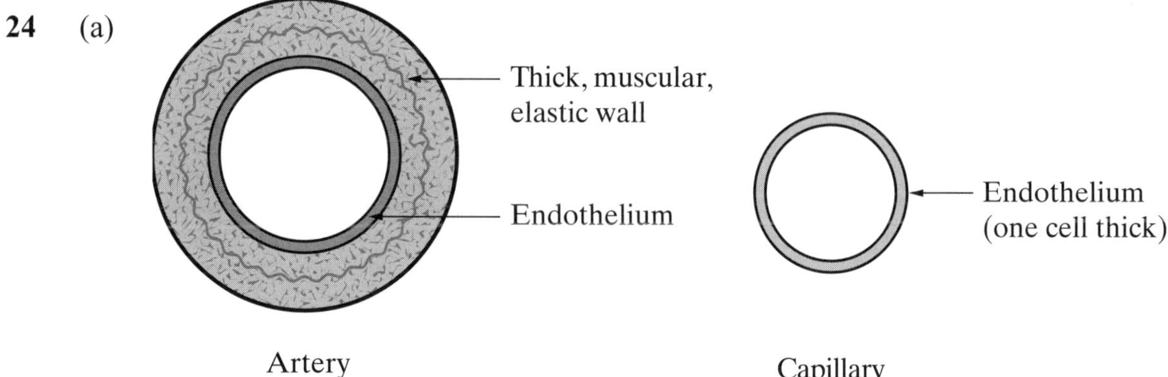

(2 marks)

(b) The capillary is only one cell thick. The resulting large surface to area ratio enables efficient (direct) diffusion of materials into and out of the capillary. *(2 marks)*

25 (a) The discovery that malaria is transmitted by a mosquito vector was a significant development in our understanding of the disease. Resulting strategies to reduce the incidence of malaria included preventing contact with mosquitos, interfering with mosquito breeding and eradicating mosquitos from populated areas by spraying with insecticide. *(2 marks)*

(b) A malaria vaccine would prevent the occurrence of the disease because vaccinated individuals that are bitten would not be affected by the plasmodium transmitted by the mosquito. Their bodies would automatically produce antibodies to fight the pathogen, preventing the individual from getting sick. Vaccination would also prevent the spread of the disease in a population because the mosquitos would not have an infected host to extract blood from. *(4 marks)*

26 Convergent evolution can be explained by the Darwin/Wallace theory of evolution, which outlines the need for variation in a gene pool and a selecting agent in the organism's environment over time. The characteristics of different environments effectively isolate species, leading to the formation of unique gene pools. Those same characteristics act on different species living in similar environments to favour particular features in their gene pool that increase the chances of survival, enabling individuals to reproduce and pass on those favourable characteristics. With similar selection pressures, the favourable characteristics are similar. For example, organisms that live/swim in water (such as fish, sharks, whales, dolphins and seals) typically have similar streamlined features including sleek, smooth bodies and a tail or fins that enable them to swim fast. *(5 marks)*

27 (a) Carbon dioxide is a product of respiration. The animals would have been exhaling carbon dioxide into the incubator, causing levels to rise. *(1 mark)*

(b)

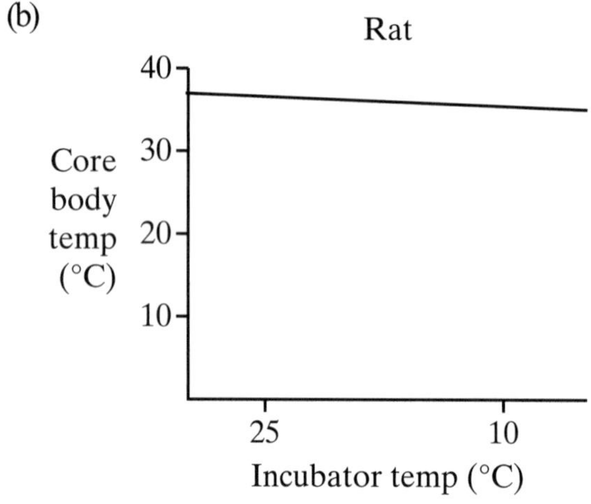

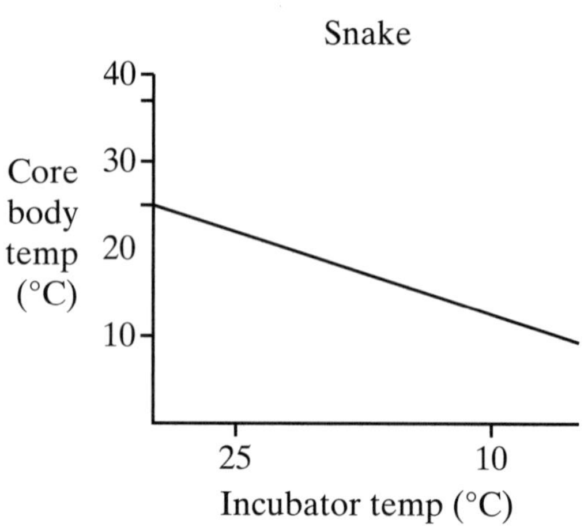

(3 marks)

(c) Answers will vary. The following table provides two examples.

Predicted observation	*Explanation*
The rat's fur might have 'fluffed up'.	The fur became erect in order to try and retain heat.
The rat may have started to shiver.	Shivering increases body heat as the rat's body would have tried to maintain a constant internal body temperature.

(3 marks)

28 (a) The average time for 300 bubbles was determined to be 203 seconds. The Trial 4 result was not included in calculating the average because it was an outlier and not considered to be statistically accurate. *(2 marks)*

(b) The results indicate that the greater the volume of carbon dioxide bubbled into the milk, the faster the rennin was able to curdle the milk. The carbon dioxide would have dissolved with the water in the milk, lowering the pH and making it more acidic. This would more closely resemble the pH in the stomach where protein digestion occurs (pH level of about 3). This caused the curdling of the milk, enabling it to remain in the stomach longer so the protease could commence protein digestion. *(4 marks)*

29 Ova are produced through the process of meiosis, which results in four daughter cells. Each daughter cell carries a haploid number of chromosomes. The chromosomes may be slightly altered by crossing over, independent assortment and random segregation of alleles. These processes all result in possible genetic variation during meiosis, so the daughter cells will not be identical. Fusing two separate ova from the same female might create a full complement of chromosomes, but each will be slightly different, resulting in a unique offspring that is not genetically identical to the original female (i.e. different genotype). As such, the offspring will not be a clone of the original female, but will be female. *(5 marks)*

30 Ideas regarding inheritance have changed considerably over time with the advent of new technologies. The work of early scientists has been verified and elaborated in light of the information these technologies have been able to provide.

For example, in the mid-1800s Mendel used garden peas to demonstrate the inheritance of 'factors' from parent plants to offspring. Using large sample sizes, Mendel controlled the fertilisation of pure-breeding pea plants and determined the 3:1 Mendelian ratio. His work was largely unnoticed due to his low profile and inability to promote his work to the scientific community. It was not until the late 19th century when technological advances in microscopy made it possible to observe mitosis and meiosis first hand. This gave rise to an understanding about cell structure and the function of the nucleus, which gave Mendel's work and theories about inheritance some support.

At the very beginning of the 20th century Mendel's work was again looked at more closely when direct observations of chromosomes became possible. The chromosome theory of inheritance was proposed, which led to the acceptance of Mendel's work. In 1903 Sutton and Boveri, working on grasshoppers and sea urchins, respectively, noted clear links between the chromosomes within nuclei and the behaviour of the inheritance factors. They identified pairs of homologous chromosomes and that the factors (now called genes) occurred as pairs, which separated during meiosis and reformed as pairs after fertilisation. With these findings Sutton and Boveri independently proposed the chromosome theory of inheritance.

Still, this theory was not well received because the evidence to support it was limited. It was not until 1910 and Morgan's work linking specific genes on a specific chromosome that Sutton and Boveri's theory became accepted. Using fruit flies Morgan was able to show how the inheritance of specific genes occurred through successive generations, and matched observed results with predicted outcomes. Furthermore, Morgan demonstrated that when a heterozygous red-eyed female was crossed with a red-eyed male, 50% of male offspring had white eyes while all females had red eyes. This did not meet the expected 3:1 Mendelian ratio. He was able to demonstrate sex linkage.

The current understanding of inheritance did not follow an easy path and was ultimately the result of a combination of independent work carried out over a longer period of time by a number of people and advances in technology that allowed discoveries to be made, verified and brought together. *(8 marks)*

Section II—Options

Question 31—Communication

(a) (i) motor neuron *(1 mark)*

(ii) stimulus → receptor → messenger → effector → response

The nervous system's response to a stimulus begins when the stimulus is detected by a sense organ. Sensory neurons transmit an impulse away from the sense organ to the effector via the central nervous system, where an association or connector neuron in the brain or spinal cord sends the impulse to a motor neuron. The motor neuron transmits the impulse away from the central nervous system to the muscle or gland (the effector) that is going to respond to the original stimulus. *(3 marks)*

(b)

Sound waves (kinetic energy)	→	Mechanical kinetic energy	→	Electrochemical energy
Outer ear		*Middle ear*		*Inner ear*
Sound is collected by the pinna. Vibrating air enters the outer ear canal. Vibration at same frequency is transmitted by the eardrum to the middle ear.		Ear ossicles transmit the vibration to the oval window.		Fluid in the inner ear vibrates, producing tension on hair cells in the organ of Corti within the cochlea. Electrochemical impulses are transmitted to the auditory nerve and are received by the brain for interpretation.

(4 marks)

(c) The eye has two types of photoreceptor cells located within the retina. The rods and cones, which both contain light sensitive pigment, contribute differently to human vision and work together to produce the clear sharp images that are possible by the human eye. The following table summarises the differences between the photosensitive cells in the eye.

Feature	*Cones*	*Rods*
Distribution	Cones are found in the middle of the retina, concentrated at the fovea region.	Rods are found across the retina but are most numerous around the periphery. They are absent in the fovea.
Structure	These cone-shaped cells contain stacks of membranes where the photosensitive photopsin pigment is found. There are three types: red, blue and green.	These rod-shaped cells contain stacks of membranes where the photosensitive rhodopsin pigment is found.
Function	Each of the three cone types is sensitive to different light intensity. This produces colour vision but relies on the stimulation of opsin by visible light.	Rhodopsin is sensitive to blue green light and works in low or dim light. Rods detect movement and shape but do not assist with detail vision, because they have only one pigment. They can only tell light from dark.

(4 marks)

(d) (i) The relationship between the position of the eyes in the head determines the visual fields of the animals. The further the eyes are positioned apart on the head, the greater the field of view. As the eyes get closer to the front of the head, the amount of overlap of the fields of view from each eye increases, giving rise to better depth perception and reducing the peripheral vision field size. *(2 marks)*

(ii) If the monkey lost one of its eyes (e.g. eye B in the diagram) its peripheral vision would be diminished. The monkey would only be able to see its surroundings through the remaining eye, so would only have vision on one side of its head (e.g. field A in the diagram). It would have no overlap of vision to the front, and so would have significantly diminished depth perception. Without depth perception the monkey would have difficulty with daily habits such as brachiating (swinging) from branch to branch and collecting food. This would greatly endanger its chances of survival in its natural environment because it would not be able to get away from predators, find food and move about in the trees safely. *(4 marks)*

(e) A successful message transfer is one where the message is sent and received without loss of content. Sound is a versatile form of communication because it can travel easily through different media (e.g. air and water) and does not require the sender or receiver to be in visual contact with one another.

Individuals of the same species have specific means of communicating by sound, such as ears in mammals, tympanic membranes in insects and the combinations of otoliths and a lateral line in fish. For example, grasshoppers detect sounds (as vibrations) using tympanic membranes on their legs that are connected to nerve fibres. Cicadas possess tympana on the base of the abdomen, which are connected to an auditory organ. Only male cicadas vocalise, yet both males and females can detect sounds. Whales detect sounds through their lower jaws, which contain a fat-filled cavity that extends back into an ear-bone complex. Here the sounds are transmitted through the ear to the auditory cortex of the brain via an auditory nerve. Fish are able to detect sound via neuromast cells in their lateral line and the stimulus is then sent to the inner ear causing the otolith bones to vibrate where an electrochemical signal is then sent to the brain for interpretation.

Organisms can detect sounds made by other species through their auditory sense organs even if they cannot be interpreted accurately. Within their own species, organisms use their auditory sense organs to interpret stimuli from their environment, and rely on their brains to identify the sounds as warnings, harmless sounds or specific communications. Within species sounds are used to warn of danger, mark territory, locate food, protect young or members of the group/family, find a mate or indicate pleasure.

Such diversity in the structures and means to detect the sounds in the organisms' environment ensures successful detection and interpretation of messages. The versatility of sound as a form of communication ultimately leads to the survival of the species.

(7 marks)

Question 32—Biotechnology

(a) (i) protein synthesis *(1 mark)*

(ii) Transcription occurs as the double-stranded DNA unwinds and a complementary messenger RNA (mRNA) strand is formed. The mRNA strand moves out of the nucleus into the cytoplasm. In the cytoplasm the strand binds to a ribosome and translation occurs. During this process transfer RNA (tRNA) molecules, each carrying an amino acid, temporarily bind to the mRNA one codon at a time. The amino acids form a polypeptide chain that codes for a specific protein, such as a functional enzyme, which is then released back into the cytoplasm. *(3 marks)*

(b)

Target sequence of DNA

DNA heated to separate strands

Primers are used to bind to the DNA strands.
(This labels the portions of DNA for amplification and the starting place for replication.)

New pieces of DNA are made that are complementary to the portions marked by primers along the strand.

This cycle is repeated multiple times, each cycle taking only a minute.
Each new segment can serve as a template for new ones. *(4 marks)*

(c)

	Traditional biotechnology: fermentation	Biotransformational biotechnology: gene therapy
Similarity	Useful application for society, which stands to gain significantly in terms of product produced (e.g. food production such as wine or bread).	Useful application for society, which stands to gain significantly in terms of medical advances, curing illness and disease (e.g. treating cystic fibrosis or Huntington's disease).
Difference	Uses naturally occurring reactants, with few impacts in terms of human evolution.	Uses well-developed technology that has the potential to alter human evolutionary patterns.

(4 marks)

(d) (i) *Thiobacillus novellus* would be suitable for the production of a human blood protein because it has a pH closest to the natural range of human blood, which is just a little higher than 7. The other species has a very low pH range by comparison. The efficiency of plasmid uptake is high for *Thiobacillus novellus*, so this would serve it well as a plasmid vector for producing blood protein. *(2 marks)*

(ii) A biotechnologist in the 1940s could Gram stain these bacteria in order to identify them under a microscope. In order to do this the bacteria would first need to be cultured. A pure culture of each bacteria would then be made by selecting each one from the original culture plate and transferring them to separate plates of sterile growth medium. A sample of the pure culture would then be transferred on to a slide and dried before applying crystal violet over the slide. Excess crystal violet would be gently rinsed off and covered with iodine, then fixed with ethanol and rinsed. The sample would then be covered with safranin, rinsed and dried at room temperature. It would then be possible to view the bacteria under a microscope.

(4 marks)

(e) Many answers are possible. This answer considers the use of transgenic technology to produce genetically modified (GM) crops.

Using transgenic technology to produce GM crops raises a number of ethical considerations for many stakeholders in the wider community. Scientists follow a rigid ethical framework for decision-making in the work they undertake; however, they are not the end users of the technology or of the products made through its use. Scientists follow strict guidelines regarding the containment of GM organisms when working in the laboratory, and take great care to ensure that their research does not cause harm to people or to the environment.

This technology gives rise to a number of concerns. For example, who determines where GM crops such as soy can be planted? Because the pollen cannot be contained GM crops can contaminate non-GM crops in surrounding farms. This makes it difficult to label such crops, and difficult for consumers who may want to buy non-GM food/products. Another concern is that these products have not been around long enough to determine their long-term effects on consumers. While agribusinesses have typically invested significantly in the research, they stand to make a lot of money from the end use of the technology. Product such as GM seed is very attractive to farmers, who stand to make a lot of money from its use by producing a high yield or superior produce that will sell well.

Despite the initial care taken in the laboratory, the use of the technology cannot be controlled once the technology is released to the marketplace. As such, the ethical framework is flawed because it does not govern what occurs once the technology has been sold to the highest bidder. *(7 marks)*

Question 33—Genetics: The Code Broken?

(a) (i) nucleotide *(1 mark)*

(ii) Copying errors can be repaired by enzymes such as DNA polymerase that can use the undamaged strand of DNA in the double helix as a template to fix and replace the incorrect damaged base sequence.

1 Endonucleases (enzymes) open up the DNA.

2 Exonucleases (enzymes) remove the damaged DNA using restriction enzymes that cut it out one base at a time.

3 Polymerase (an enzyme) replaces the damaged/incorrect bases.

4 Ligase (an enzyme) joins the new section to the old. *(3 marks)*

(b)

Select gene	→	Cut gene out of DNA	→	Insert gene into new organism	→	Make copies of the gene
Human insulin gene		Restriction enzymes are used to cut out the insulin gene		Gene is stitched into plasmid using ligase		Plasmid is inserted into bacteria and copies are made

(4 marks)

(c) Blood groups and highly variable genes can be very effective in paternity testing as they provide clear markers or reference points when comparing DNA samples. Inheritance of blood type follows simple Mendelian inheritance patterns. In cases of paternity identification, ABO blood typing can be used to exclude a man from being a child's father rather than proving paternity. For example, a child with type O blood cannot be fathered by a man who has type AB blood, because he would pass on either the A or the B allele to all of his offspring. In recent times specific genetic markers have been used more often as they are more reliable and inclusive than using blood typing. In the 1970s, testing for human leukocyte antigens (HLAs) made it possible to rule out men as fathers with 80% effectiveness. Since then electrophoresis of various biochemical markers has become available. In this process proteins from blood or other tissues are separated by an electric current that is run through a gel containing the tissue sample. Different forms of the proteins separate according to their electrical charge or size. Similarities in the patterns produced by samples from both the child and potential fathers are compared. This has been shown to be a better indicator of genetic relationships or parentage than earlier testing methods. *(4 marks)*

(d) (i) There are 20 different amino acids used to make proteins, but there are only four different nucleotides in DNA and RNA. The protein-coding sequence of a gene is composed of three-letter triplets that make up codons. There are also special start and stop codons that mark the beginning and end of a gene. Multiple codons can code for the same amino acid, giving rise to the same protein and making this part of the code redundant (i.e. most of the amino acids have at least two different codons). Most living things use this exact code to make proteins from DNA. The data shows these sequences in grey, with only a few breaks that are redundant, resulting in the high percentage of correlation in the protein type. *(2 marks)*

(ii) One strength in using this data to determine evolutionary relationships is that the DNA sequence can be specifically identified and mapped. The closest correlation between the species can then be identified. For example, species 2 appears to have the highest correlation with the mouse data, followed by species 3. The closest correlation for eyes in mice occurs with Drosophila, an insect, rather than another mammal.

However, a chromosome map based on recombination frequencies does not fully represent a chromosome. As a result, comparisons based on these segments can be erroneous because the rate of recombination will vary from species to species. Such comparisons can be limiting as they provide relative order but not position on the chromosomes. *(4 marks)*

(e) These experiments are very significant for medical technology and assisting patients with diseases that can be linked to specific genes. In particular, the XIST gene may provide a treatment for those that suffer from trisomy/multiple X diseases associated with the X chromosome. If the inserted XIST gene silences the X chromosome that it is inserted into then the symptoms that manifest themselves with the extra X chromosome may be diminished.

Such treatment is called gene therapy, which has been successfully carried out on patients with cystic fibrosis. In these patients the genes necessary for healthy lung function are inserted into the lung cells using adenoviruses as vectors, delivered as inhalants. There has been varied success with this treatment due to problems associated with the size of the gene, lung cells being replaced relatively frequently and sometimes the carrier virus causes inflammation of the lung tissue. However, there has been success with cancer suppressor cells administered via adenovirus. In conjunction with anti-cancer drugs these treatments have resulted in regression of tumours.

The use of retroviruses for gene therapy has had limited success but research in this area continues due to the possible benefits if the technology improves, which is only a matter of time. However, such treatment may pose ethical considerations. If carried out as somatic gene therapy, the functional gene would be inserted into the appropriate body cells, and as such would not be passed on. If the treatment involved germ line therapy (manipulation of gene in a gamete) the change could be passed on. This gives rise to ethical concerns, with questions arising regarding the nature of treatment and the possible implications for the individual in the future and the impact on the wider community. *(7 marks)*

Question 34—The Human Story

(a) (i) Large, well-developed brains and the ability to think logically classifies humans into the genus Homo. *(1 mark)*

(ii) Interpreting data from fossils is difficult if the conditions under which the fossils formed are not appropriate. The best fossils are formed when organisms with hard parts are buried quickly and remain undisturbed, preventing decay by decomposing organisms. Another difficulty is that conflicting dates can occur depending on the type of dating technique used. The fossil record also has gaps, with scarce evidence in some areas, so different interpretations of the same evidence can occur.

(3 marks)

(b)

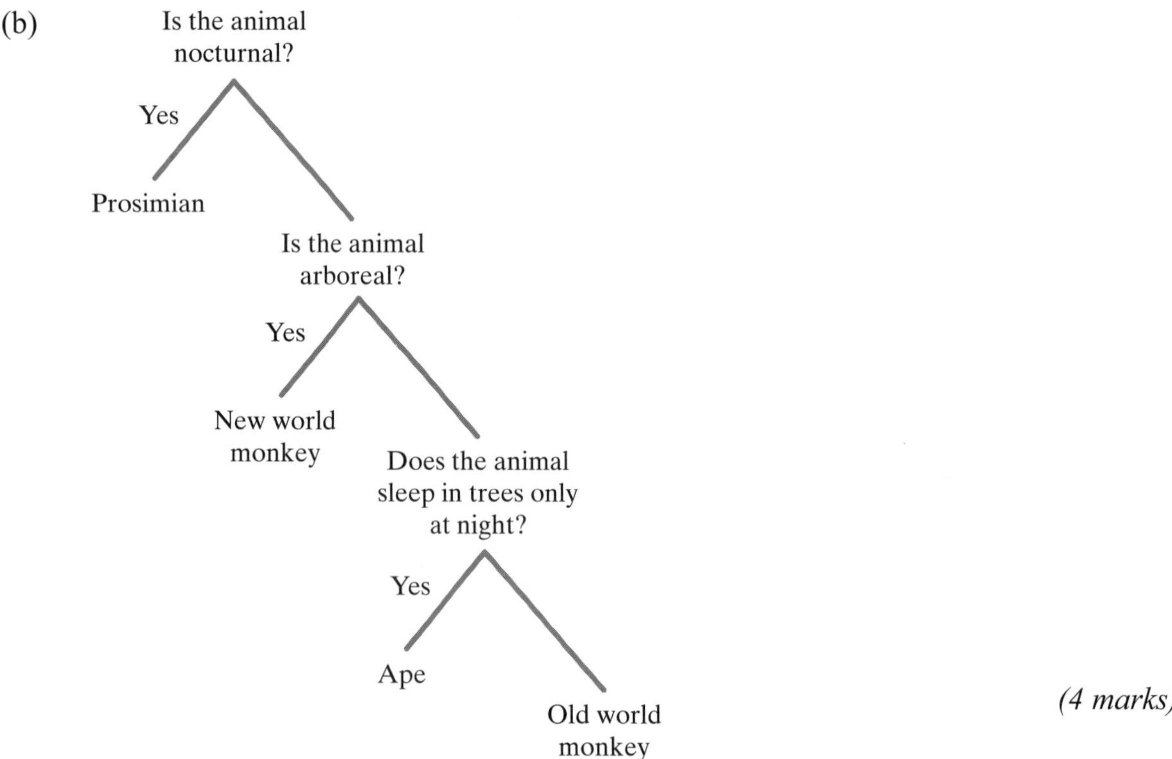

(4 marks)

(c) The Human Genome Project has provided a map of the position of every gene on each human chromosome and the nucleotide sequence of every gene. This knowledge, combined with the advances in genetic technology, can lead to significant changes within a population.

Individuals can be screened for disease or potential defective genes that can make them genetically unfit, as far as their potential health is concerned. Based on this information they could be refused employment or insurance, or prevented from participating in certain activities. It may become possible to genetically modify individuals and effectively breed defective genes out of the population or create a manifestation of selected genes over less desirable ones.

Defective genes that have been identified may be removed from the gene pool with gene therapy or by cloning healthy individuals. This could change the trend in biological evolution in the future. For example, cystic fibrosis is caused by a defective gene in the lungs. With gene therapy it is possible to insert a defect-free gene into the lungs thereby preventing death. This is an example of somatic cell manipulation, so it is not passed on. If manipulation of germ line genes is carried out for illnesses such as cystic fibrosis, Huntington's disease or certain bone marrow disorders, the resulting embryo is free of the disease and would not pass it on. This certainly changes the gene pool permanently and subsequent biological evolution. *(4 marks)*

(d) (i) The two general trends presented in the graph are that the average amount of skin pigmentation in humans decreases with distance from the equator, and that the average amount of skin pigmentation in humans living in the Southern Hemisphere tends to be greater than for those living in the Northern Hemisphere. *(2 marks)*

(ii) Answers will vary. The following is one example.

If more than one form of a particular gene is present in a population it is said to be polymorphic. One example of polymorphism in humans is sickle cell anaemia. This is a genetic disease that reduces the oxygen-carrying capacity of blood vessels, causing damage to body organs. Those individuals that carry the sickle cell allele (that is, are heterozygous) do not suffer from the disease, but are resistant to malaria. This is advantageous for heterozygous carrier individuals who tend to live in areas where malaria is common, such as Asia and western and Central Africa. *(4 marks)*

(e) The article describes the discovery of new fossil sites that show how hominins may have migrated west across Europe and further south-east to Papua New Guinea. The recent findings add more ideas to the current arbitrary theories, which each tend to lack conclusive evidence and in all probability will incorporate mutual similarities based on the interpretation of similar evidence.

Scientists have used advanced dating techniques to determine the ages of fossils more accurately, leading to the proposal of the multiregional model of human origins. Various biochemical processes to determine biological similarities between specimen samples, including the tracing of mitochondrial DNA, have given much support to the 'Out of Africa' model of human origins. The transition of the hominins around the globe was complicated and may very well be a combination of the two models. Modern humans may have evolved from an isolated area of Africa and then interbred with local groups after migration away from Africa resulting in the phenotypic variation in existence today.

As new evidence comes to light, and old evidence is re-examined with new technology, past and present models are continually being modified as the information gained is interpreted. *(7 marks)*

CHAPTER 14

2015 HIGHER SCHOOL CERTIFICATE EXAMINATION

Biology

General Instructions

- Reading time – 5 minutes
- Working time – 3 hours
- Write using black pen
- Draw diagrams using pencil
- Board-approved calculators may be used

Total marks – 100

Section I

75 marks

This section has two parts, Part A and Part B

Part A – 20 marks

- Attempt Questions 1–20
- Allow about 35 minutes for this part

Part B – 55 marks

- Attempt Questions 21–31
- Allow about 1 hour and 40 minutes for this part

Section II

25 marks

- Attempt ONE question from Questions 32–35
- Allow about 45 minutes for this section

Section I
75 marks

Part A – 20 marks
Attempt Questions 1–20
Allow about 35 minutes for this part

Use the multiple-choice answer sheet for Questions 1–20.

---

1 What is the name of the theory which describes evolution as patterns of rapid first appearances or extinctions followed by periods of little or no change?

(A) Gradualism

(B) Convergence

(C) Adaptive radiation

(D) Punctuated equilibrium

2 What are alternative forms of genes called?

(A) Alleles

(B) Autosomes

(C) Chromatids

(D) Chromosomes

3 Which adaptation would decrease water loss from a plant in a region with low rainfall?

(A) Broad leaves

(B) Surface roots

(C) Sunken stomates

(D) Loosely packed epidermal cells

4 Which of the following does NOT play a role in maintaining the health of an organism?

(A) Mitosis

(B) Cell death

(C) Gene expression

(D) Sexual reproduction

5 The diagram shows a model for enzyme action.

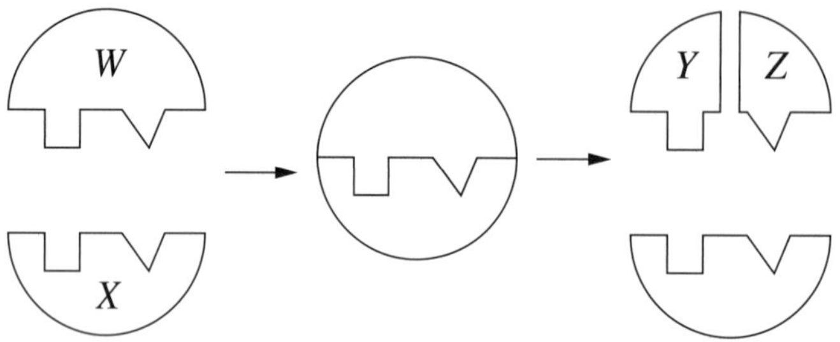

Which letter indicates the substrate in this diagram?

(A) *W*

(B) *X*

(C) *Y*

(D) *Z*

6 An extremely high concentration of carbon dioxide is undesirable in active muscle tissue because it will

(A) increase the pH.

(B) cause enzymes to denature.

(C) increase cellular reaction rates.

(D) cause haemoglobin to release oxygen.

Refer to the following information to answer Questions 7 and 8.

The diagram shows a homeostatic mechanism in a mammal.

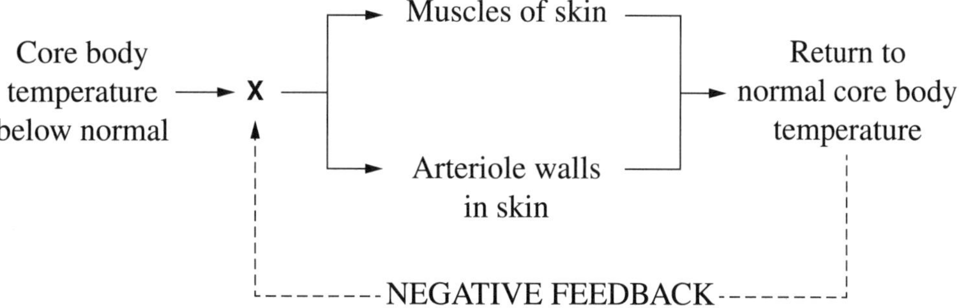

7 What does **X** represent in the diagram?

(A) The heart

(B) The brain

(C) A thermoreceptor in the skin

(D) A pressure receptor in a blood vessel

8 Which of the following describes what happens to the muscles and the arteriole walls in the skin when the core body temperature is below normal?

	Muscles of skin	*Arteriole walls in skin*
(A)	Relax to lower epidermal hairs	Expand
(B)	Contract to raise epidermal hairs	Contract
(C)	Relax to raise epidermal hairs	Expand
(D)	Contract to lower epidermal hairs	Contract

9 A pathogen and a red blood cell are drawn to the same scale, with some features indicated.

Pathogen

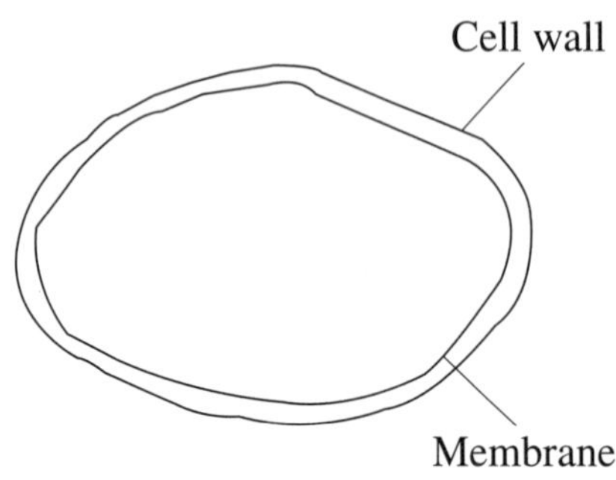

Red blood cell

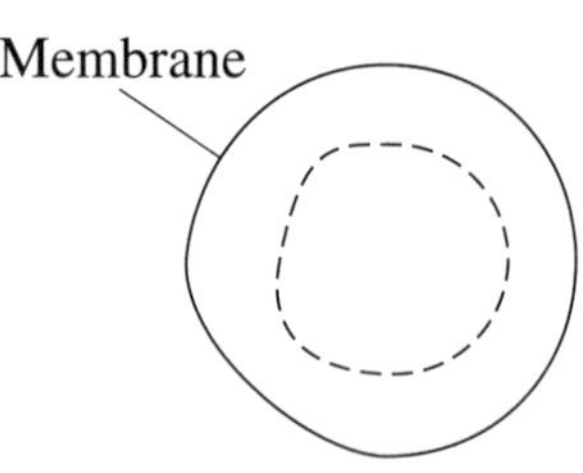

What type of pathogen is this?

(A) A virus

(B) A prion

(C) A fungus

(D) A bacterium

10 ATP is an organic molecule common to the metabolism of every species on Earth.

Which of the following statements explains why this fact is considered evidence for evolution?

(A) All species have evolved to be suited to the environment of Earth.

(B) ATP molecules were transferred between organisms early in the history of life.

(C) The gene for ATP developed in the earliest organisms and has been retained by subsequent species.

(D) All species on Earth developed the same gene for ATP as a consequence of convergent evolution.

11 Environment can affect phenotype by altering the sequence of bases in DNA.

Which of the following is an example of this?

(A) High protein diets making children taller than their parents

(B) Stress causing the expression of one set of genes instead of another

(C) Language and music lessons improving intelligence in young children

(D) Nuclear fallout from atomic bombs increasing birth defects in populations

Refer to the following information to answer Questions 12 and 13.

> The larvae of fruit flies damage fruit in Western Australia. To control the problem, growers are advised to spray fruit trees with pesticides. Any already damaged fruit is boiled and disposed of as chicken food or landfill. Another control measure is the release of genetically engineered infertile flies of this species.

12 Which reason best explains why the corresponding control measure reduces this problem?

	Control measure	*Reason*
(A)	Release of infertile flies	Infertile flies do not eat the fruit
(B)	Release of infertile flies	The number of flies in the next generation is decreased
(C)	Boiling fruit and feeding to chickens	Chickens are unaffected by the damaged fruit
(D)	Boiling fruit and feeding to chickens	The amount of waste for landfill is reduced

13 The following measures could be used to prevent the spread of this fruit fly across Australia.

1. Australia-wide release of infertile fruit flies
2. Aerial spraying of orchards throughout the country
3. Spot spraying of newly affected orchards in Western Australia
4. Stopping the transport of fruit from Western Australia to other states

To prevent the spread of this fruit fly across Australia, which combination of measures would be most practical to use?

(A) 1 and 2

(B) 1 and 4

(C) 2 and 3

(D) 3 and 4

14 The table shows the base triplets in mRNA for amino acids.

From the table, the amino acid Serine (Ser) can be coded for by the base triplet UCG.

Base triplets found in messenger RNA

First base	Second base: U	C	A	G	Third base
U	Phe	Ser	Tyr	Cys	**U**
	Phe	Ser	Tyr	Cys	**C**
	Phe	Ser	Stop	Stop	**A**
	Phe	Ser	Stop	Trp	**G**
C	Leu	Pro	His	Arg	**U**
	Leu	Pro	His	Arg	**C**
	Leu	Pro	Gln	Arg	**A**
	Leu	Pro	Gln	Arg	**G**
A	lle	Thr	Asn	Ser	**U**
	lle	Thr	Asn	Ser	**C**
	lle	Thr	Lys	Arg	**A**
	Met	Thr	Lys	Arg	**G**
G	Val	Ala	Asp	Gly	**U**
	Val	Ala	Asp	Gly	**C**
	Val	Ala	Glu	Gly	**A**
	Val	Ala	Glu	Gly	**G**

Which base triplet could code for the amino acid Tyrosine (Tyr)?

(A) CCU

(B) CAU

(C) UAA

(D) UAC

15 In a certain plant species, individual plants have either yellow, red or orange flowers.

Two plants, each with a different flower colour, were crossed in a breeding experiment like those carried out by Mendel. The F2 results were: 6 red, 11 orange and 5 yellow flowered plants.

What were the genotypes of the original parent plants?

(A) RY and RY

(B) RR and rr

(C) RR and YY

(D) Rr and RY

16 Why might epidemiology be considered more essential for the study of non-infectious diseases than for the study of infectious diseases?

(A) The causes of infectious diseases have already been determined.

(B) Only non-infectious diseases are affected by patterns of behaviour.

(C) Epidemiology cannot be used to find the causes of infectious diseases.

(D) Koch's postulates are not useful in finding the causes of non-infectious diseases.

17 The diagram shows an interaction between cells of the immune system.

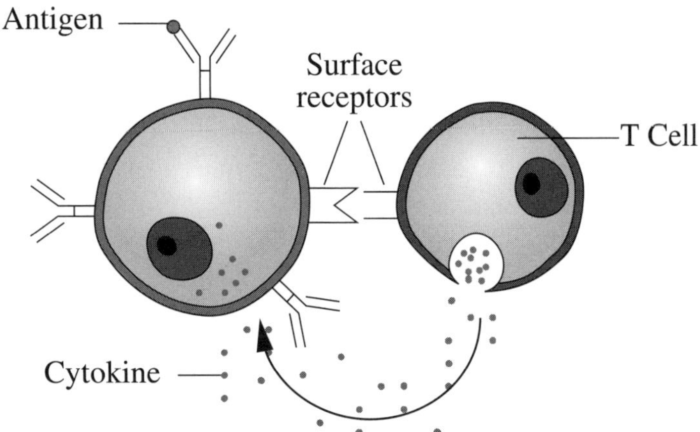

What specific process is shown in the diagram?

(A) B cell encountering an antigen

(B) Activation of a macrophage by a helper T cell

(C) Stimulation of a B cell to become a plasma cell

(D) Cytotoxic T cell destroying a virus-infected cell

18 Students conducted a large first-hand investigation into enzyme activity.

The aim in the report is shown.

Aim: To determine the optimum pH of four different enzymes.

How many independent variables were in this first-hand investigation?

(A) 1

(B) 2

(C) 4

(D) 5

Refer to the following information to answer Questions 19 and 20.

> The intestinal tract of a human foetus is sterile.
>
> After birth, microflora from the mother are transferred to the baby's mouth through close contact. After a year, the microflora of the baby is similar to the mother's, with the baby's immune system ignoring these microbes.
>
> Also during the first year of life, breast milk from the mother provides antibodies to the baby for any disease the mother has already experienced. When breastfeeding ceases, these antibody levels in the baby start to fall.
>
> After the first year, any new species of invading bacteria is treated as a pathogen by the baby's immune system.

19 A medical consequence for six-month-old babies that have only been bottle-fed with formula milk and not breastfed is that

(A) they will not develop microflora.

(B) their immune system will be damaged.

(C) their consumption of milk cannot be quantified.

(D) they will be at increased risk of infectious disease.

20 Strict hygiene practices are followed in the care of newborns, whereas hygiene practices in the care of older babies are less emphasised.

Which of the following is the best reason for this difference?

(A) Vaccinations render personal hygiene unnecessary for older babies.

(B) Procaryotic cells are not identified as antigens in early development.

(C) Antibiotic treatments kill bacterial populations in the digestive system.

(D) Early exposure to pathogens helps to build a strong immune system.

2015 HIGHER SCHOOL CERTIFICATE EXAMINATION

Biology

Centre Number

Student Number

Section I (continued)

Part B – 55 marks
Attempt Questions 21–31
Allow about 1 hour and 40 minutes for this part

Answer the questions in the spaces provided. These spaces provide guidance for the expected length of response.

Write your Centre Number and Student Number at the top of this page.

Question 21 (2 marks)

Explain ONE control measure for malaria. **2**

....

....

Question 22 (2 marks)

Explain why insects excrete uric acid as their principal nitrogenous waste. **2**

....

....

....

....

Question 23 (3 marks)

The diagram shows a model involving DNA.

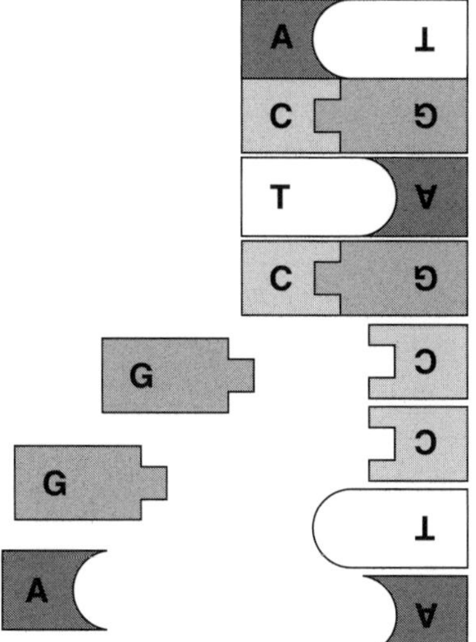

(a) What process is being modelled? 1

....

(b) Identify TWO structural features of the DNA molecule which are NOT shown in this model. 2

....

....

....

....

Question 24 (7 marks)

Data can be provided by a pulse oximeter pegged to a person's finger, as shown in the diagram.

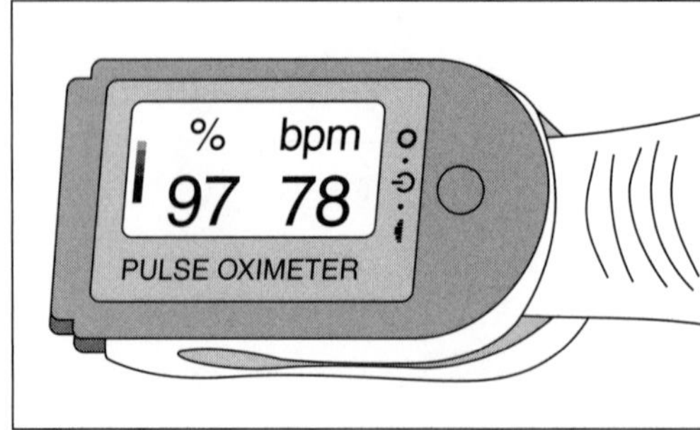

(a) What is the oxygen saturation for this person? **1**

....

(b) Outline TWO limitations of using only the information provided in the diagram to determine the 'health' of a person. **2**

....

....

....

....

(c) Explain TWO advantages in using a pulse oximeter to measure oxygen saturation compared to using another named technology in a specific setting. **4**

....

....

....

....

....

....

....

....

....

Question 25 (5 marks)

A group of students wanted to test whether water purifying tablets were effective in making creek water free from bacteria.

They conducted an experiment using a water sample collected from the creek and found that the tablets were effective.

(a) Describe a means of addressing ONE identified hazard relevant to this investigation. **2**

....

....

....

....

(b) Illustrate the results of this experiment in diagrammatic form. **3**

Use labels to clearly identify the data collected.

Question 26 (5 marks)

Sugar is transported in vascular tissues in plants and animals. **5**

Contrast the structure and workings of ONE named plant tissue and ONE named animal tissue used to transport sugar.

....

....

....

....

....

....

....

....

....

....

Question 27 (5 marks)

(a) Outline TWO differences between whole blood and plasma. **2**

....

....

....

....

(b) The steps below show the preparation and use of blood products in the treatment of Ebola Virus Disease. This disease is characterised by significant blood loss. **3**

World Health Organisation Protocol

1. A patient recovers from Ebola Virus Disease.
2. The same patient is disease-free for 28 days.
3. Blood is taken from the patient and screened for transmissible diseases.
4. Plasma is separated from the whole blood.
5. Plasma is transfused into another person with early signs of Ebola Virus Disease.

Explain why this protocol produces an effective treatment for Ebola Virus Disease.

....

....

....

....

....

....

Question 28 (5 marks)

A group of students hypothesised that the height of plants decreases with increased elevation.

The students planted ten plant cuttings from the same plant at each of five locations. The locations were at varying elevations in the same mountain range. All the cuttings were provided with the same volume of water on planting, and no fertiliser was applied. The students returned after the same growth period and measured the height of the plants.

The cross-section shown indicates the average height of the plants in metres after the growth period at each location in the mountain range.

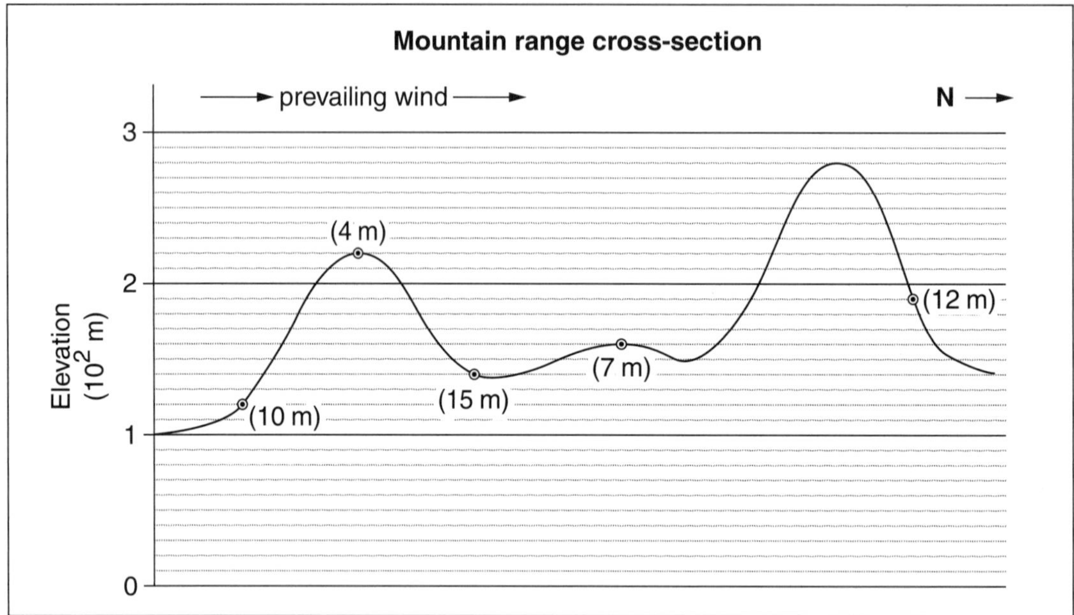

(a) Evaluate the validity of the experiment. **3**

....

....

....

....

....

....

Question 28 continues

Question 28 (continued)

(b) Complete both columns of the table to best present the data for the analysis of any trend. **2**

End of Question 28

Question 29 (6 marks)

'The application of modern reproductive techniques in plant and animal breeding limits genetic diversity.' **6**

Discuss this statement.

Question 30 (7 marks)

The graph shows the history of the relative numbers of three varieties of bird (*X*, *Y* and *Z*) within a bird species, on a remote island in the Pacific Ocean.

The bird species arrived on the island in a migration event. Before migration, the bird species was not present on the island.

The graph record of bird numbers on the island is divided into two sections (**1** and **2**). Over the time data were recorded, the environment of the island did not change.

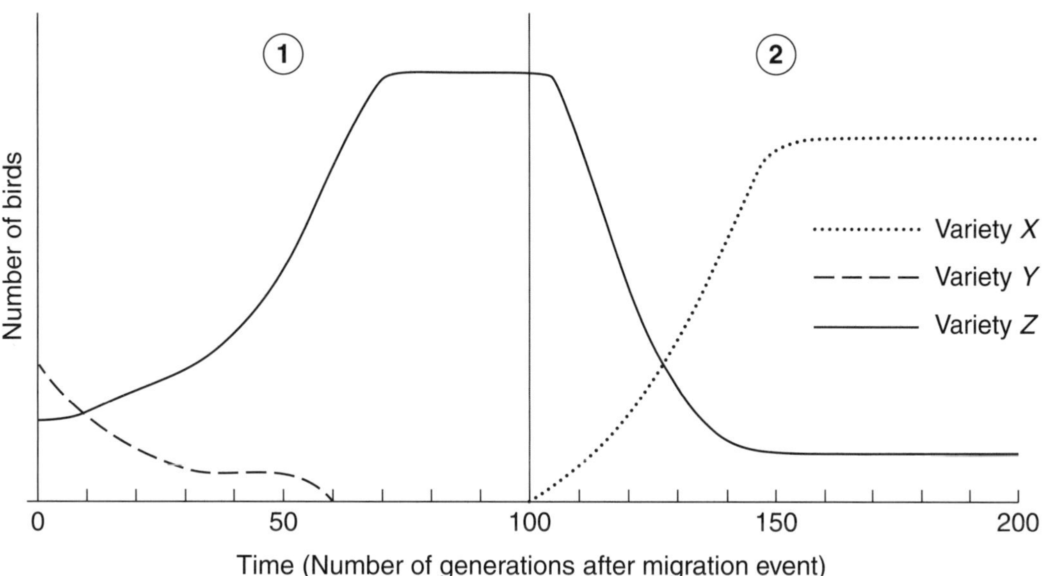

(a) What bird varieties originally migrated to the island? **1**

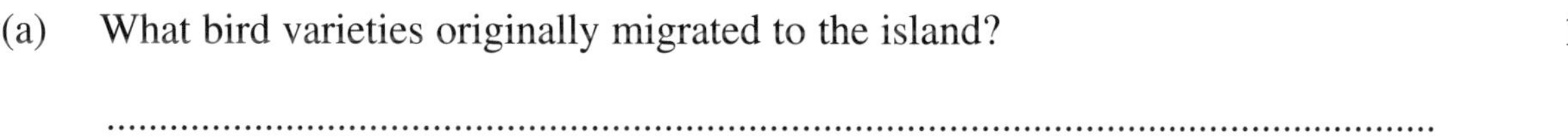

Question 30 continues

Question 30 (continued)

(b) Using the Darwin/Wallace theory of evolution, and making reference(s) to the data in the graph, explain the changes to the population of each variety of bird in

(i) section (1) of the graph. **2**

....

....

....

....

....

(ii) section (2) of the graph. **4**

....

....

....

....

....

....

....

....

....

....

....

....

....

....

End of Question 30

Question 31 (8 marks)

'Renal dialysis and kidney transplants are very different treatments for the same medical condition. Each treatment was developed from a new application of biological knowledge.' **8**

Justify these statements.

2015 HIGHER SCHOOL CERTIFICATE EXAMINATION

Biology

Section II

25 marks
Attempt ONE question from Questions 32–36
Allow about 45 minutes for this section

Answer parts (a)–(e) of one question in the Section II Writing Booklet. Extra writing booklets are available.

Question 32 Communication

Question 33 Biotechnology

Question 34 Genetics: The Code Broken?

Question 35 The Human Story

Question 36 Biochemistry *(Not included in this reproduction)*

Question 32 — Communication (25 marks)

Answer parts (a), (b) and (c) of the question on pages 2–4 of the Section II Writing Booklet. Start each part of the question on a new page.

(a) The diagram shows changes in membrane potential in a neurone after a stimulus is applied at time *t*.

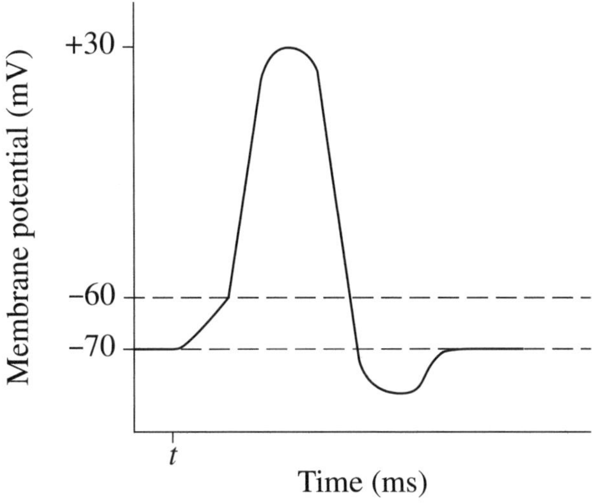

(i) What is the resting membrane potential of this neurone? **2**

(ii) With reference to this diagram, explain why some stimuli would not generate an action potential in a neurone. **2**

(b) Different species of animals can see the same environment differently. **4**

Explain this in terms of the wavelengths of the electromagnetic spectrum that animals can detect. Use specific examples in your answer.

(c) Account for the diversity of senses in humans, in terms of receptors and corresponding regions of the brain. Support your answer with TWO examples. **4**

Question 32 continues

Question 32 (continued)

Answer parts (d) and (e) of the question on pages 6–8 of the Section II Writing Booklet. Start each part of the question on a new page.

(d) A sound shadow is produced when the wavelength (λ) of a sound is shorter than the length of the human head (ℓ). When a sound of wavelength longer than the head is used, no sound shadow occurs. See Figure 1.

Figure 1

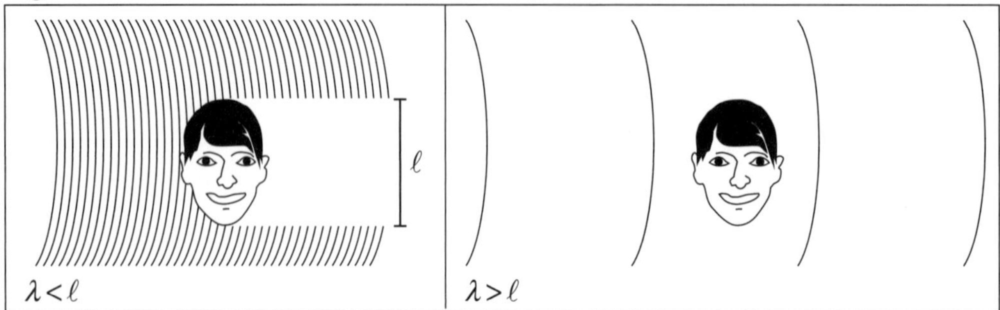

Figure 2

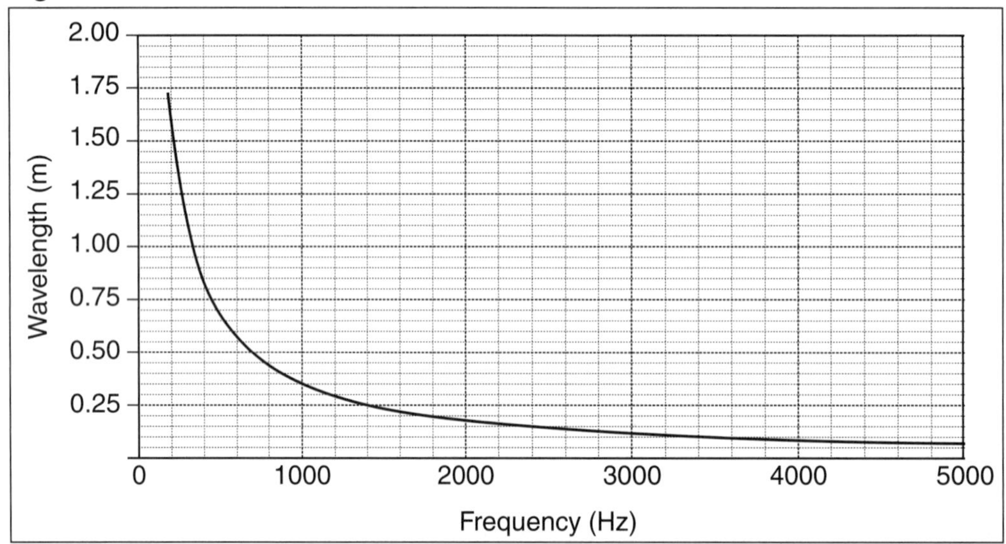

(i) Predict a difference in the loudness of the sound in each ear if the sound has a wavelength shorter than the length of the head. Justify your answer. **2**

(ii) A person's head is 25 cm long. Using Figures 1 and 2, and showing all the steps, determine the frequency range of sound for which a sound shadow occurs for this person. **4**

(e) 'Science is used to solve problems for the benefit of society.' **7**

Justify this statement with reference to the scientific knowledge used to solve ONE problem in hearing and ONE problem in visual accommodation.

End of Question 32

Question 33 — Biotechnology (25 marks)

Answer parts (a), (b) and (c) of the question on pages 2–4 of the Section II Writing Booklet. Start each part of the question on a new page.

(a) The fermentation reaction shown occurs as yeast biomass increases.

$$\text{Glucose} \longrightarrow X + Y$$

(i) Identify the products X and Y. **2**

(ii) Describe how, after the 18th century, the fermentation of glucose was modified to increase the yield of yeast biomass. **2**

(b) DNA can be extracted in a school laboratory. **4**

Describe separately the method used to get the DNA out of the cells, and the method used to isolate the DNA from other cell components in this setting.

(c) Explain why the domestication of species for agriculture is described as a biotechnology. In your answer, refer to ONE specific domesticated organism. **4**

Question 33 continues

Question 33 (continued)

Answer parts (d) and (e) of the question on pages 6–8 of the Section II Writing Booklet. Start each part of the question on a new page.

(d) (i) Why is lactic acid fermentation used in the transformation of milk into cheese? **2**

(ii) Several groups of students in a class investigated the optimum conditions for the transformation of milk into cheese using a common bacterial starter culture. **4**

Each group conducted experiment 1 and the class obtained average values of their results. The class then designed experiment 2 and each group conducted the experiment, again obtaining average values of their results. The average values for each experiment are shown below.

Experiment 1 – The effect of temperature on lactic acid production in milk at pH 5

Time after bacteria added (hours)	*Average lactic acid concentration in substrate* (%)		
	Fermentation temperature 2°C	Fermentation temperature 25°C	Fermentation temperature 35°C
0	0.1	0.1	0.1
10	0.1	0.2	0.2
20	0.1	0.3	0.2
30	0.2	0.8	0.4
40	0.3	1.8	0.7
50	0.4	2.2	1.0
60	0.6	2.3	1.3

Question 33 continues

Question 33 (continued)

Experiment 2 – The effect of pH on lactose conversion in milk at 25°C

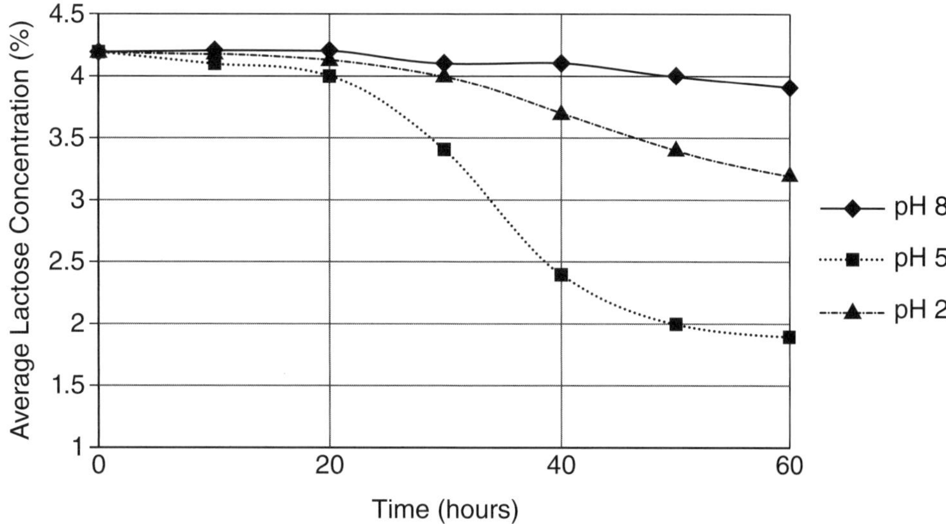

The students concluded that the results were contradictory and that they could not define the optimum conditions for their cheese production.

Evaluate the students' conclusion with reference to their data from both experiments.

(e) 'Science is used to solve problems for the benefit of society.' **7**

Justify this statement with reference to the scientific knowledge behind DNA technology, using ONE example from medicine and ONE from forensics.

End of Question 33

Question 34 — Genetics: The Code Broken? (25 marks)

Answer parts (a), (b) and (c) of the question on pages 2–4 of the Section II Writing Booklet. Start each part of the question on a new page.

(a) The diagram shows two cells taken from the same organism.

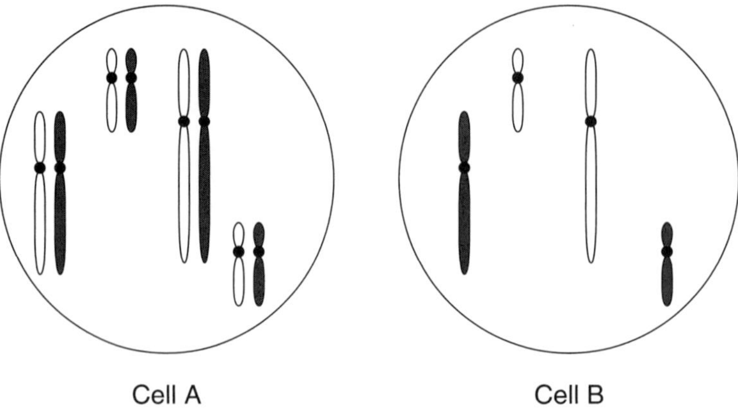

(i) What type of cell is Cell B? **1**

(ii) Explain the potential effects on this species of a disadvantageous mutation in each of these cell types. **3**

(b) Describe how the current models of the structure of tRNA and mRNA have helped us understand the production of a sequence of amino acids. **4**

(c) Describe how the action of genes can explain features of embryonic development. **4**

Question 34 continues

Question 34 (continued)

Answer parts (d) and (e) of the question on pages 6–8 of the Section II Writing Booklet. Start each part of the question on a new page.

(d) A biology class searched the internet for published results of cross-breeding experiments on the same three traits controlled by genes known to be linked.

They found two breeding experiments on the same species of flowering plant, involving the genes for stamen length, flower colour and fruit colour.

Each experiment investigated two of the three genes. The experimental results were presented in different formats.

Experiment 1: AaEe × aaee

Raw data: Phenotypic ratios in offspring
- 27 long stamen, red flowers
- 3 long stamen, yellow flowers
- 3 short stamen, red flowers
- 27 short stamen, yellow flowers

Experiment 2: BbEe × bbee

After data analysis: Result indicated on a reference graph.

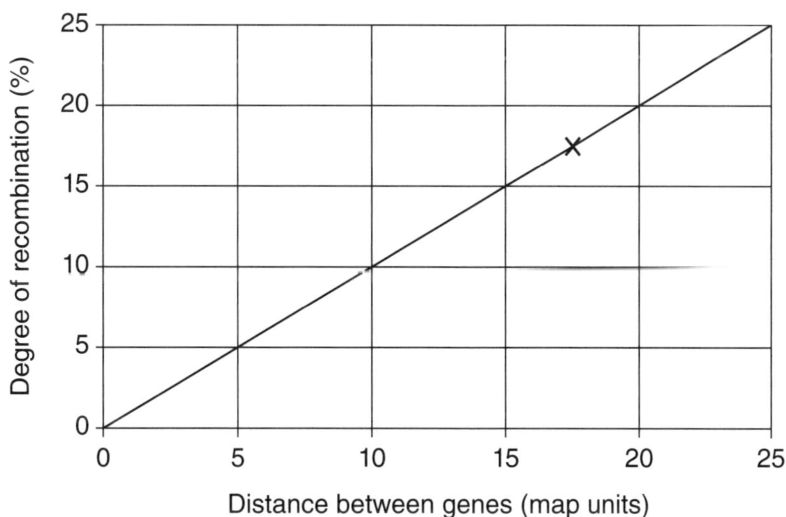

Question 34 continues

Question 34 (continued)

The class concluded that the distance between Gene A and Gene E is less than the distance between Gene B and Gene E.

(i) Show, with working, that Gene A is 10 map units from Gene E. **2**

(ii) Evaluate the conclusions drawn by the class. **2**

(iii) What further data would need to be collected to draw a quantitative map of the chromosome showing the relative positions of all these genes? Justify your answer. **2**

(e) 'Science is used to solve problems for the benefit of society.' **7**

Justify this statement with reference to the scientific knowledge behind DNA technology, using ONE example from medicine and ONE from forensics.

End of Question 34

Question 35 — The Human Story (25 marks)

Answer parts (a), (b) and (c) of the question on pages 2–4 of the Section II Writing Booklet. Start each part of the question on a new page.

(a) (i) The diagram shows two primates. **2**

Figure 1 **Figure 2**

Provide the word that describes the tail for the classification of each primate.

(ii) Spider monkeys and baboons are members of the same order. Use the hierarchical classification system to explain whether they would necessarily be in the same phylum and the same genus. **2**

(b) Explain ONE piece of evidence for, and ONE piece of evidence against, the theory of regional continuity. **4**

(c) Account for identified differences in the cultural development of *Australopithecines* and *Homo neanderthalensis* in relation to their cranial capacities. **4**

Question 35 continues

Question 35 (continued)

Answer parts (d) and (e) of the question on pages 6–8 of the Section II Writing Booklet. Start each part of the question on a new page.

(d) (i) Distinguish between relative dating and absolute dating of fossils. **2**

(ii) Students found three fossils (**A**, **B** and **C**) at an archaeological site. **4**

The depth of each fossil below the surface was recorded and the strata of rocks at the site were mapped.

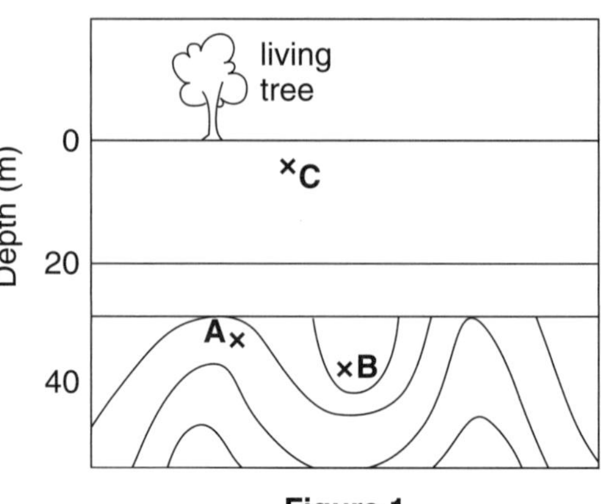

Figure 1

The fossils were analysed in the laboratory for ratios of C^{14} to C^{12}, and the results were presented on a C^{14} half-life curve.

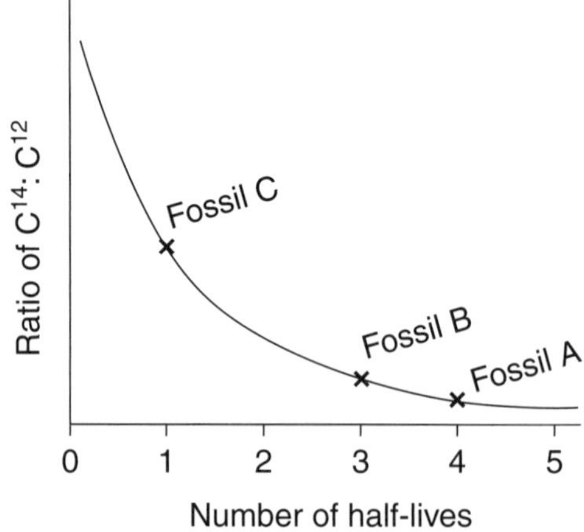

Figure 2

The students concluded that their data were conflicting and they could not determine the relative ages of the fossils.

Evaluate the students' conclusion with reference to the data presented.

(e) 'Science has been used to solve problems in the investigation of evolutionary relationships between humans and other primates, and so has provided information of interest to society.' **7**

Justify this statement in terms of the scientific knowledge behind DNA–DNA hybridisation AND karyotype analysis.

End of Question 35

End of paper

2015 HSC Examination Paper

Sample Answers

Section I Part A *(Total 20 marks)*

1 D This term describes short bursts of change.

2 A This is the only term related to variation in a gene. The other terms relate to chromosome structure.

3 C Sunken stomata prevent the loss of water.

4 D Sexual reproduction is not associated with the health of an organism.

5 A *W* is the substrate as it does change in the reaction.

6 A Carbon dioxide will form carbonic acid in the blood and lower the pH, thus increasing the acidity.

7 B The brain sends messages to the effectors in order to promote a response to a stimulus.

8 B If body temperature is low, the hairs stand on end and blood vessels constrict to prevent heat loss.

9 C The size and structure of the pathogen correlate to a fungus.

10 C This statement concurs with evolutionary theory regarding common ancestry.

11 D Radiation is a common cause of mutation in genes.

12 B Fewer flies in subsequent generations will eventually reduce or eliminate the population altogether.

13 D By not taking affected fruit carrying the flies to other states and eliminating the new populations, the problem may be contained to local areas only.

14 D This option contains the correct possible combination for tyrosine.

15 C The F2 parents were both heterozygous, so the original parents were both homozygous.

16 D Koch's postulates will only assist in identifying infectious disease.

17 B The helper T cell is releasing cytokine to assist in activation of the macrophage.

18 C pH is the dependent variable. The optimum pH for four different enzymes is being tested, so there are four independent variables.

19 D The babies will have a vulnerable immune system without the benefit of the immunity that is conferred through breast milk from the mother.

20 B The immune systems of newborns is immature and will not work as effectively as in older babies.

Section I Part B

21 An example of a control measure for malaria is spraying breeding grounds with pesticides. This reduces the number of vectors, thereby decreasing the likelihood of disease spread. *(2 marks)*

22 Insects excrete uric acid as their principal nitrogenous waste because it requires less water to produce this type of waste and enables the organism to minimise its water intake. *(2 marks)*

23 (a) The model represents DNA replication and shows unzipping of a double helix, with complementary base pairs moving apart. *(1 mark)*

(b) The model does not represent the sugar-phosphate backbone of the nucleotide attached to each base, nor does it show the helical structure of DNA. The model also does not show the hydrogen bonds that connect the base pairs. *(2 marks)*

24 (a) The oxygen saturation for the person is 97%. *(1 mark)*

(b) One limitation is that only oxygen saturation and pulse data are depicted in the diagram; however, there are additional factors that contribute to overall health of an individual. Another limitation is that these measurements do not take into consideration the person's overall body mass, blood pressure or their cholesterol. These factors have consequences for circulation of blood and can significantly affect the individual's health. *(2 marks)*

(c) The pulse oximeter can give fairly accurate data immediately and is very portable and compact. Two advantages in using an oximeter to measure oxygen saturation in a hospital, ambulance or surgery are that it produces the data required very quickly and it does not require invasive procedures.

Another method that can be used to obtain this information is arterial blood gas determination. This involves the collection of blood from an artery by inserting a syringe. This can be difficult to do at times for various reasons including the location of blood vessels and the hydration of the patient. This method also has an increased risk of infection and is time consuming. The results are not immediate as the blood has to be transported to a pathology lab and analysed before the results can be communicated back to the doctor. This can mean a delay in diagnosis in a hospital emergency ward. *(4 marks)*

25 (a) One potential hazard that students could encounter in this investigation is accidentally ingesting the water-borne pathogens. One way to minimise this hazard is to wear gloves when handling the water samples and to thoroughly wash hands after the task. *(2 marks)*

(b)

Effectiveness of water purifying tablets on creek water bacteria			
Trial	Number of tablets	Quantity of bacteria present on culture (agar in petri dish)	
		Water sample 1	Water sample 2
A (Control)	0	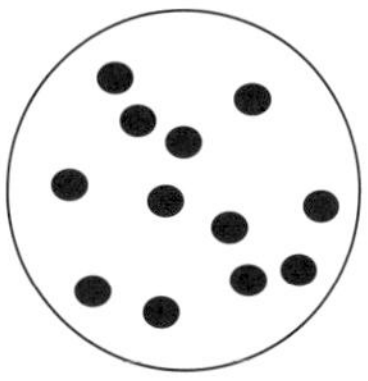	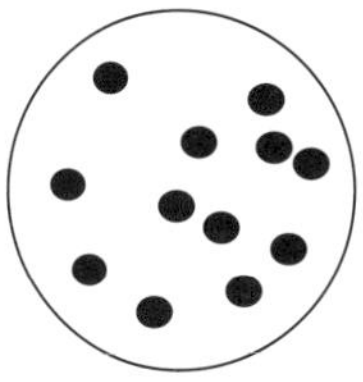
B	1	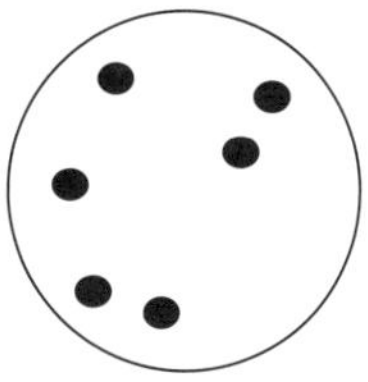	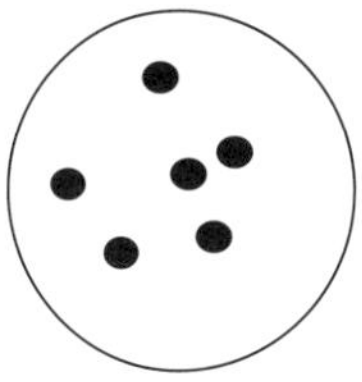
C	2	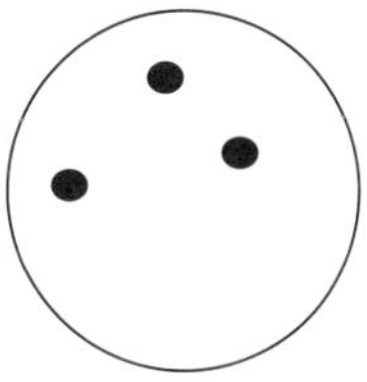	
D	3	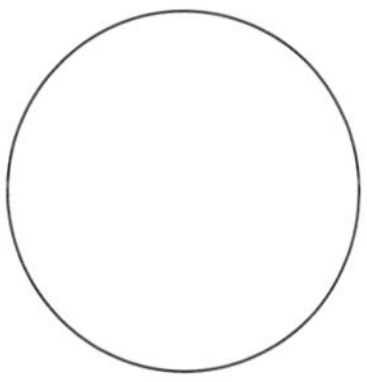	

Note to students: diagrams could include agar plates, beakers, flasks, test tubes, tables or graphs. In this example results are represented as a table with diagrams of agar plates. *(3 marks)*

26 Sugars are transported by blood vessels in animals and by phloem in plants.

Arteries are blood vessels that have thick muscular walls which maintain a pulse. The contractions of the muscular walls of the arteries ensure that the sugar, absorbed from digested nutrients in the alimentary canal, are effectively transported around the body.

Phloem tissue transports the products of photosynthesis (glucose) from the source of production in the leaves to other parts of the plants for use in respiration (sink cells in roots). The living phloem tissue is made up of elongated cells arranged end to end, separated by sieve elements which have companion cells alongside. *(5 marks)*

27 (a) Whole blood contains multiple blood cell types (including red blood cells, white blood cells and platelets) suspended in a straw-coloured liquid called plasma.

Plasma is the liquid part of the blood, which is 95% water. It has had the blood cells removed from it and is a mixture of proteins, enzymes, nutrients, wastes, hormones, gases and globulins. *(2 marks)*

(b) The protocol described represents possible effective treatment of Ebola patients because the globulins transported in plasma can act as antibodies. The original patient with Ebola recovered and was able to produce the antibodies which can then be transferred to the new patient and assist their immune system to fight the disease. *(3 marks)*

28 (a) While some variables such as volume of water, no fertiliser and taking cuttings from the same plant were controlled, other factors such as exposure to prevailing winds and direction with respect to the sun were not controlled. As both of these factors will affect plant growth, the experiment is not valid. *(3 marks)*

(b)

Elevation (10^2 m)	*Height (m)*
1.2	10
1.4	15
1.6	7
1.9	12
2.2	4

(2 marks)

29 The application of some modern reproductive techniques in plant and animal breeding can limit the genetic diversity, while other techniques enhance genetic diversity.

If plants or animals are cloned (sheep and pigs) or grafted (fruit trees) then genetic diversity is reduced, as you are manifesting a limited amount of the genes from a species genetic pool. This can have negative consequences if the environment in which the organisms live is a changing one. However, if the sole purpose of breeding organisms in this way is to produce organisms with a limited set of desirable characteristics for agricultural purposes then it can be beneficial to society.

In selective breeding or in-vitro fertilisation (IVF) programs, the diversity is not reduced in the same way because new genetic combinations are being produced and passed, thereby maintaining genetic diversity. Artificial pollination is the transfer of pollen from one plant to the stigma of another plant. Artificial insemination is collecting semen from one animal and transferring it to the reproductive organs of another organism. Examples of where this technology has been used include developing cows with high butter fat in milk or leaner meat, and sheep with better wool quality.

When organisms are manipulated genetically, such as through the use of recombinant DNA, new genes are introduced into a species' gene pool. The manipulated or inserted genes are passed to the next generation through the transgenic organisms, thereby increasing the genetic diversity within that species' genetic pool. This technology has been used to create frost-resistant strawberries and disease-resistant Bt cotton and corn.

As such, it cannot be said that modern reproduction techniques are limiting genetic diversity; it is quite the opposite effect, as genetic diversity is being created through the use of the reproductive technology. *(6 marks)*

30 (a) Bird varieties *Y* and *Z* originally migrated to the island. *(1 mark)*

(b) (i) Section 1 of the graph demonstrates Darwin's theory of evolution by the reduction in number of bird variety *Y*. These birds were not able to survive because their genetic combination was not sufficient for them to maintain the population, that is, they were not 'fit' enough to survive and breed beyond 60 generations. Variety *Z* must have had a more diverse genetic pool. The birds were able to survive and reproduce, increasing in number until levelling off after about 60 generations. *(2 marks)*

(ii) In section 2 of the graph, variety *Z* drops dramatically in the 110th generation, which coincides with the appearance of variety *X*. The two varieties, *X* and *Z*, must have been in competition with each other for resources on the island because the environmental conditions had not changed in 200 years. It appears that variety *X* was better suited to the environment (i.e. had more favourable characteristics), and so was better able to survive on the island than variety *Z*, which declined rapidly with the arrival of variety *X*. For this reason, variety *X* survived to pass on this combination of successful genes to the subsequent generations over time. *(4 marks)*

31 Renal dialysis and kidney transplants are used to treat patients with chronic kidney diseases that prevent the purification of their blood. These diseases include glomerulonephritis, polycystic kidneys and diabetes mellitus.

Renal dialysis has been relatively readily available to patients for about the past 50 years. This treatment is based on knowledge of kidney function and osmotic processes in the body. Two distinctly different types of dialysis are possible: haemodialysis, which filters the blood outside the body; and peritoneal dialysis, which operates within the body. Dialysis works on the principles of the diffusion of solutes and ultrafiltration of fluid across a semi-permeable membrane, effectively replacing the work done by the glomerulus in the nephron. Nitrogenous wastes (such as urea) diffuse across a concentration gradient from a high concentration in the blood to a low concentration in the dialysate solution in canister. Today patients can have their own dialysis machines at home and treat themselves with minimal disruption to work and home life.

With advances in technology and biological understanding of kidney function, it has also become possible to undergo a kidney transplant for the most severe cases. This involves a patient being the recipient of a healthy kidney from a donor that has had the tissue type matched. A foreign tissue contains antigens which stimulate cytotoxic T cells to attack the transplanted organ. Increased understanding of immuno-suppressor drugs, which must be given to the recipient of the transplanted organ, and effective tissue matching techniques has increased the chances of success for transplant patients. The number of people who are willing to donate kidneys has also increased with the knowledge that the procedure will not impact on their lives detrimentally.

Dialysis and transplant options have developed for the same purpose from the increase in biological knowledge over time, with significantly increasing success in each of the treatment options. *(8 marks)*

Section II—Options

Question 32—Communication

(a) (i) The resting membrane potential for this neurone is –70 mV. *(2 marks)*

(ii) In order to generate a response to a stimulus, the depolarisation must reach a threshold that is at least 15 mV more than the resting potential, which on the graph is 70 mV. Depolarisation occurs below this. It is an all or nothing event. *(2 marks)*

(b) Different animals see the environment differently because they detect different parts of the electromagnetic spectrum. Bees can detect ultraviolet (UV) radiation (short wave radiation) as well as green/blue light (longer wavelength than UV). Snakes see their environment through the infrared spectrum. Mammals tend to be bi-chromatic: they have two types of cones and are sensitive to blue/green frequencies of the visible spectrum. Primates are tri-chromatic with cones sensitive to blue, green and red, while aquatic mammals tend to have sensitivity primarily to red and green. The red wavelength is shorter than wavelengths at the blue and green end of the visible spectrum. As a result of the various combinations that are possible with their particular set of photosensitive cones, each of these animals will have a slightly different 'view' of their surrounds. *(4 marks)*

(c) Humans are fortunate to have five distinct sensory receptors: photoreceptors on the retina detect light; mechanoreceptors and hair cells in the organ of Corti detect sound; chemoreceptors in the nose and tongue detect smell and taste; and mechanoreceptors in the skin detect pressure or touch. These receptors are connected to the brain by a series of sensory neurones. The brain is divided up into distinct sensory areas. The back of the cerebrum, directly above the cerebellum, is the visual cortex; smell is processed on the underside of the frontal lobe of the cerebrum; and hearing on the temporal lobe. *(4 marks)*

(d) (i) If the sound has a wavelength shorter than the head, the sound will be detected by the ear facing the sound directly and not by the ear facing away from the sound. Sound at this ear will be significantly softer and possibly with a delay due to the wavelength being blocked by the head between the two ears. *(2 marks)*

(ii) Figure 1 shows that the head will block frequencies with a wavelength shorter than the head length. If the person's head is 25 cm, then the head will block frequencies above 1400 Hz as these frequencies have a wavelength shorter than the head length (as indicated on Figure 2). Figure 2 indicates that frequencies below 1400 Hz have a wavelength above 0.25 m and as such will be able to avoid the sound shadow as depicted in Figure 1. The human ear can detect frequencies between 20 Hz and 20 kHz, so the person should be able to detect frequencies in the range 20–1400 Hz. *(4 marks)*

(e) Answers may vary. One example is given below.

Science is used to solve problems for the benefit of society by providing technology to assist people to overcome certain physical and medical conditions.

Hearing defects can be overcome in varying degrees through the use of hearing aids or, if the auditory nerve is still functioning, cochlear implants.

Hearing aids are used to facilitate the transmission of sound waves from the outer ear to the inner ear by magnifying them so that the organ of Corti can receive the signal and pass it on to the auditory nerve.

People with visual impediments have also benefited from the application of scientific knowledge through various technologies. These include the use of spectacles, and surgical procedures such as laser surgery, lens/corneal replacement and, more recently, the development of 'bionic' eye technology.

Spectacles have been used for many centuries, since people developed an understanding of light transmission through lenses and the ability to shape lenses so they refract light in order to focus it on the retina. Accommodation in the eye is achieved through the use of either convex lenses that extend the focal length to overcome short-sightedness, or concave lenses that decrease the focal length to achieve a focused image on the retina for people with long-sightedness. *(7 marks)*

Question 33—Biotechnology

(a) (i) *X* is ethanol, *Y* is carbon dioxide. *(2 marks)*

(ii) Since the 18th century it has become possible to modify the fermentation of glucose to produce alcohol-based products on a larger scale, primarily by using different carbohydrates and microorganisms. These products included glycerol, lactic acid, citric acid and yeast biomass. *(2 marks)*

(b) The following steps describe how to extract DNA from a strawberry in a laboratory.

1. Crush the fruit using a mortar and pestle or blender, to break open the cell walls of the cells.
2. Filter the crush into a beaker to remove the pulp.
3. Add detergent to the filtered liquid to break down the nuclear membrane.
4. Add salt to the beaker to enable the DNA strands to clump together, isolating the DNA from the other cell components.
5. Add methylated spirits to make the DNA strands visible as they float to the top of the beaker. *(4 marks)*

(c) Domestication of animals for agriculture is considered to be a form of biotechnology because it introduced an intervention in the process of natural selection, rather than the natural environment being the selective agent.

Domestication is a form of artificial selection because it involved the deliberate selection of animals such as sheep, cows, goats and pigs for specific characteristics that were desirable for human consumption or production of animal products for human use. The animals were kept close to where people lived, creating a readily available source of food such as meat and milk. The deliberate breeding of particular animals gave rise to higher yields, and the availability of materials such as leather or wool enabled durable fibres and other products to be made. Today selective breeding is used to create cows that produce milk with higher butter fat content and leaner meats with greater muscle density. *(4 marks)*

(d) (i) Lactic acid is used in the transformation of milk into cheese because it gives rise to a bacterial fermented product that sours faster than other acids and has a rich flavour/aroma. Fermentation is achieved by adding bacteria such as *S. thermophilus* or *L. bulgaricus* to heat-treated milk. The growing bacteria produce the lactic acid, giving rise to the creamy consistency and flavour. *(2 marks)*

(ii) The students' experimental conclusion is not valid because the two experiments are investigating two different variables: temperature and pH. Experiment 1 investigated the effect of temperature on the production of lactic acid, and showed that the most effective temperature for lactic acid production was 25°C. Experiment 2 investigated the effect of pH on lactose conversion at 25°C, and indicated that the optimum pH is 8. *(4 marks)*

(e) Science is used to solve problems for the benefit of society by providing DNA technology to assist with medical and forensic applications.

Since the mapping of the human genome, DNA technology has been used in medicine to identify specific genes and chromosomes that are responsible for genetic conditions such as Tay-Sachs disease and Down syndrome. DNA technology is also used for the identification of compatible tissue that can be used for organ transplants or tissue donations for treating patients with injuries or blood cancers. Recombinant DNA can be used to produce insulin for treatment of diabetes, human growth hormone to overcome stunted growth, and even proteins to dissolve clots in the treatment of heart attack patients. New applications of this technology are being developed all the time. The technology has assisted greatly in the interpretation and study of cloned genes, identifying their position on a chromosome, base sequence and difference in individuals, all of which can assist in the study of hereditary diseases.

In forensics there have also been significant problems that have been solved using science for the benefit of society. DNA fingerprinting has become widely used to solve crimes. It is used to isolate an individual's DNA sequence to identify body tissue or cells. This can be used to trace both suspects and victims in a crime. DNA fingerprinting is also used to identify the parents of children in family law matters. This scientific knowledge is very accurate compared to the traditional blood sampling and matching, and has greater scope. It also does not require large samples of tissue, fibre or hair, as the technology is very precise. *(7 marks)*

Question 34—Genetics: The Code Broken?

(a) (i) Cell *B* is a gamete. *(1 mark)*

(ii) A potential mutation in each of these cell types can have significant effects on a species. If cell *A* was to have a disadvantageous mutation, the organism might die due to the inability of the affected cell/tissue to function correctly. It could manifest itself as a cancer or possibly organ failure if it is localised. This could mean that the affected individual would not be able to reproduce, which could lead to the demise of the species. If cell *B* was to have a disadvantageous mutation it could be passed on to the offspring. This could result in the organism being sick or unable to look after itself, resulting in decreased numbers of the species. *(3 marks)*

(b) The current models of tRNA and mRNA have assisted us in understanding the production of a sequence of amino acids because they are much shorter than DNA strands, they are single stranded and are very specific in their function. In DNA adenine pairs with thymine, but mRNA uses uracil instead of thymine. (Cytosine pairs with guanine in both DNA and mRNA.) tRNA will only attach to one specific amino acid. During transcription an enzyme builds a molecule of mRNA with the corresponding complementary base in triplets/codons and this molecule moves away from the DNA in the nucleus to the ribosomes in the cytoplasm. Here translation occurs with the anticodons being temporarily matched in the form of tRNA along the mRNA. Another enzyme forms the peptide bonds between the amino acids which then break off, leaving a chain of amino acids which are then folded into the correct shape forming a protein. This process enables us to see how DNA is replicated and the correct sequence produced. *(4 marks)*

(c) All sexually reproducing organisms pass on a haploid set of chromosomes to their offspring, which each have their own genetic information. These combine with a haploid set from a parent of the opposite sex to form an embryo that has a unique set of genetic information, compared to either parent. Mitosis occurs, resulting in undifferentiated cells. As the embryo develops, cells differentiate and become specialised in their size, shape and function according to the genes that regulate these developments. All cell activities, and production of proteins and enzymes, at different times of development are regulated by genes. A variety of proteins, activators and repressors link up with different parts of DNA to cause interaction between other proteins and genes. The result is that some genes are affected and expressed, while others are repressed or inhibited. Genes are turned on or off according to their position in the embryo, function and age of the developing embryo. *(4 marks)*

(d) (i) The frequency of recombination (shown in Experiment 1) is 3/30 which equals 10%. In Experiment 2, we see that a 10% degree of recombination (reading across the graph) results in a separation in gene distance of 10 map units. *(2 marks)*

(ii) The students' conclusion is accurate. The greater the percentage of recombination, the larger the distance that the genes are apart. There is a 10% degree of recombination because A–E = 10 map units and B–E is 18 map units. As such the distance between A–E is less than genes B–E, so the class is correct. *(2 marks)*

(iii) A chromosome map that has been produced by percentages of recombination of genes during crossing over does not completely represent a whole chromosome. Chromosome maps based on recombination can only be used to produce the relative positions of genes, not the absolute ones. You would need to compare the map for Experiment 1 as well as the recombination of the offspring to get a true indication of separation, by conducting another set of breeding experiments to show the AaBb × aabb back cross. Then the percentage of recombinants can be used to determine the relative distance of B–A as either 8 or 28 map units. *(2 marks)*

(e) Science is used to solve problems for the benefit of society by providing DNA technology to assist with medical and forensic applications.

Since the mapping of the human genome, DNA technology has been used in medicine to identify specific genes and chromosomes that are responsible for genetic conditions such as Tay-Sachs disease and Down syndrome. DNA technology is also used for the identification of compatible tissue that can be used for organ transplants or tissue donations for treating patients with injuries or blood cancers. Recombinant DNA can be used to produce insulin for treatment of diabetes, human growth hormone to overcome stunted growth, and even proteins to dissolve clots in the treatment of heart attack patients. New applications of this technology are being developed all the time. The technology has assisted greatly in the interpretation and study of cloned genes, identifying their position on a chromosome, base sequence and difference in individuals, all of which can assist in the study of hereditary diseases.

In forensics there have also been significant problems that have been solved using science for the benefit of society. DNA fingerprinting has become widely used to solve crimes. It is used to isolate an individual's DNA sequence to identify body tissue or cells. This can be used to trace both suspects and victims in a crime. DNA fingerprinting is also used to identify the parents of children in family law matters. This scientific knowledge is very accurate compared to the traditional blood sampling and matching, and has greater scope. It also does not require large samples of tissue, fibre or hair, as the technology is very precise. *(7 marks)*

Question 35—The Human Story

(a) (i) Spider monkeys have a 'prehensile' tail. The baboon does not have the same gripping tail. *(2 marks)*

(ii) The broadest grouping in the hierarchical classification system is kingdom, followed by phylum, class, order, family, genus and species. Members of any grouping in this hierarchy will share the same broader grouping (i.e. all members of a particular class are in the same phylum and kingdom) but not necessarily any subgrouping (i.e. all members of a particular family do not necessarily belong to the same genus).

Spider monkeys and baboons are both members of the same order, Primates. For this reason, they are also both in the same phylum, Chordata (as they have a backbone). However, they are in different genus groups. *(2 marks)*

(b) Advocates for regional continuity will argue that a second migratory wave did not occur and that populations were dispersed and isolated leading to diversity of the gene pool in regional settings. This is evidenced by the mitochondrial DNA found in Mungo man. The case against regional continuity is based on fossil evidence that implies an earlier migration out of Africa had occurred and that the Aboriginals migrated to Australia via Asia 50 000–60 000 years ago. This remains an area of contention in the evolution of man. *(4 marks)*

(c) *Homo neanderthalensis* lived at the same time as early *Homo sapiens*. Tall and strongly built, with better defined muscles than modern humans, *H. neanderthalensis* had a smaller forehead/large eyebrow ridge and prominent jaw with little or no chin. The cranial size was typically larger than modern humans (1400–1750 cc). *H. neanderthalensis* used tools and hunted large animals in socially organised groups with cultural aspects such as jewelry, clothing and burial. This corresponds to a larger cranial capacity.

Australopithicines are much older than hominids. *Australopithecus africanus* had a cranial capacity of 420–490 cc and *A. afarensis* had a cranial capacity of 375–550 cc. Both were much smaller than *H. neanderthalensis,* with upright posture and unspecialised teeth. They appear to have lived in groups, and did not use simple tools, indicating a gatherer habit associated with a smaller cranial capacity. *(4 marks)*

(d) (i) Relative dating compares rock and/or fossil sequence or strata to estimate their age. Using this method you can determine which is older or younger than the other, but not exactly how old they are. Absolute dating uses specific technology, such as radiometric dating, to determine a numerical age of a fossil or rock strata. *(2 marks)*

(ii) The students' data is not conflicting. They have shown that the fossil *C* is the youngest with a half-life of 1, and indeed it was found at approximately 4 m depth, much closer to the surface than fossils *A* and *B*. Absolute dating indicates that fossil *A* (half-life of 3) is older than fossil *B* (half-life of 4); however, the two were found very close in depth, at around 36–38 m, with fossil *A* slightly nearer the surface.

This apparent contradiction can be explained by the folding of the strata in the past, causing uplift and depression. As such, the data is not conflicting but might require further analysis to more accurately determine the geology of the area. *(4 marks)*

(e) Biochemical characteristics are used to analyse evolutionary relationships between organisms: the higher degree of similarity between two species, the closer the organisms are on the evolutionary tree. In particular, two methods of analysis have been used: DNA–DNA hybridisation and karyotype analysis. Scientific knowledge has enabled us to understand the information gained from such technology to bring about the relationship studies that provide the evidence required to make the links between the species. This information is of particular interest to society as it provides evidence for evolution and gives rise to an understanding of the world around us.

DNA–DNA hybridisation attempts to determine the similarity of DNA from different species and hence their genetic relationship. Understanding of the structure of DNA has made this technology possible.

Karyotype analysis involves the study of the chromosome map. This is a picture taken after cells are removed from an organism, grown outside the body and specially stained prior to photographing them. This allows the size, shape, structural markings and number of chromosomes to be compared, once again giving rise to the ability to deduce evolutionary relationships between organisms. For example, the karyotypes of chimpanzees and humans indicate a great deal of similarity with at least eight chromosomes that are alike, but vary slightly in number of chromosomes.

Certainly both these forms of analysis indicate that we are more closely related to chimpanzees than other primates such as gorillas and orangutans, which is of great interest to humans and worth investigating further. *(7 marks)*

CHAPTER 15

2016 HIGHER SCHOOL CERTIFICATE EXAMINATION

Biology

General Instructions

- Reading time – 5 minutes
- Working time – 3 hours
- Write using black pen
- Draw diagrams using pencil
- Board-approved calculators may be used

Total marks – 100

Section I

75 marks

This section has two parts, Part A and Part B

Part A – 20 marks

- Attempt Questions 1–20
- Allow about 35 minutes for this part

Part B – 55 marks

- Attempt Questions 21–31
- Allow about 1 hour and 40 minutes for this part

Section II

25 marks

- Attempt ONE question from Questions 32–36
- Allow about 45 minutes for this section

Section I
75 marks

Part A – 20 marks
Attempt Questions 1–20
Allow about 35 minutes for this part

Use the multiple-choice answer sheet for Questions 1–20.

---

1 The diagram shows the shape of an enzyme and its substrate, before and after heating.

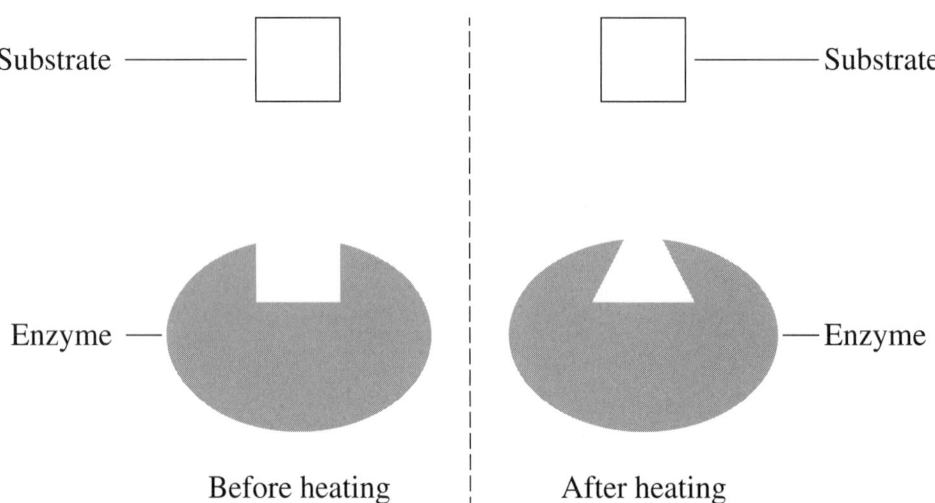

As a result of the temperature increase, enzyme activity will

(A) cease.

(B) increase.

(C) slow down.

(D) stay the same.

2 Normal microflora are considered to be pathogenic when they

(A) increase in size.

(B) are resistant to antibiotics.

(C) cause symptoms in a person.

(D) are transferred to another person.

3 Which of the following scientists proposed the chromosomal theory of inheritance?

(A) Walter Sutton

(B) Gregor Mendel

(C) Macfarlane Burnet

(D) Thomas Hunt Morgan

4 Beadle and Tatum's 'one-gene-one-protein' hypothesis was modified to the 'one-gene-one-polypeptide' hypothesis.

What does this modification tell us about their work?

(A) Their results were inaccurate.

(B) Their experiments were poorly designed.

(C) Their results were not analysed properly.

(D) Their conclusions were provisional in nature.

5 The graph shows the relationship between substrate concentration and reaction rate in an enzyme-catalysed reaction.

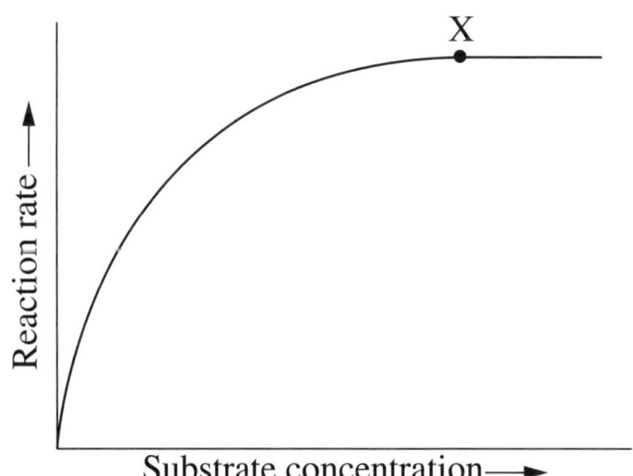

Why does the reaction rate NOT continue to increase after point X?

(A) The enzyme has been denatured.

(B) The substrate has been denatured.

(C) The enzyme is no longer acting as a catalyst.

(D) All the enzyme active sites are being occupied.

6 Which of the following best describes *metabolism*?

(A) The optimum activity of an organism

(B) All of the enzymes in a living organism

(C) An enzyme-catalysed reaction in an organism

(D) All the chemical reactions inside cells of an organism

7 The diagram shows a simplified model of a mammalian nephron and three processe labelled *1*, *2* and *3*.

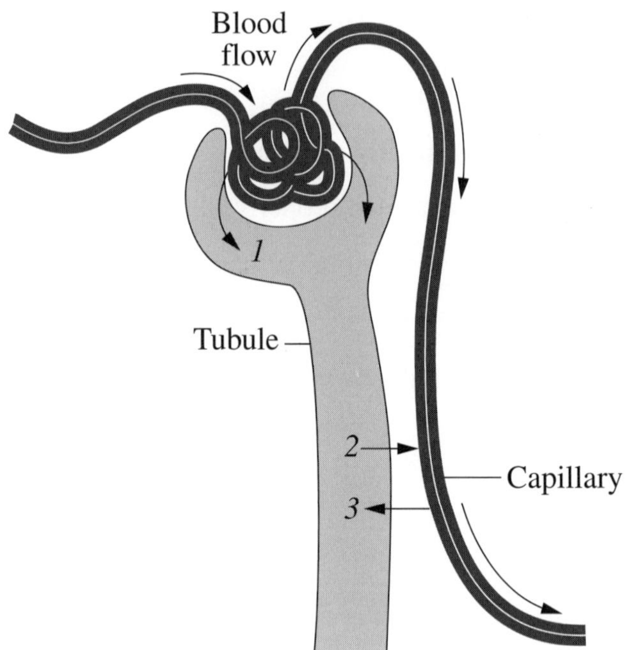

Which row of the table correctly identifies each of the processes?

	1	*2*	*3*
(A)	Reabsorption	Filtration	Secretion
(B)	Filtration	Reabsorption	Secretion
(C)	Secretion	Reabsorption	Filtration
(D)	Filtration	Secretion	Reabsorption

The model of a mammalian nephron in Question 7 is from Bob Constantin, et al., *Inquiry into Biology*, McGraw-Hill Ryerson, Whitby, Ont., 2006, p. 311, fig. 9.4, and is reproduced with the kind permission of McGraw-Hill Education.

8 The pedigree shows the inheritance of a characteristic.

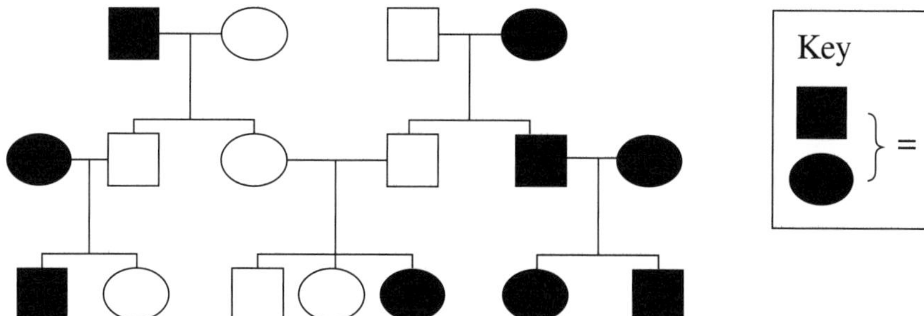

What pattern of inheritance is shown?

(A) Dominant and sex-linked

(B) Recessive and sex-linked

(C) Dominant and not sex-linked

(D) Recessive and not sex-linked

9 Why is passive transport alone inadequate for the production of urine that is high in nitrogenous wastes?

(A) Osmosis cannot target specific solutes.

(B) Solutes cannot move against a concentration gradient.

(C) Solute transport increases at low concentration gradients.

(D) Osmosis moves water from a low concentration to a high concentration.

10 A model of DNA is shown.

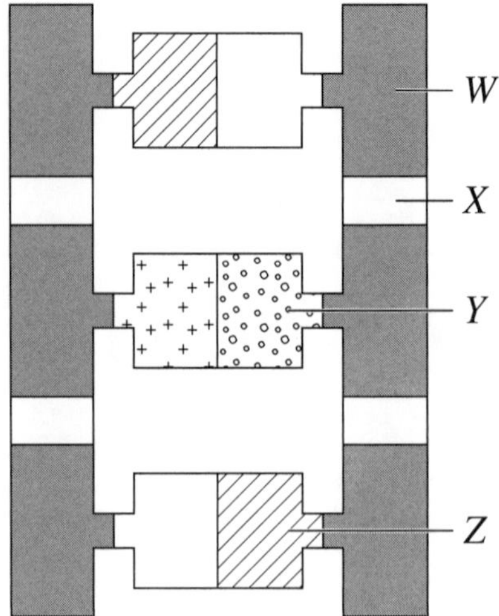

Which row of the table correctly identifies the different components of the model?

	W	*X*	*Y*	*Z*
(A)	Sugar	Phosphate	Adenine	Guanine
(B)	Phosphate	Sugar	Guanine	Cytosine
(C)	Sugar	Phosphate	Adenine	Thymine
(D)	Phosphate	Sugar	Guanine	Thymine

11 Antibodies are molecules released by

(A) memory B cells.

(B) memory T cells.

(C) specialised B cells.

(D) specialised T cells.

12 Some students investigated the response of a plant cutting to an increase in ambient air temperature. They used the apparatus shown.

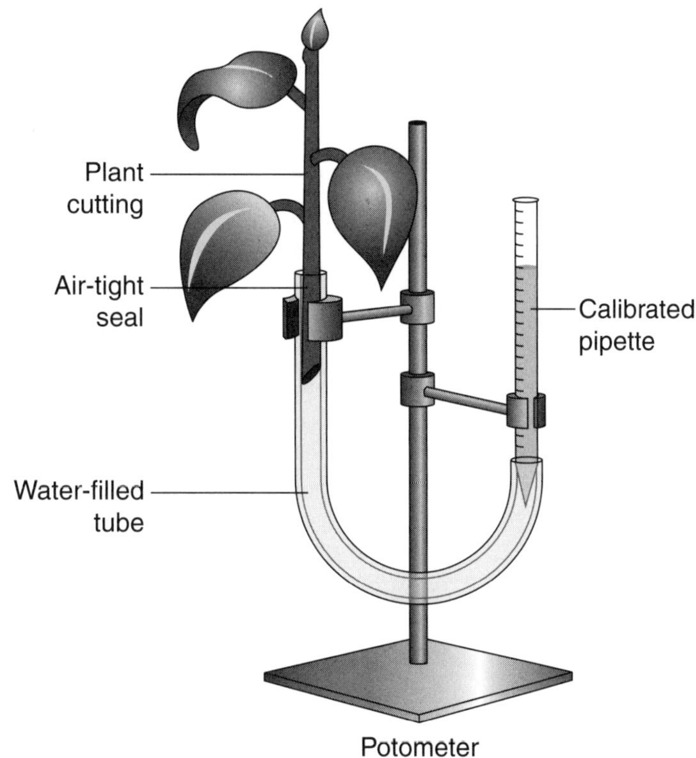

Potometer

Their data are shown below.

Air temperature (°C)	*Water lost in 60 minutes* (mL)
10	1
18	3
21	4
25	5
30	6

The benefit for the plant of this response would be a decrease in the

(A) rate of transpiration.

(B) rate of photosynthesis.

(C) temperature of the plant.

(D) amount of water in the plant.

Refer to the following information to answer Questions 13 and 14.

The diagram shows some chromosomes during some stages of meiosis.

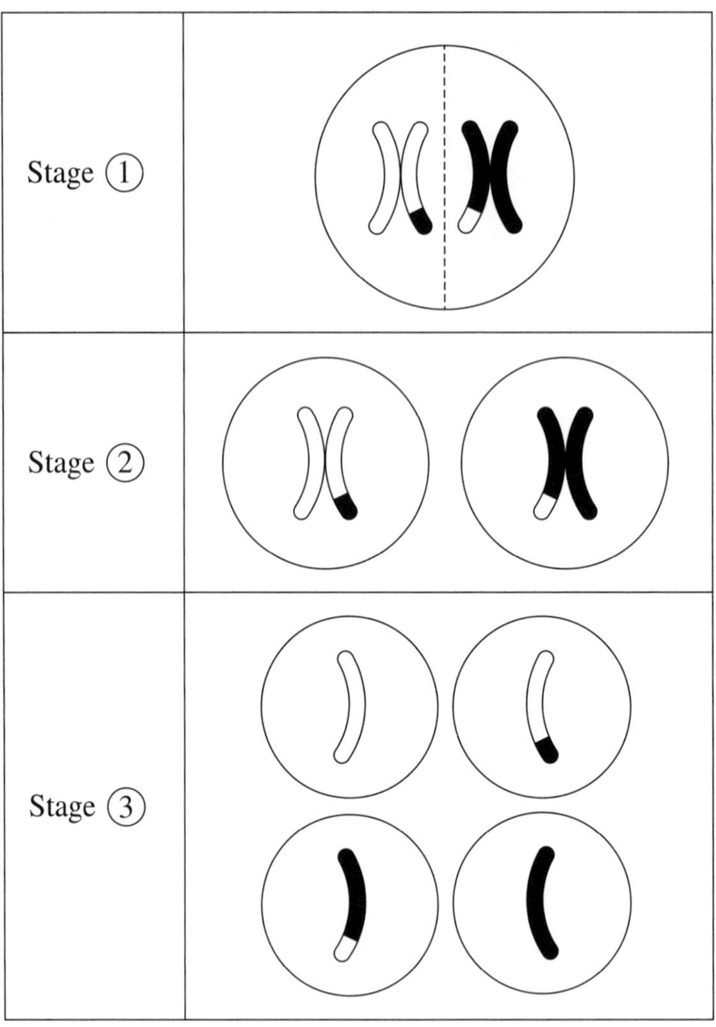

13 When does the segregation of homologous chromosomes occur?

(A) Before stage ①

(B) Between stages ① and ②

(C) Between stages ② and ③

(D) Between stages ① and ② and again between stages ② and ③

14 The chromosomes shown carry

(A) different genes and different alleles.

(B) different genes and the same alleles.

(C) the same genes and different alleles.

(D) the same genes and the same alleles.

15 The table shows four features that can be used to distinguish between types of pathogens.

Which row of the table is correct?

	Feature	*Prion*	*Protozoan*
(A)	Contains DNA	✓	✗
(B)	Cellular	✗	✓
(C)	Can reproduce	✓	✗
(D)	Composed of protein	✗	✓

16 Ringbarking is the removal of a thin strip of bark from the entire circumference of a tree.

The tree will initially survive, but the roots will eventually die because ringbarking stops

(A) photosynthesis.

(B) the transport of water.

(C) the transport of oxygen.

(D) the transport of dissolved nutrients.

17 Capillaries have thin walls that help them perform their main function.

The best explanation for this is that

(A) capillaries are the smallest vessels in the body.

(B) thin walls are an adaptation that help diffusion.

(C) blood flow in capillaries is under very low pressure.

(D) thin walls maximise the surface area available for gas exchange.

18 How does the production of a new transgenic species have the potential to alter the path of evolution?

(A) The creation of new genes increases biodiversity.

(B) The removal of genes from a species decreases biodiversity.

(C) The transfer of genes within a species increases biodiversity.

(D) The transfer of genes between two species increases biodiversity.

Refer to the following information to answer Questions 19 and 20.

> Melanomas are characterised by uncontrolled cell division caused by mutations that continue to occur once the tumour has developed. Scientists have discovered that vaccines produced using antigens extracted from the patient's own melanoma cells can be useful in treating melanoma. When injected, the vaccines stimulate an immune response.

19 What can be inferred from the scientists' discovery?

(A) Cancer cells carry unique antigens.

(B) Self-antigens are not present on cancer cells.

(C) The melanoma patient has a dysfunctional immune system.

(D) The body cannot mount an immune response against cancer cells.

20 The effect of the melanoma vaccine is to stimulate

(A) T cells which produce antibodies.

(B) cytotoxic T cells which activate B cells.

(C) cell division to produce more lymphocytes.

(D) production of B cells which destroy melanoma cells.

2016 HIGHER SCHOOL CERTIFICATE EXAMINATION

Biology

Centre Number

Student Number

Section I (continued)

Part B – 55 marks
Attempt Questions 21–31
Allow about 1 hour and 40 minutes for this part

Answer the questions in the spaces provided. These spaces provide guidance for the expected length of response.

Show all relevant working in questions involving calculations.

Write your Centre Number and Student Number at the top of this page.

Question 21 (2 marks)

Give TWO potential benefits of artificial blood. **2**

....

....

Question 22 (2 marks)

Name an Australian ectotherm and describe its response to a decrease in ambient temperature. **2**

....

....

....

....

Question 23 (5 marks)

(a) Explain ONE reason why the concentration of water in cells should be maintained within a narrow range for optimal cell function. **2**

....

....

....

....

(b) A person has consumed large amounts of water. Complete the table to show the effect on each of the variables listed. **3**

Variable	*Effect*
Urine volume	
ADH secretion	
Salt concentration in blood	

Question 24 (3 marks)

Name an infectious disease and explain how ONE host response is a defence adaptation. **3**

....

....

....

....

....

....

Question 25 (6 marks)

Rabies is a disease caused by a virus that affects mammals.

In 1880 Louis Pasteur investigated dogs that were suffering from rabies in order to find the cause. He believed rabies was caused by a microorganism but could not culture it in broth nor observe it under the light microscope. However, he could cause the disease in healthy dogs by injecting them with saliva from infected dogs. He was able to repeat the disease cycle in this way.

(a) Why was Pasteur NOT able to observe the rabies virus? **2**

....

....

(b) Explain why Pasteur needed to identify and culture the microorganism in order to meet the scientific standards for establishing the cause of rabies. **4**

....

....

....

....

....

....

....

....

Question 26 (7 marks)

Students conducted preliminary experiments across different species to analyse their DNA base composition.

The table shows the experimental data collected.

Species	*% Adenine*	*% Guanine*
A	38	12
B	26	22
C	8	40
D	20	32
E	33	18

(a) On the grid below, plot the % Adenine vs % Guanine of the species analysed AND draw a suitable line of best fit. **3**

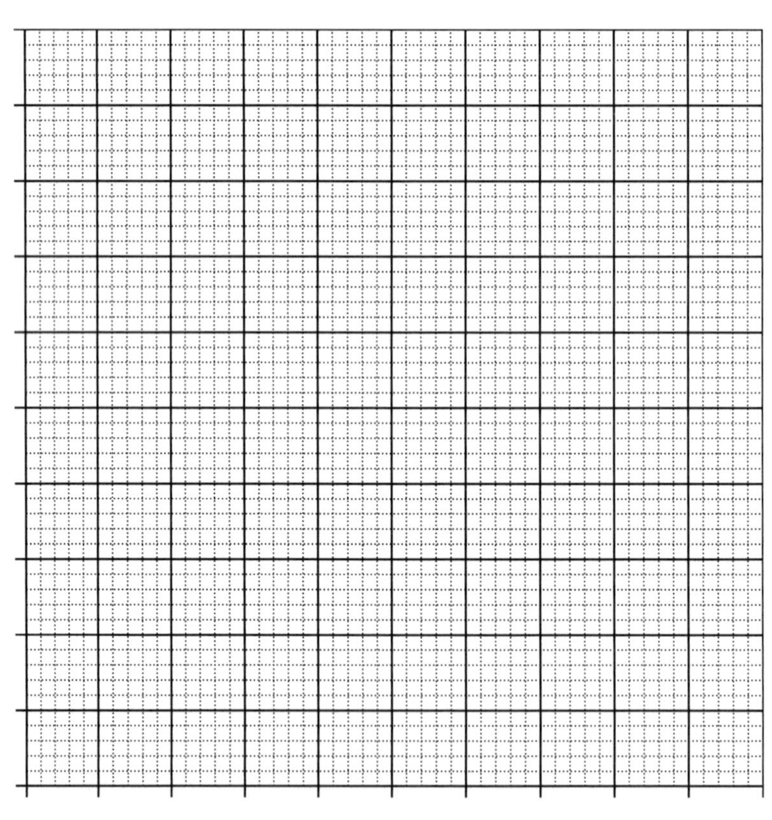

Question 26 continues

Question 26 (continued)

(b) Identify the relationship shown by the data.

....

....

(c) Explain the relationship shown by the data.

....

....

....

....

....

....

End of Question 26

Question 27 (4 marks)

The diagram shows a vascular bundle from a flowering plant.

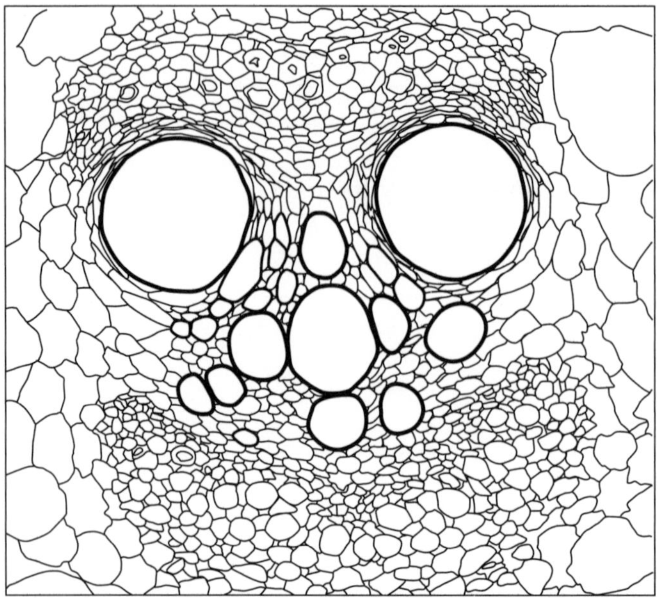

(a) On the diagram, clearly label a xylem vessel. **1**

(b) A mutation in this species causes the walls of the xylem vessels to be significantly reduced in thickness. **3**

Explain why the leaves of these mutant strains will wilt more easily.

....

....

....

....

....

....

Question 28 (6 marks)

Bacteriophages are viruses that use bacteria as host cells. One type of bacteriophage infects only one species of bacterial cell. The infected cells are destroyed as the bacteriophage particles burst out of them, completing their life cycle.

Bacteriophages can be used to fight bacterial infections. As bacteria evolve, bacteriophages also evolve.

(a) Explain TWO advantages of using bacteriophage treatment compared to antibiotic treatment for bacterial infections. **4**

....

....

....

....

....

....

....

....

....

....

(b) Describe a possible disadvantage of using bacteriophage treatment. **2**

....

....

....

....

Question 29 (7 marks)

The diagram shows the parts of a pea plant flower, similar to those used by Mendel in his experiments to investigate the inheritance of characteristics.

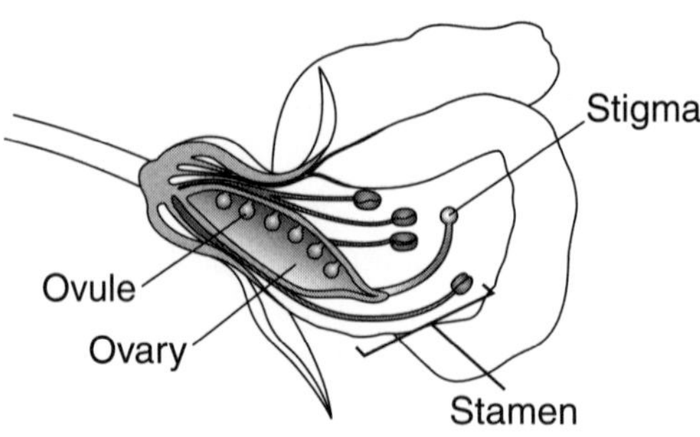

(a) Describe TWO ways in which Mendel's use of flower parts contributed to the success of his experiments. **3**

....

....

....

....

....

....

....

....

(b) How did Mendel's experimental results advance our knowledge of the inheritance of characteristics? **4**

....

....

....

....

....

....

....

....

....

....

Question 30 (5 marks)

Explain why the combined use of quarantine and vaccination programs is a more effective way of controlling disease than using only one of these strategies. **5**

....

....

....

....

....

....

....

....

....

....

....

....

Question 31 (8 marks)

As altitude increases, the partial pressure of oxygen (pO_2) in air decreases. **8**

Species A and B are closely related endotherms that live in different habitats in Asia. The minimum pO_2 required for 100% blood oxygen saturation differs in these species because of differences in their haemoglobin structure. Data related to these two species are shown below.

Endotherm species	*Habitat altitude*	*Minimum pO_2 for 100% Hb saturation*
A	High	54
B	Low	80

Explain how the differences in these species could have arisen, using the Darwin/Wallace theory of evolution and your understanding of the adaptive advantage of haemoglobin.

....

....

....

....

Question 31 continues

Question 31 (continued)

End of Question 31

2016 HIGHER SCHOOL CERTIFICATE EXAMINATION

Biology

Section II

25 marks
Attempt ONE question from Questions 32–36
Allow about 45 minutes for this section

Answer parts (a)–(e) of one question in the Section II Writing Booklet. Extra writing booklets are available.

Show all relevant working in questions involving calculations.

Question 32 Communication

Question 33 Biotechnology

Question 34 Genetics: The Code Broken?

Question 35 The Human Story

Question 36 Biochemistry (*Not included in this reproduction*)

Question 32 — Communication (25 marks)

Answer parts (a), (b) and (c) of the question on pages 2–4 of the Section II Writing Booklet. Start each part of the question on a new page.

(a) The diagram shows a cross-section of the part of the human body responsible for voice production.

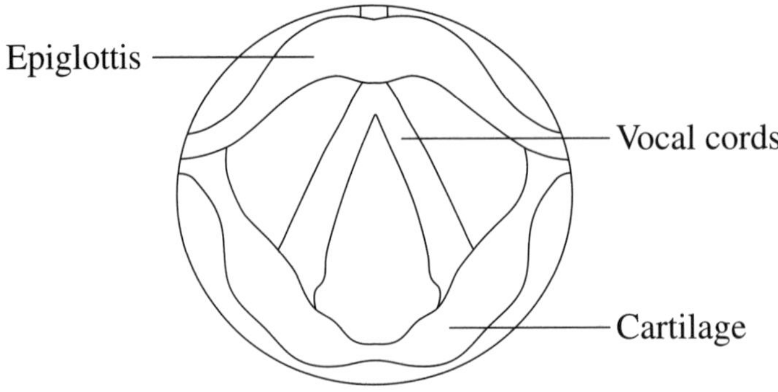

(i) Name the part of the human body shown. **1**

(ii) Describe how vocal cords produce sounds of different frequencies. **2**

(b) A student conducted an experiment to model the process of accommodation. **5**

Describe the experiment and how the first-hand data collected could contribute to an understanding of how accommodation works in human vision.

(c) (i) Describe how electrochemical changes in the membrane of a neurone allow information to be transferred. **2**

(ii) The brain includes the cerebrum, cerebellum and brain stem. Draw an outline diagram of the human brain, and shade and label the area of the cerebrum involved in the perception of sound. **2**

Question 32 continues

Question 32 (continued)

Answer parts (d) and (e) of the question on pages 6–8 of the Section II Writing Booklet. Start each part of the question on a new page.

(d) Total colour blindness is a condition that causes people to see only black, white or grey. Colours all appear as grey.

An eyeborg is a new technology that can be worn externally on the head to help these people perceive differences in colour.

The diagram shows an eyeborg. It has a receiver which points to an object. Light reflected from the object causes an electronic message to be sent along a wire to a speaker which produces a sound of a specific frequency that is detected by the ear. Different colours produce sounds of different frequencies.

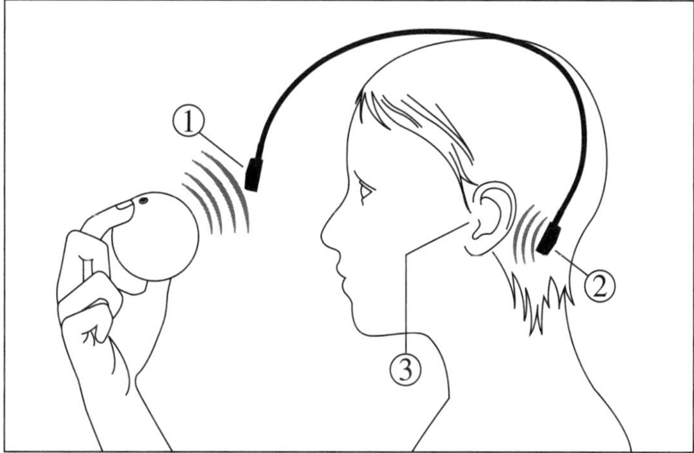

(i) Describe the energy transformations occurring at locations 1, 2 and 3 in the diagram. **2**

(ii) Other technologies also improve human communication. Name ONE such technology and compare it to an eyeborg in terms of the impact it has on the user. **4**

(e) Compare mechanisms in the human body for detection and perception of a range of frequencies in visual and auditory communication. **7**

End of Question 32

Question 33 — Biotechnology (25 marks)

Answer parts (a), (b) and (c) of the question on pages 2–4 of the Section II Writing Booklet. Start each part of the question on a new page.

(a) The diagram shows a step in protein synthesis.

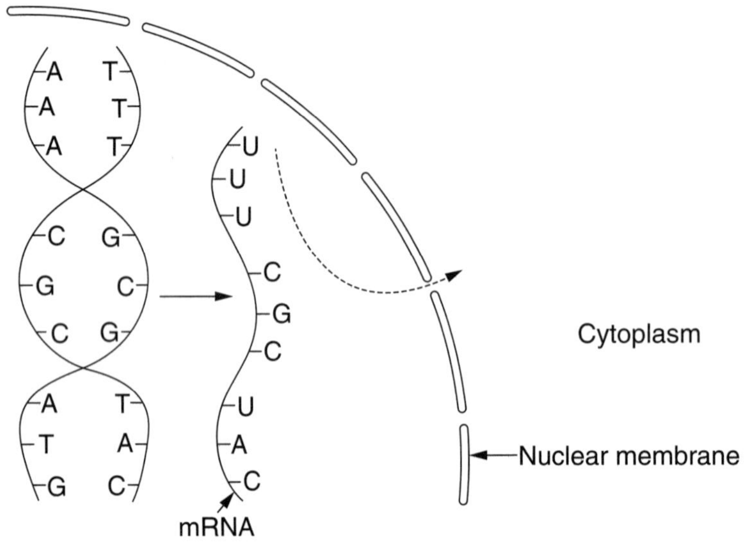

(i) Name the process by which mRNA is produced. **1**

(ii) Explain why the mRNA moves in the direction shown. **2**

(b) A student conducted an experiment to investigate the rate of enzyme activity. **5**

Relate the understanding gained from performing this experiment to an application of biotechnology.

(c) (i) Explain why the Polymerase Chain Reaction (PCR) is used in biotechnology. **2**

(ii) Draw a flow chart to show the sequence of events that results in the formation of recombinant DNA. **2**

Question 33 continues

Question 33 (continued)

Answer parts (d) and (e) of the question on pages 6–8 of the Section II Writing Booklet. Start each part of the question on a new page.

(d) (i) Explain a benefit of strain isolation in biotechnology. **2**

(ii) Assess the impact of our understanding of restriction enzymes on developments in biotechnology. **4**

(e) 'Artificial selection and animal cloning are similar biotechnologies, used to create organisms with improved characteristics.' **7**

Evaluate this statement. In your response, include reference to the processes involved in the two biotechnologies.

End of Question 33

Question 34 — Genetics: The Code Broken? (25 marks)

Answer parts (a), (b) and (c) of the question on pages 2–4 of the Section II Writing Booklet. Start each part of the question on a new page.

(a) The diagram shows a process used in agriculture.

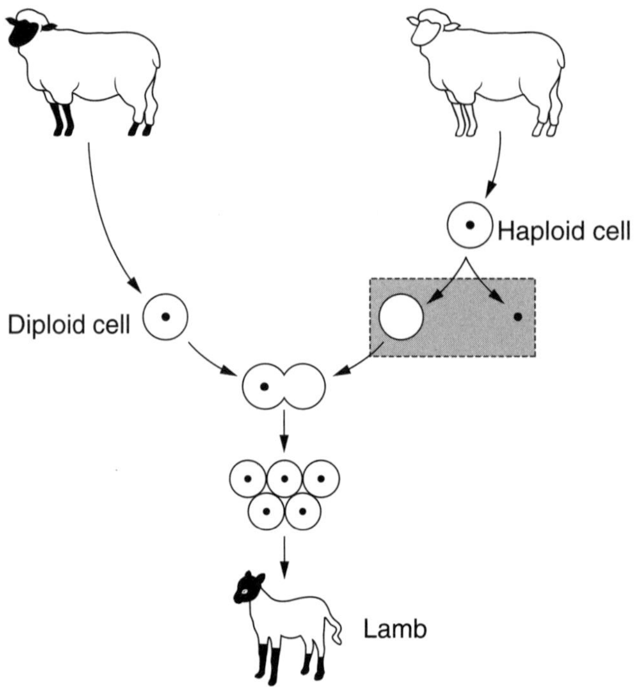

(i) Name the process shown. **1**

(ii) Give TWO reasons for what occurs in the shaded area. **2**

(b) Explain how selective breeding has resulted in a series of changes in an agricultural species. **5**

(c) (i) Provide a human and a non-human example of multiple allelism controlling a trait. **2**

(ii) Sketch a graph that shows the variation of a named polygenic trait within a population. **2**

Question 34 continues

Question 34 (continued)

Answer parts (d) and (e) of the question on pages 6–8 of the Section II Writing Booklet. Start each part of the question on a new page.

(d) (i) Using an example, describe what is meant by *gene cloning*. **2**

(ii) Assess the impact of advances in our understanding of mutations on the development of technologies. **4**

(e) 'Genes influence proteins and proteins influence genes.' **7**

Evaluate this statement with reference to the structure and function of genes and proteins.

End of Question 34

Question 35 — The Human Story (25 marks)

Answer parts (a), (b) and (c) of the question on pages 2–4 of the Section II Writing Booklet. Start each part of the question on a new page.

(a) The diagram shows rock layers containing a number of fossils. A dating technique was used on this data alone to determine that the youngest fossil was in layer *D*.

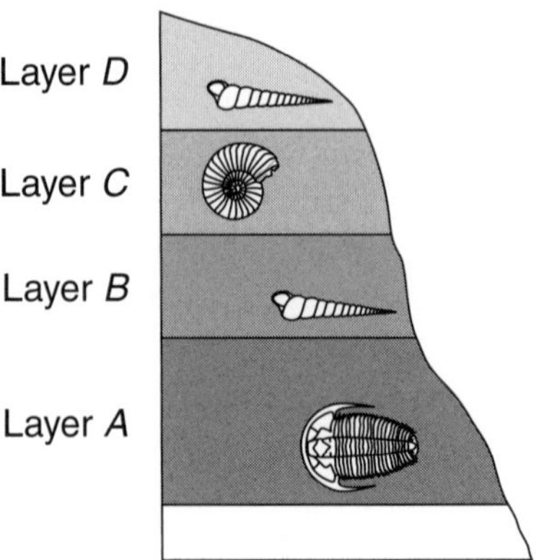

(i) Name the dating technique used. **1**

(ii) Name a different dating technique and describe how it could be used to verify this finding. **2**

(b) The table compares some of the amino acids present in a particular protein in different primates. **5**

Human	PRO	GLY	GLY	ALA	ASN	ALA	THR	ARG
Ape (chimpanzee)	PRO	GLY	GLY	ALA	ASN	ALA	THR	ARG
Ape (gorilla)	PRO	GLY	GLY	ALA	ASN	ALA	THR	LYS
New world monkey	PRO	GLY	GLY	ASN	ASN	ALA	GLN	LYS
Prosimian	PRO	GLY	SER	HIS	ASN	ALA	GLN	LYS

Using these data and your knowledge of the characteristics of primate groups, explain why using different types of data improves the reliability of estimated evolutionary relationships.

Question 35 continues

Question 35 (continued)

(c) (i) What are TWO features of chromosomes used in karotype analysis? **2**

(ii) Sketch a graph that depicts changes in human population over the last 10 000 years. Identify a point on the graph that shows the effect of a technological advance on world population. **2**

Answer parts (d) and (e) of the question on pages 6–8 of the Section II Writing Booklet. Start each part of the question on a new page.

(d) A new fossil form was recently found in South Africa. This fossil shares characteristics with both the genus Australopithecus and the genus Homo.

There has been debate as to whether this new fossil form should be classified in the genus Australopithecus or in the genus Homo.

(i) Describe a key difference between fossils classified as the genus Australopithecus and those classified as the genus Homo. **2**

(ii) Explain how DNA sequencing technology could be used to determine which genus the new fossil belongs to. In your answer, refer to relevant hominid species. **4**

(e) 'Biological evolution influences cultural development and cultural development influences biological evolution.' **7**

Evaluate this statement with reference to examples in hominid evolution.

End of Question 35

2016 HSC Examination Paper

Sample Answers

Section I Part A

(Total 20 marks)

1 A Enzyme active sites are denatured and hydrogen bonds broken.

2 C If normal conditions change, microflora can increase in number thereby causing symptoms.

3 A Walter Sutton, together with Theodore Boveri, is attributed with the chromosomal theory of inheritance.

4 D Experimental conclusions tend to be provisional in nature and subject to further testing, particularly as technology develops.

5 D Increasing concentration when all active sites are occupied cannot increase activity as the enzyme is functioning at capacity.

6 D *Metabolism* is a general term relating to the chemical reactions within an organism.

7 B Filtration, reabsorption and secretion is the only correct sequence of processes.

8 D Sex-linked traits are located on the 'X' and so are either present or not.

9 B B is the best of the options given. D doesn't make sense as osmosis moves water from a high to low concentration (or from low to high solute) and the other options are incorrect.

10 A Sugars (*W*) attach to bases and the two bases cannot be complementary.

11 C Specialised B cells or plasma cells produce antibodies.

12 C The benefit of increased ambient temperature is a decrease in the temperature of the plant as increased water loss will help regulate the temperature of the plant.

13 B Homologous chromosomes are only in stage 1 so segregation occurs during meiosis 1 (between stages 1 and 2).

14 C Given the chromosomes are homologous, by definition they carry the same genes but there has been an exchange of alleles.

15 B Protozoans are cellular organisms capable of reproduction.

16 D Ringbarking prevents the transport of photosynthetic products.

17 D Thin walls assist in the increase of the surface area for diffusion.

18 D Transgenesis relates to the transfer of genes between species.

19 B If self-antigens are not present in the cancer cells then the vaccine can create an immune response against them.

20 C The vaccine would promote development of lymphocytes.

Section I Part B

21 Two potential benefits of artificial blood would be a longer storage or shelf life and overcoming the problems associated with compatibility and tissue matching that is experienced with donated blood. (*2 marks*)

22 The Eastern Brown snake is an ectotherm. If the ambient temperature drops below the optimum, the snake will bask in the sun to gain body heat. (*2 marks*)

23 (a) The concentration of water in cells should be maintained within a narrow range as water is a universal solvent and many substances are dissolved or carried in it. Water forms the basis of the cytoplasm in cells and is crucial for metabolic reactions in cells. These reactions are under the control of specific enzymes and so the cells' water levels need to remain at an appropriate level for optimal cell function. Deviations from the narrow range of water concentration results in enzymes denaturing as pH levels or substrate concentrations change. (*2 marks*)

(b)

Variable	**Effect**
Urine volume	increases
ADH secretion	reduces
Salt concentration in blood	reduces

(*3 marks*)

24 Responses can vary but must name an infectious disease.

An infectious disease caused by *Vibrio cholerae* is cholera. The host response is to produce antibodies to the toxin. If the patient survives, future exposure to the same strain results in resistance. This is an example of antibody-mediated immunity and results in increased resistance which is of adaptive advantage as it can result in the survival of the individual. (*3 marks*)

25 (a) The rabies virus is far too small to be seen with the naked eye or even a simple light microscope. The technology available to Pasteur was not as advanced as it is today; he needed an electron microscope. (*2 marks*)

(b) Pasteur needed to culture the microorganism so that he could identify it and link it directly to the cause of symptoms in individuals. The scientific standard for identifying a disease is based on Koch's postulates:

- the disease must be present in every host with the disease
- the organism needs to be isolated from the host and made into a pure culture
- a healthy host without symptoms must develop the same symptoms after inoculation with the pure culture
- the microorganism needs to be isolated from the new host and a second pure culture made that is the same as the original culture.

So by culturing the rabies virus in unaffected dogs Pasteur was able to determine the cause and produce a vaccine to prevent others from developing the disease if exposed to the same virus. (*4 marks*)

26 (a)

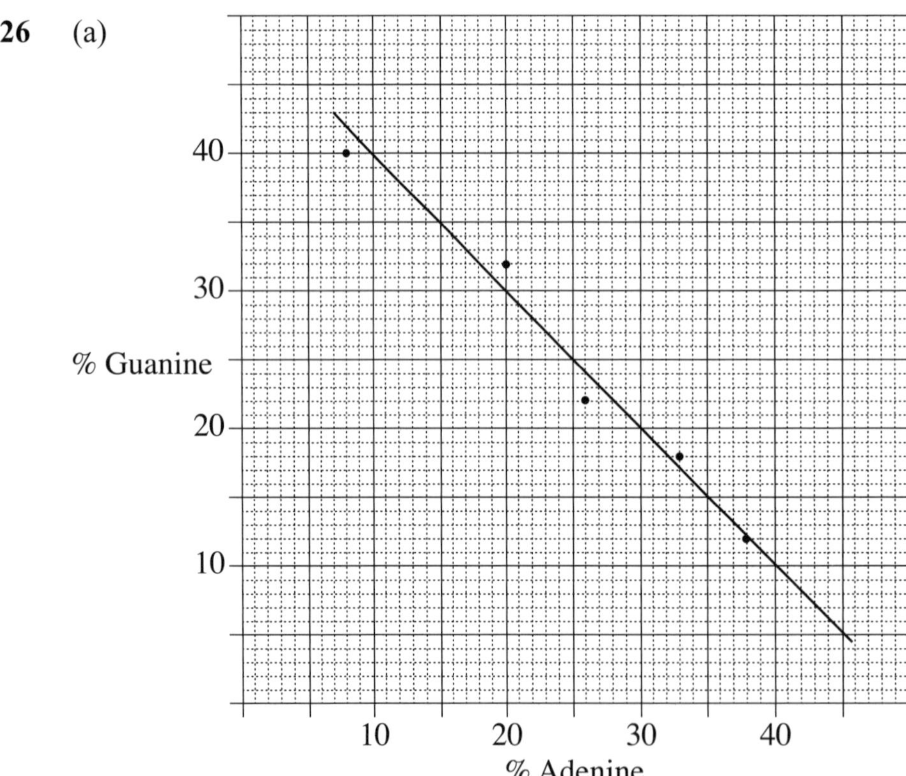

(*3 marks*)

(b) As the percentage of adenine increases, the percentage of guanine decreases. (*1 mark*)

(c) As the percentage of adenine increases, you would expect the percentage of thymine to increase as these are a complementary base pair. Guanine and cytosine are complementary, not adenine and guanine. So the graph indicates that some species have a higher adenine/thymine proportion of base pairs than cytosine/guanine. (*3 marks*)

27

(a)

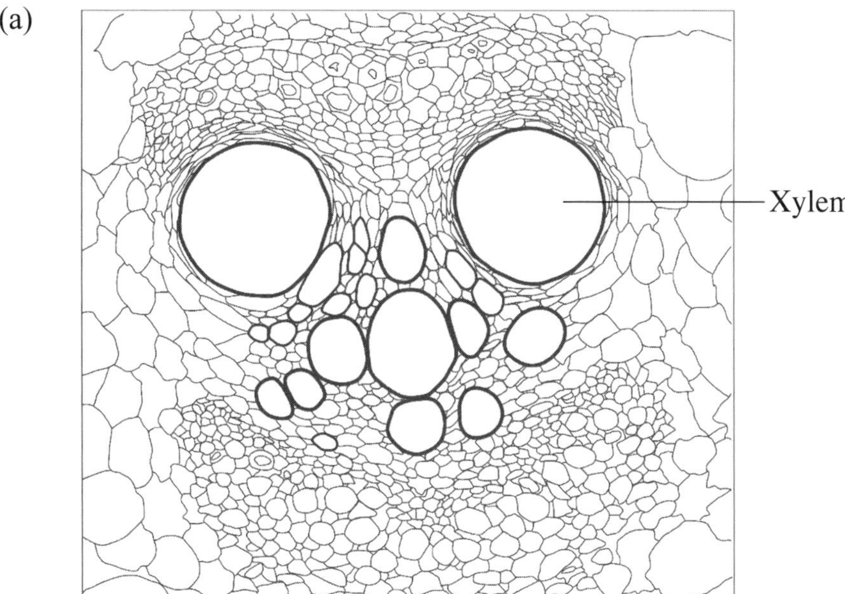

(1 mark)

(b) If the walls of the xylem are significantly reduced in thickness then the mutant strain will wilt more easily as it is the xylem vessel that enables the plant to be more rigid, thus providing support for the plant. The lignin that would normally be deposited in the xylem cells in ring/spiral patterns strengthens these cells, making them impermeable to water. If they are significantly thinner then water will be lost from the vascular bundles and wilting of the leaves will result. If the vessels were to collapse it may reduce the transpiration stream (movement of water to the leaves) leading to wilting. (*3 marks*)

28 (a) An advantage of using bacteriophage treatment is that antibiotic resistance could be avoided, as no antibiotics are used in the treatment of the disease, thereby preventing overuse or misuse of antibiotics. A second advantage of using bacteriophage treatment is that each bacteriophage is specific to a particular bacteria. If introduced into an infected patient the bacteriophage can destroy the pathogenic bacteria as they use them to reproduce. Once the pathogenic bacteria are destroyed the bacteriophage would also reduce in number as they would no longer have a suitable host to complete their life cycle. (*4 marks*)

(b) A possible disadvantage of using bacteriophage treatment is the unknown effect the residual bacteriophage may have on the body until it completely disappears and the possibility that it may evolve to use other host cells other than the original bacteria it was used to destroy. (*2 marks*)

29 (a) Mendel successfully manipulated pea flowers in two ways. First, he removed the stamens from one breeding pair to ensure that self-pollination did not take place. Second, he transferred pollen by hand using a brush from the stamen (male part) of another pure-bred plant flower to the stigma (female part) of the original flower with stamens removed. (*3 marks*)

(b) Mendel's work advanced our understanding of the inheritance of characteristics as he identified that organisms pass on characteristics or 'factors' from parent to offspring. The 'factors' he identified we now refer to as traits or genes. These will determine the phenotype of an individual offspring depending on the type of inheritance. Each individual carries two forms of each gene known as alleles in a paired combination referred to as genotypes. These can be dominant or recessive. In the heterozygous combination the dominant allele will mask the recessive allele. An individual must be homozygous of the recessive allele combination of the gene in order to exhibit that phenotype. (*4 marks*)

30 The combined use of quarantine and vaccination programs is a more effective way of controlling disease than using either exclusively because the herd immunity resulting from effective vaccination programs can be backed up by isolation of discrete occurrences of the disease. An effective vaccination program can ensure that individuals that are weak, such as the very young with under-developed immune systems that cannot be immunised or the frail and elderly with weakened immune systems, can be protected by the herd immunity. This means that there is enough of a barrier in the population or community to prevent the disease from being passed on should an infected individual be introduced. If infected individuals are identified, they should be immediately quarantined to prevent them mixing with the community/population so that they do not come into contact with the vulnerable individuals and the disease then is less likely to be spread. A good recent example of this was the isolation of Ebola victims with the simultaneous research into possible vaccinations to limit the spread of the extremely infectious disease. (*5 marks*)

31 Haemoglobin is a pigment in red blood cells which increases the oxygen-carrying capacity of red blood cells by up to 100 times. Haemoglobin contains four active sites to which oxygen molecules attach and this gives mammals a significant survival advantage. As oxygen concentrations decrease at higher altitudes, the adaptive advantage of mammals to increase the quantity of haemoglobin in the blood is significant as it increases the ability of the blood to transport oxygen around the body tissues. The selective advantage of having a larger heart to circulate larger volumes of blood rich in red blood cells allowing organisms to breathe more deeply enables organisms' survival at higher altitudes. The Darwin/Wallace theory would explain that those organisms that possessed these features—larger heart and higher concentration of red blood cells—were able to breathe more deeply, supplying oxygen to the body tissues and thereby enabling survival in the thinner atmosphere. These characteristics are then passed onto the offspring, manifesting the features in the population and species. Therefore species A survived better at higher altitudes with these more suitable features while species B was not able to do as well and did not venture to the higher altitudes as a result. (*8 marks*)

Section II—Options

Question 32—Communication

(a) (i) the larynx (*1 mark*)

(ii) The larynx is in the throat that contains the vocal cords and is a series of cartilages joined by ligaments and membranes. Sound is produced when the folds of cartilage constrict or relax as air passes through them. Higher frequencies or pitch are produced as the vocal cords vibrate rapidly when air is pushed through them from the lungs while they are tighter and shorter. If the cords are longer or thicker and vibrate slowly then a lower pitched sound is produced. (*2 marks*)

(b) Accommodation is the process by which the lens of the eye changes shape to focus an image onto the retina, irrespective of the distance of the object from the eye. Accommodation can be modelled using the following equipment: a candle (light source), a piece of white paper (retina), a series of different bi-convex lenses of different thicknesses (independent variable) and a stand on which to place the lenses.

- Place the thickest lens in the stand at a distance of 10 cm from the paper.
- Move the candle to the other side of the lens until the flame comes into focus on the paper.
- Record the distance between the lens and the candle (dependent variable).
- Repeat this for the thinner lens ensuring the distance between the paper and lens remains 10 cm (controlled variable).

The data obtained can be used to depict the functioning of the eye and show how the lens gets thicker if the object is closer so that the light is refracted more significantly than when it is further away and the lens is thinner causing less refraction in order to have the image focused onto the retina. In short, the more curved the lens the greater the refractive power and the shorter the focal length. (*5 marks*)

(c) (i) Electrochemical changes in the membrane of a neurone allow information to be transferred through a series of polarity reversals between the inside and outside of the cell's plasma membrane. A resting potential difference exists between the inside and outside of the membrane resulting in a cellular voltage, the resting membrane potential being –70 millivolts (mv). Sodium ions move within, having a net negative charge and the membrane is therefore polarised. As a stimulus passes, the sodium/potassium and chlorine ions are able to pass through the membrane allowing the impulse to pass. This is called the sodium/potassium pump. When the resting potential of –70 mv changes to +30 mv an action potential is generated in which electrical impulses are generated and transferred between nerve cells. (*2 marks*)

(ii)

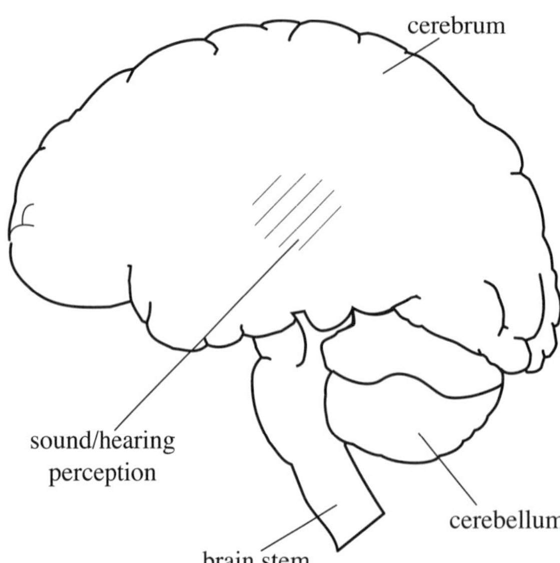

(*2 marks*)

(d) (i) 1: The electromagnetic transverse wave (visible light) reflected from the object is converted into an electrical signal.

2: The electrical signal is converted to sound energy causing air particles to vibrate.

3: The kinetic energy from the vibrating air particles is transferred as a mechanical vibration through the ear to the cochlea where it is converted into an electrochemical signal and travels to the brain. (*2 marks*)

(ii) Note: answers can vary as there are a few different examples that students can use to answer the question. One example is:

The use of cochlear implants has enabled the profoundly deaf to be able to interpret sound, by bypassing the outer and middle ear. Similar to the eyeborg, the cochlear implant is a device attached to the head surgically. It basically bypasses the damaged cochlea in the recipient. The device picks up sound waves with a microphone which are then converted by a speech processor into an electrical code. Similar to the eyeborg, which relays electrical energy into electrochemical impulses that are transferred to the brain, this is then sent to the internal implant within the inner ear that transforms it into an electrical pulse that stimulates the nerve endings, sending the message to the brain for decoding.

The use of these devices has not only made it possible for people to hear if their cochlea is damaged, but for young children born profoundly deaf to be able to hear. This is important for speech development and cognition of information. The impact on the user is immense as they have a whole new sensorial world to explore and the implant will help them survive independently. It could be argued that the cochlear implant has a more significant impact on an individual than simply being able to see colour, a disability which may not necessarily hinder daily existence. (*4 marks*)

(e) Note: answers can vary as there are a few different examples that students can use to answer the question. One example is:

The human body has a range of mechanisms to be able to detect and perceive visual and auditory communication. The ear is able to detect a variety of sound frequencies from 20–20 000 Hz. Although this tends to be a narrower range than that of other mammals such as cats or dogs, it is still significant and relevant to the human lifestyle. For example, humans don't need to communicate over long distances like whales, who communicate below the 20 Hz frequency. The Organ of Corti within the cochlea is responsible for converting physical/mechanical vibrations into an electrochemical message that is passed via the auditory nerve to the brain for interpretation. Similarly the photoreceptors on the retina convert a narrow range of the electromagnetic spectrum or the 'visible light' (380–720 nm) component into an electrochemical message sent via the optic nerve to the brain for interpretation. Within the human eye the rod and cone cells are located on the retina. There are 6 to 7 million cone cells on the macula of which there are three types, each stimulated by a different region of the visible light spectrum: blue—450 nm; green—540 nm; and red—570 nm. When stimulated in different amounts we perceive the colour spectrum. There are about 120 million rod cells on the fovea which are used for black-and-white perception. Within the brain the visual cortex located at the back of the brain perceives the message and we 'see' the message as a series of images. The auditory messages are perceived by a smaller part in the centre of the brain. The perception of the relevant auditory/visual stimulus is then interpreted, based on the individual's past experience and levels of specific function in these areas. If someone has damage to any part of this message transfer, be it sight or sound, they may have difficulty in getting the same message as the next person. *(7 marks)*

Question 33—Biotechnology

(a) (i) transcription (*1 mark*)

(ii) mRNA moves across the nuclear membrane out of the nucleus into the cytoplasm as it carries the DNA base template to the ribosomes located within the cytoplasm for translation to occur. (*2 marks*)

(b) Experiments utilising enzymes can demonstrate that enzymes have a limited range of activity. For example:

Aim: To investigate how temperature affects enzyme activity

Method: Dissolve a junket tablet in 25 mL of water in a measuring cylinder. Junket tablets contain an enzyme, rennin, which acts on the milk protein, casein.

1. Measure 10 mL of full-cream milk into each of four large test tubes. Mark these 'A', 'B', 'C' and 'D'.
2. Place test tubes B, C, D in a water bath at 37 °C.
3. Place test tube A in a beaker of ice.
4. Treat each of the test tubes in the following way:

Test tube A: When the temperature of the **milk** reaches approximately 5 °C add 4 mL rennin solution and observe any changes in the milk by gently shaking the test tube at regular intervals.

Test tube B: Leave as the control.

Test tube C: When the temperature of the **milk** reaches approximately 37 °C add 4 mL rennin solution and observe any changes in the milk by gently shaking the test tube at regular intervals.

Test tube D: When the temperature of the **milk** reaches approximately 37 °C add 4 mL of **boiled rennin** and observe any changes in the milk by gently shaking the test tube at regular intervals.

Results:

Test tube A	Test tube B (control)	Test tube C	Test tube D
No curdling	No curdling	Milk curdled	No curdling

The results indicated that the enzyme rennin will operate in a temperature range around 37 °C. The boiled rennin was denatured and the test tube in ice did not clot. Similar investigations can be carried out by varying the pH or the substrate concentration. Each of these factors will also affect the rate of enzyme activity.

Understanding as a result of experiments using enzymes include:

- Enzymes are biological catalysts that promote a specific biochemical reaction within a range of conditions including temperature range, pH and substrate concentration, e.g. rennin works best at 37 °C.
- Enzymes can be denatured and will no longer work if exposed to extreme conditions such as high temperatures, e.g. rennin will no longer work if boiled.

Processes such as fermentation have been used in biotechnology to produce beer, wine, cheese, yoghurt and bread. The microorganisms that carry out the fermentation processes rely on particular enzymes that act on particular substrates to produce the desired products. A key to successful production is the maintenance of appropriate conditions of temperature and pH for the enzymes in the microorganism to operate. Bakers' yeast requires a specific temperature range to be active and thereby 'prove' the dough.

Another application of biotechnology is that particular enzymes have been incorporated into laundry detergents to break down organic stains in clothes to make them appear cleaner. The experiments show why these enzymes need to work at appropriate temperatures and pH to give maximum benefit. (*5 marks*)

(c) (i) PCR is used in biotechnology to produce larger amounts of DNA from very small amounts of DNA or from impure samples. This is beneficial when larger amounts of DNA are needed, e.g. in forensic analysis of samples of tissue from a crime scene. (*2 marks*)

(ii)

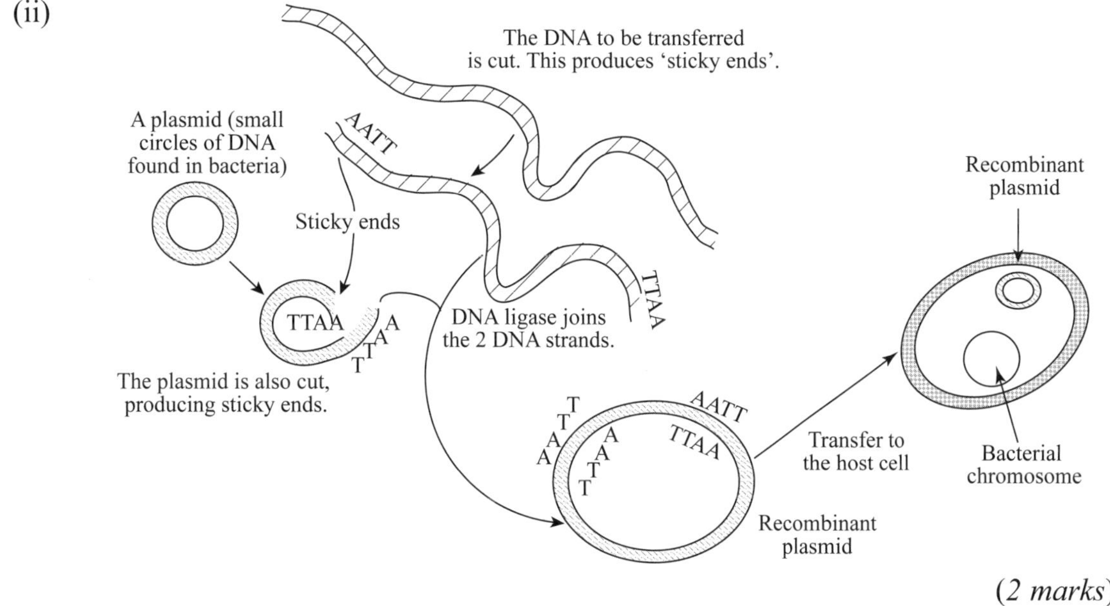

(*2 marks*)

(d) (i) A benefit of using strain isolation in biotechnology is that, once higher yielding organisms have been identified, they can be cultured in a pure form and mass produced for commercial use, e.g. the isolation of *Penicillium chrysogenum* discovered growing in a melon, which yielded much higher quantities of penicillin than the original strain. (*2 marks*)

(ii) Restriction enzymes have the ability to cut open sections of DNA or, in other words, produce gene fragments by gene splicing. This process is not random but particular restriction enzymes break DNA segments at specific sequences of bases. If the base sequence of a particular gene is known, selection of the appropriate restriction enzyme can help to splice it out of the remaining DNA. Without restriction enzymes, recombinant DNA could not be produced and gene cloning would not be possible. The use of biotechnology to produce human insulin, recombinant vaccines for disease prevention and the production of genetically modified organisms would not be possible without our understanding and use of restriction enzymes. (*4 marks*)

(e) Artificial selection and animal cloning are both biotechnologies that have been utilised in agriculture in particular. Artificial selection has been a common practice for a very long time; it involves the selection of organisms that have desirable characteristics and ensures their manifestation by assisting or facilitating the breeding of the organism, e.g. taking the prize bull or semen bought from an agricultural show back to the farm and using it to inseminate the cows. This can increase the butter-fat content in dairy cows or the marbling and muscle development of beef cattle. It is done without much interference and there is no manipulation of the DNA.

Cloning is similar in that desirable characteristics are identified in organisms, but these are then physically manipulated to ensure that embryos will be identical to the desirable organism. We have come a long way since the first animal clone 'Dolly the sheep'.

Whole animal cloning converts the nucleus from a mature organism into a zygote that develops into an organism genetically identical to the single parent organism (somatic cell nuclear transfer). It is a form of manipulation that is a result of manipulation of asexual reproduction (versus sexual reproduction used in artificial selection). Some unintended outcomes have included the early ageing and death of the cloned progeny. Success rates of producing offspring from animal cloning are low and the technique could be still regarded as experimental and not widely accessible. Both techniques produce animals with desirable qualities and reduce genetic diversity in the population, but animal cloning is more successful in that respect as it creates genetically identical clones. (*7 marks*)

Question 34—Genetics: The Code Broken?

(a) (i) Cloning (*1 mark*)

(ii) The cell in the shaded area represents the removal of the nuclear material from a haploid body cell of an animal. This is to provide a clear cell, free of any nuclear material, which can then grow into an exact copy of the parent cell once the diploid nuclear material from the parent animal is inserted. (*2 marks*)

(b) Cattle (*Bos taurus)* have been domesticated and selectively bred since the original wild type were used in south-eastern Turkey and Pakistan about 10 500 years ago. Their selection has resulted in recognised breeds that are used for their meat (beef and veal), milk and dairy products or for their strength as draught animals. They have also been selected for suitability to specific environments.

Initially cattle were bred for particular characteristics, e.g. quantity or quality of milk production in dairy breeds led to the Jersey and Friesian breeds. Size and meat quality were selected in beef cattle. Cattle have also been selected for their temperament and ease of handling. The best cattle within a strain were selected for breeding.

Artificial insemination and embryo transfer have now been used to promote the qualities of the more desirable animals. More recently hybrid vigour (cross-breeding different breeds) and disease resistance has been recognised as important.

Originally different species of cattle were recognised from Europe and India. They arc now considered to belong to the same species as so much interbreeding between different strains has occurred. European breeds were initially introduced to Australia and it has been found that cross-breeding with Indian strains has resulted in greater resistance to heat, drought and ticks.

In Japan particular selection for the marbling of fat in beef has resulted in four strains of cattle producing Wagyu beef. These cattle produce beef that is highly prized not only because of the marbling of the fat but also because it contains more omega-3 and omega-6 fatty acids and a higher ratio of monosaturated fats to saturated fats.

Cattle nowadays are widespread, produce half the quantity of meat consumed and are quite diverse in their breeds. (*5 marks*)

(c) (i) Multiple alleles can be found in humans in the form of blood groups: three alleles, A, B and O, result in four possible blood types. In non-human organisms the 'Tabby Cat' is determined by three alleles: the blotched, Abyssinian and mackerel patterns. (*2 marks*)

(ii)

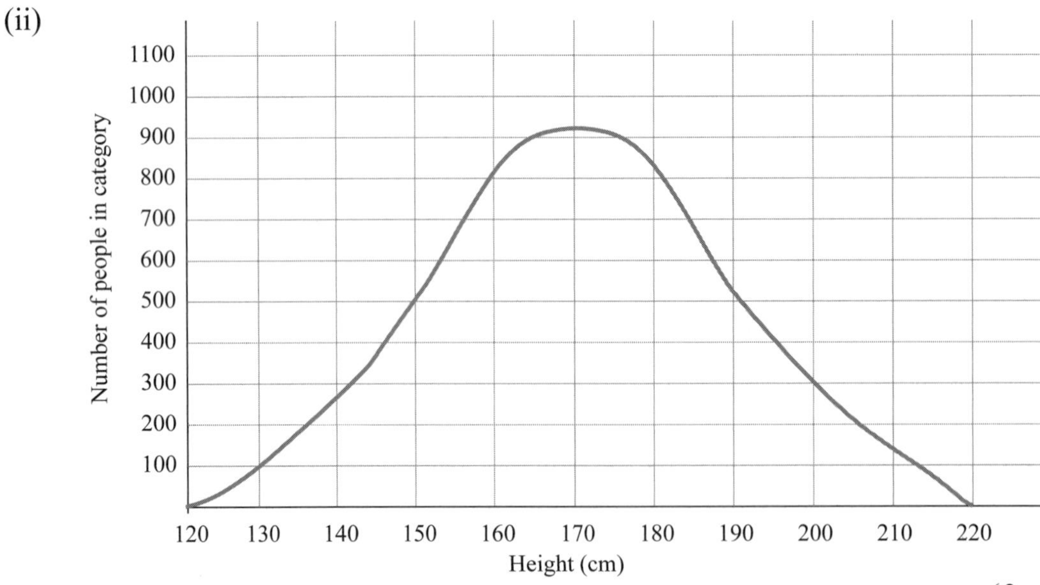

(*2 marks*)

(d) (i) Gene cloning is the process in which a gene of interest is located and copied (cloned) out of DNA extracted from an organism. Large amounts of DNA are needed for genetic engineering. Multiple copies of a piece of DNA can be made either by using polymerase chain reaction (PCR) or by cloning DNA in cells. This is then used for a variety of biotechnological applications including investigations into:

- the function of a gene
- gene characteristics such as size or expression
- how mutations may affect a gene's function. (*2 marks*)

(ii) Mutations result from a change in DNA. Mutation may be the result of a substitution of one nitrogenous base for another, and insertion or deletion of bases, broken and inverted DNA sequences, DNA duplication or breaks and rearrangements of whole chromosomes.

The mutations can be caused by a number of factors including temperature, radiation or exposure to certain chemicals. Many mutations can be detrimental while others are advantageous and are the basis of evolution within a species. The understanding of the causes of mutation has enabled the study of genetics to proceed, with many new applications resulting. Mutagens have been used to create organisms that carry new alleles. Where these characteristics are desirable they can be used extensively to propagate new varieties of crops and vegetables. Geneticists have been very successful in developing technologies that are able to produce new combinations of genes to further our understanding of inheritance and the economic viability of the agricultural industry.

The production of mutant organisms such as cancer-susceptible mice has been used to trial new cancer treatments.

Mutations may result from changes in chromosomes, e.g. Down's syndrome is a condition of trisomy of chromosome 21. Technologies that have resulted from the understanding of this type of mutation include embryo screening and IVF. (*4 marks*)

(e) Genes are discrete segments of the macromolecule deoxyribonucleic acid (DNA) that are found on chromosomes and code for polypeptides. DNA is a double helical molecule consisting of two strands of sugar and phosphate groups that are joined by pairs of nitrogenous bases (adenine A, thymine T, guanine G and cytosine C). Protein synthesis is the process by which the gene is transcribed and translated within the cell to result in the production of a chain of amino acids that comprise a polypeptide. The specific pairing of the DNA bases (A with T, G with C) in sets of three bases results in triplets that each specifically code for a particular amino acid in the polypeptide sequence. Proteins are macromolecules that are made up of one or more polypeptide chains. Each polypeptide chain becomes folded in a specific manner and there are several polypeptide chains in a particular protein (e.g. haemoglobin consists of four polypeptides each with an iron-containing haem group embedded). The actual structure of the protein determines its function e.g. the haemoglobin can carry oxygen molecules in the blood in a loose association so that the oxygen is released where it is in low concentration in the body. Other proteins act as enzymes catalysing specific chemical reactions within the organism. Enzymes bind with their specific substrate at their active sites to promote the reaction with less energy requirements so that the reactions can occur at sufficient speed to support life. If the active site of the enzyme does not have the correct structure it cannot carry out its role. This may occur if there has been a mutation in the DNA of the gene coding for that specific enzyme protein. So there is a high level of dependence of proteins, whether they be structural or functional within an organism on the gene/s that coded for their structure. Genes work in a coordinated manner, especially in embryonic development where cascades of genes need to operate in sequence to produce limbs and other skeletal and neurological structures. The switching off and on of genes may be a response to particular gene products, proteins. Hormones produced from the activation of genes during puberty may in turn promote changes that result in the activation of other genes. Growth hormone will prompt the action of genes to result in the coordinated changes of a spurt of growth. While it is obvious that proteins depend on genes, there is also some evidence of examples of when genes depend on proteins. (*7 marks*)

Question 35—The Human Story

(a) (i) Stratigraphic correlation (*1 mark*)

(ii) Absolute dating utilises radiometric technology to provide an estimation of the age of rock strata based on the radioactive decay of certain specific elements. For example, uranium 238 decays slowly into lead; it has a half-life of 4.5 billion years and can be used to date the oldest rocks. The fossils shown are contained in layers of sedimentary rock. Because these are made of fragments of older rocks, they cannot directly be dated using radiometric dating, but igneous rocks in the vicinity may provide clues. For example, if there was a volcanic intrusion into the lower layer but not the higher layers you would know that it occurred after the lower-layer sedimentary rock formed but before the upper layers were deposited. Another technique for absolute dating is optically stimulated luminescence which dates when a sedimentary rock was last exposed to light. Care has to be taken to ensure the sample is collected without exposure to light. (*2 marks*)

(b) The data depicted compares the amino acids present in proteins of different primates; as such it shows clear links between 'Great Apes' and some deviation between them and the New world monkeys as well as the Prosimians with the number of differences increasing in the latter two. As classification is so arbitrary in nature it is important to use different types of data to increase reliability of estimated evolutionary relationships. Such additional data can come in the form of phylogenetic trees that are based on fossil evidence and radiometric dating data or biochemical analysis of proteins such as haemoglobin on Cytochrome C. The more data that can be gathered and compared, the more accurate the knowledge about the evolutionary pathways will become. (*5 marks*)

(c) (i) Two features of chromosomes used in a karotype analysis include:

- the physical number of chromosomes possessed by individual species
- the location of the centromere within the chromosome pair varies between species. (*2 marks*)

(ii)

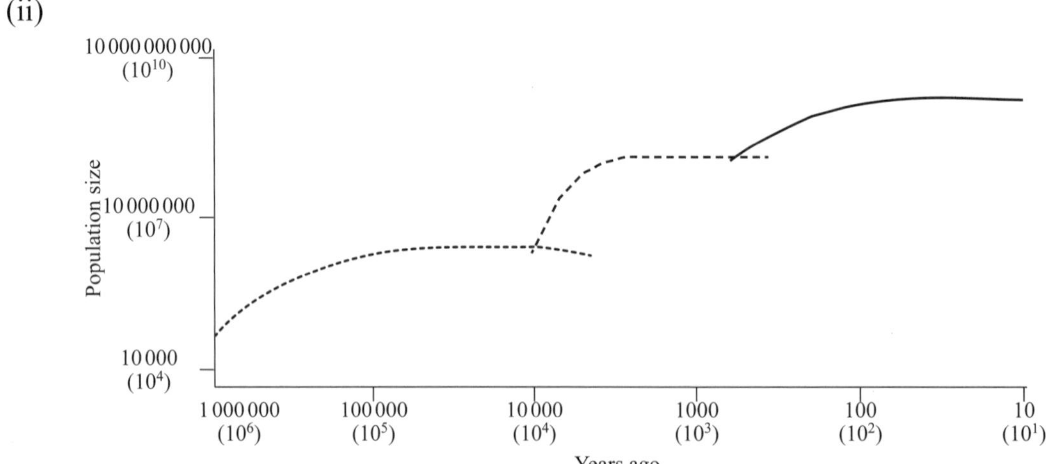

The rate change at 10 000 years was associated with the development of agriculture and the change at 400–500 years ago with urbanisation and industrialisation. (*2 marks*)

(d) (i) Examples of key differences between fossils classified as genus *Australopithecus* and genus *Homo* include: *Australopithecus* has a smaller brain capacity/volume of about 450 mL, substantial browridge, long projecting snout, large teeth and bigger jaw than that of the genus *Homo* which has larger body size, reduced jaw and molar teeth and a larger brain of up to about 650 mL. (*2 marks*)

(ii) Classification of fossils to both genus and species level is hindered by a lack of evidence. Comparison of the DNA sequences or genomes of extant species is a useful tool in determining evolutionary relationships. DNA deteriorates in heat, acidity and alkalinity so the ability to extract and use DNA from fossils of extinct species (remembering that the only extant species of *Homo* is *Homo sapiens*) is limited. Most DNA analysis of human evolution has come from early modern humans and Neanderthals (*Homo neanderthalensis*). The modern human and Neanderthal lineage is believed to have separated between 500 000 and 400 000 years ago and only one in every 600 base pairs differs between modern humans and Neanderthals. All non-African modern humans have between 2 and 5% of genes coming from Neanderthals. Over 50 000 years ago some interbreeding occurred between Neanderthals and modern humans. Interbreeding between species can only occur if they are closely related—probably from the same genus.

DNA extracted from a finger bone found near Siberia demonstrated that the female owner was probably a Denisovan who descended from *Homo erectus* and shared a last common ancestor with humans and Neanderthals between onc and one and a half million years ago.

If DNA could be extracted from the 'unknown' fossil and it showed a very high level of difference between DNA from modern humans, and analysis of mitochondrial DNA showed a really long period since they shared a common ancestor, there would be the suggestion that the fossil belonged to a species in a different genus to the *Homo* genus. When technology improves and DNA can be successfully analysed from fossils of the genus *Australopithecus* then DNA analysis could confirm whether a fossil is likely to belong to that genus. (*4 marks*)

(e) That biological evolution influences cultural development and vice versa can be seen with respect to hominid evolution. As the hominid brain capacity increased, the use of tools developed. This development of tools naturally opened up a new dimension of application with the environment which then, like a chain reaction, spread to further development/application, and so forth. With the development of tools and culture a snowball effect resulted, causing the build-up of culture and interdependence between individuals, tribes and then populations/civilisations. If we apply this to human development we can begin with the rudimentary use of simple tools (e.g. stones, sticks) by the *Australopithecus ramidus* which progressed to digging in the *A. afarensis* and *A. africanus. Homo habilis* was thought to use stone tools to remove meat from bones and there was elaborate use of stone hand axes by *Homo ergaster*. Along the way the cranial capacity and mobility of these hominids was increasing. They were mastering their environment and developing community and culture. *Homo erectus* discovered fire and hunted larger animals with stone axes, *H. heidelbergenesis* prepared core tools and moved into caves, using animal hides for shelter and clothing. *H. neanderthalensis* further developed these skills and added adornments and jewellery to their culture and ceremonial burial. The *H Sapiens* used complex tools such as bone needles to sew, blade tools to hunt and developed art as a means to communicate events.

A major turning point in the evolution of human culture began about 12 000 years ago with the beginning of agriculture. Since then the more reliable food supply and industrialisation has led to major changes in the environmental factors that are selecting humans in the process of evolution. Selection or survival of the fittest is really about the fittest for reproduction. Increased life spans means that modern humans with disability or other impairments are likely to survive and maybe even reproduce to pass on their genes but humans are developing diseases such as type 2 diabetes, suffering from obesity and its associated diseases in higher proportions. The pace of cultural change has been so great that our lifestyles are at a mismatch with our biological evolution. Biological evolution and cultural development depend on one another, influence one another and evolve in parallel. (*7 marks*)

CHAPTER 16

2017 | HIGHER SCHOOL CERTIFICATE EXAMINATION

Biology

General Instructions

- Reading time – 5 minutes
- Working time – 3 hours
- Write using black pen
- Draw diagrams using pencil
- NESA approved calculators may be used

Total marks: 100

Section I – 75 marks

This section has two parts, Part A and Part B

Part A – 20 marks

- Attempt Questions 1–20
- Allow about 35 minutes for this part

Part B – 55 marks

- Attempt Questions 21–31
- Allow about 1 hour and 40 minutes for this part

Section II – 25 marks

- Attempt ONE question from Questions 32–36
- Allow about 45 minutes for this section

Section I
75 marks

Part A – 20 marks
Attempt Questions 1–20
Allow about 35 minutes for this part

Use the multiple-choice answer sheet for Questions 1–20.

---

1 What is the name of the process that enables organisms to maintain a relatively stable internal environment?

A. Osmosis

B. Adaptation

C. Homeostasis

D. Active transport

2 Which of the following body systems is involved in detecting and responding to environmental changes?

A. Circulatory

B. Digestive

C. Excretory

D. Nervous

3 Which scientist contributed to our understanding of the immune response?

A. Robert Koch

B. Louis Pasteur

C. Maurice Wilkins

D. Frank Macfarlane Burnet

4 What is the role of lymphocytes in the body?

A. They fight infection.

B. They initiate blood clotting.

C. They transport oxygen around the body.

D. They transport carbon dioxide around the body.

5 Four vertebrate forelimbs are shown.

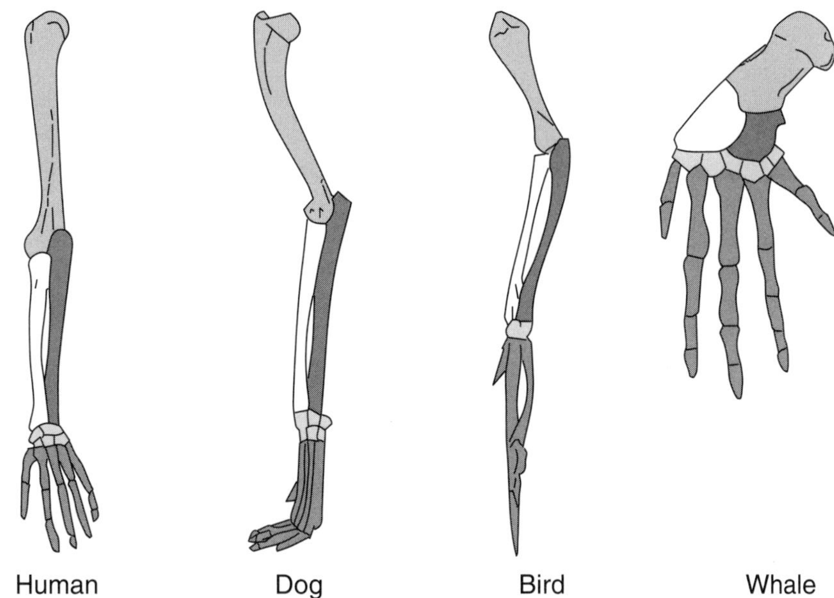

In which area of study do these forelimbs support the theory of evolution?

A. Biogeography

B. Comparative anatomy

C. Comparative embryology

D. Palaeontology

6 What name is given to the process whereby a white blood cell engulfs a microorganism?

A. Infection

B. Inflammation

C. Phagocytosis

D. Vaccination

7 Which waste product does renal dialysis remove?

A. Urea

B. Urine

C. Lipids

D. Vitamins

8 Which row of the table shows the effects of dissolving carbon dioxide in water?

	pH	*Acidity*
A.	Increases	Increases
B.	Decreases	Decreases
C.	Decreases	Increases
D.	Increases	Decreases

9 An experiment was planned to investigate the effect of the enzyme, amylase, on starch. The following combination of test tubes was considered.

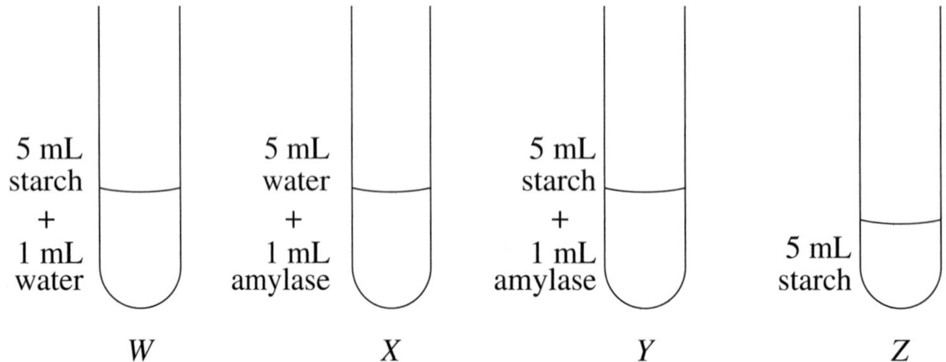

Two drops of iodine will be added to each test tube.

Which combination of test tubes would ensure that the experiment is valid?

	Experiment	*Control*
A.	*Y*	*W*
B.	*Y*	*Z*
C.	*X*	*W*
D.	*X*	*Z*

10 A scaled outline of a cell is shown.

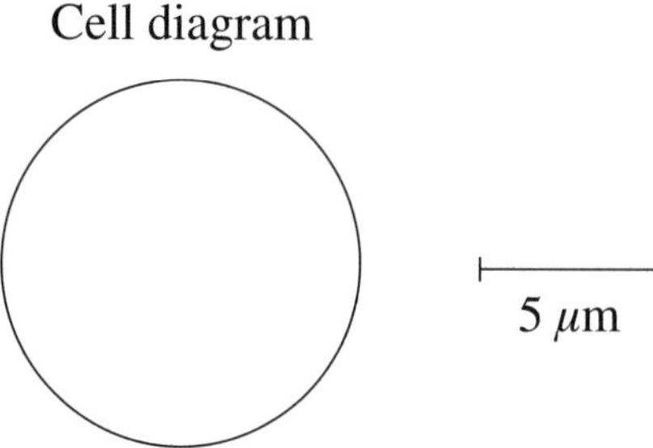

What is the diameter of the cell?

A. 7 μm

B. 10 μm

C. 12 μm

D. 15 μm

11 A student was asked to complete a table showing whether T cells and B cells have particular characteristics.

Which row did the student complete correctly?

	Characteristic	*T cell*	*B cell*
A.	Produces plasma cells	✓	✓
B.	Produces antibodies that are released in body fluids	✓	✗
C.	Cell surface receptor can recognise a specific antigen	✓	✓
D.	Forms clones once stimulated	✗	✓

12 What is the probability of producing a tall pea plant when a heterozygous tall pea plant is crossed with a homozygous short pea plant?

A. 0%

B. 50%

C. 75%

D. 100%

13 A section of DNA has the following nucleotide sequence.

AGG TCT CAG ATC

What is the nucleotide sequence of the newly-made strand following DNA replication?

A. AGG TCT CAG ATC

B. AGG UCU CAG AUC

C. UCC AGA GUC UAG

D. TCC AGA GTC TAG

14 Which statement correctly describes fungi and protozoans?

A. Fungi and protozoans are unicellular.

B. Fungi and protozoans have chloroplasts.

C. Fungi have a cell wall and protozoans do not.

D. Fungi are procaryotic and protozoans are eucaryotic.

15 A plant breeder crosses two plants of the same species. They are both pure-breeds for flower colour. The colour of the flowers of all the offspring is the same, but different to that of the parents.

What type of inheritance does this result show?

A. Sex-linked

B. Recessive

C. Dominance

D. Co-dominance

16 Which row in the table correctly identifies the features of the named transport mechanism?

	Transport mechanism	*Is energy required?*	*Molecules transported*	*Direction of movement (concentration of transported molecules)*
A.	Diffusion	No	Gases	From low to high
B.	Osmosis	No	Water	From high to low
C.	Active transport	Yes	Ions	From high to low
D.	Passive transport	No	Sugars	From low to high

17 Which feature of DNA was discovered as a result of Rosalind Franklin's work?

A. Double helix shape

B. Long stranded molecule

C. Sugar-phosphate backbone

D. Complementary nucleotides

18 A student used a microscope to estimate the size of blood cells. Two types of cells were observed. The student estimated one type to be about 50% larger than the other.

Which of the following could be used to assess the accuracy of the student's findings?

A. The size of other body cells

B. The sizes of blood cells estimated by other students

C. The expected sizes of blood cells quoted in scientific literature

D. The average size of blood cells from three repetitions of the investigation

19 A zebronkey hybrid is the result of crossing a male zebra which has 44 chromosomes with a female donkey which has 62 chromosomes.

How many chromosomes will the zebronkey have?

A. 53

B. 75

C. 84

D. 106

20 A student performed a valid enzyme-substrate experiment. At the end of each 10-minute period, the quantity of the gaseous product formed was collected, removed and measured. The graph shows the results of this experiment.

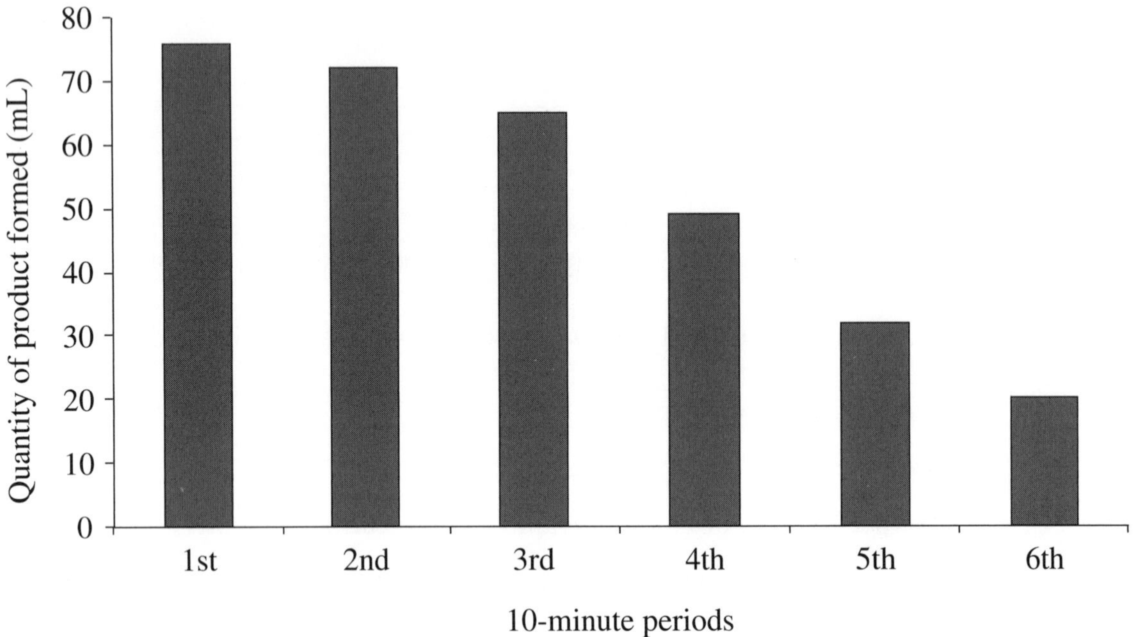

Which of the following statements explains the trend shown in the graph?

A. The rate of enzyme activity is decreasing.

B. The concentration of the product is decreasing.

C. The concentration of the enzyme is decreasing.

D. The concentration of the substrate is decreasing.

2017 HIGHER SCHOOL CERTIFICATE EXAMINATION

Centre Number

Student Number

Biology

Section I Part B Answer Booklet

55 marks
Attempt Questions 21–31
Allow about 1 hour and 40 minutes for this part

Instructions

- Write your Centre Number and Student Number at the top of this page.
- Answer the questions in the spaces provided. These spaces provide guidance for the expected length of response.
- Show all relevant working in questions involving calculations.

Please turn over

Question 21 (2 marks)

A model of enzyme activity is shown. **2**

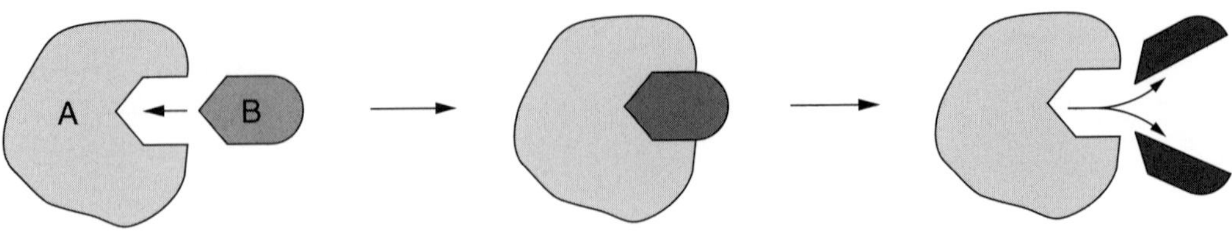

Name the TWO labelled components.

A: ....

B: ....

Question 22 (6 marks)

(a) For each type of disease in the following table, name a specific disease and its cause. **2**

Type of disease	*Name of disease*	*Cause of disease*
Infectious disease		
Non-infectious disease		

(b) Explain how TWO different methods used to treat drinking water reduce the risk of infection. **4**

....

....

....

....

....

....

....

....

Question 23 (5 marks)

Complete the table with reference to the two types of blood vessel shown. **5**

Diagram of vessel	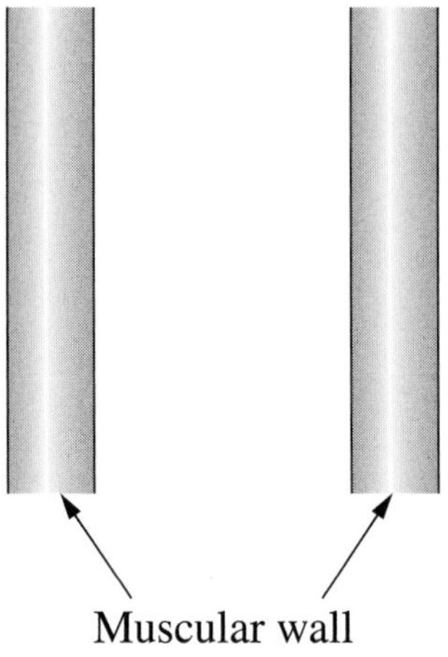 Muscular wall	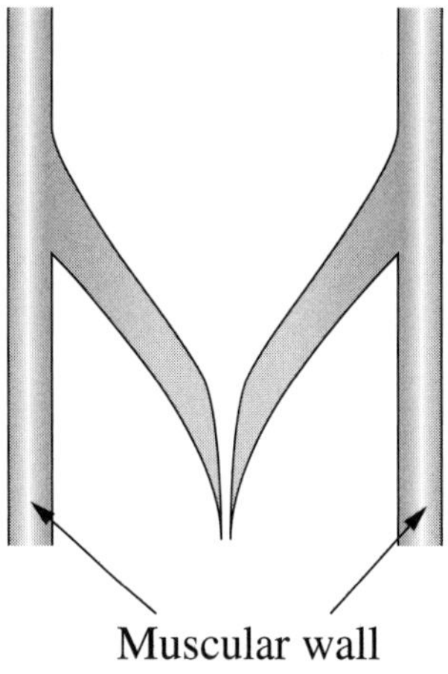 Muscular wall
Name the vessel	....	....
Explain how ONE structural feature of the vessel enables it to carry out its function.	....	....

Question 24 (7 marks)

(a) Three genes are arranged along a homologous pair of chromosomes as shown.

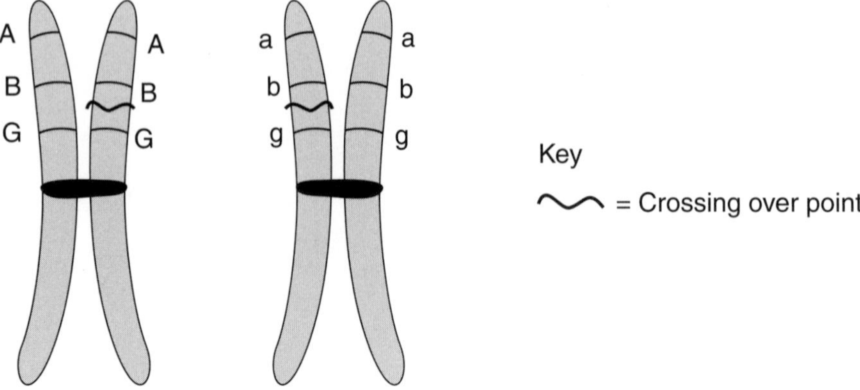

(i) What is the individual's genotype before crossing over occurs? **1**

....

(ii) Label, on the diagram below, the alleles after crossing over has occurred. **1**

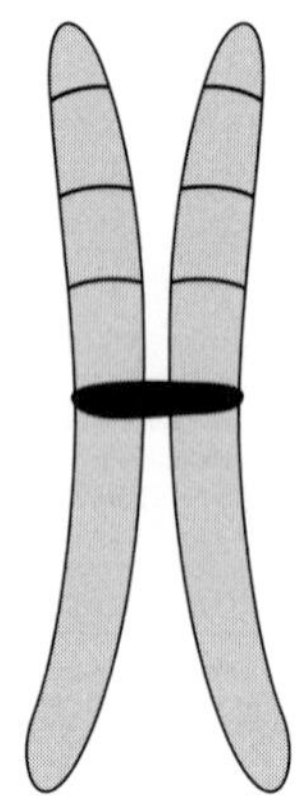

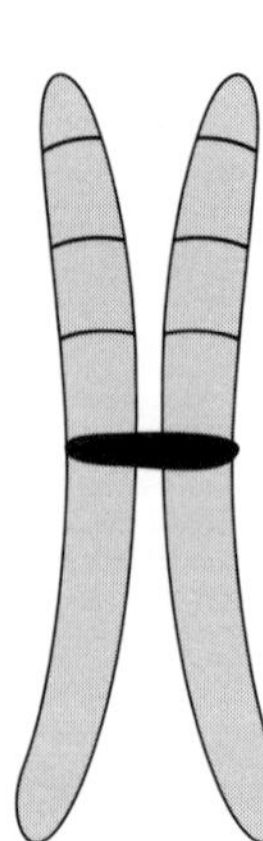

(b) Explain the effect of independent assortment of chromosomes on the genotype of the offspring. **2**

....

....

....

....

Question 24 continues on the following page

Question 24 (continued)

(c) Explain the role of isolation in the process of evolution. **3**

....

....

....

....

....

....

....

....

....

....

Question 25 (4 marks)

Explain the difference in the urine concentration of marine fish and freshwater fish. **4**

....

....

....

....

....

....

....

....

....

....

Question 26 (4 marks)

A controlled experiment was performed to investigate the effect of substrate concentration on the rate of an enzyme-catalysed reaction. Data were collected and are presented in the graph.

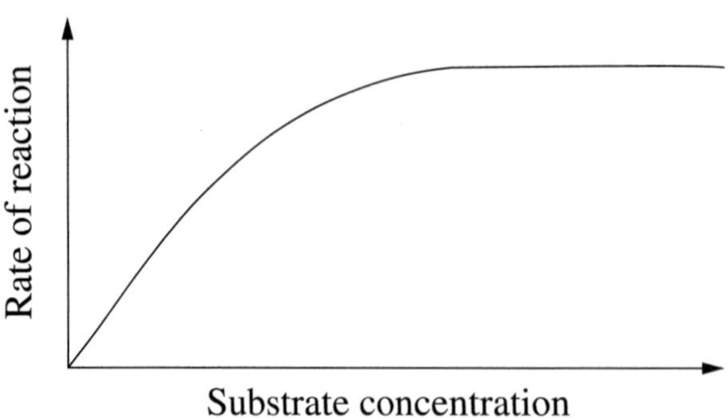

(a) What is the independent variable in this experiment? **1**

....

(b) Explain the trends shown in the graph. **3**

....

....

....

....

....

....

Question 27 (5 marks)

(a) Draw labelled diagrams to distinguish between transverse sections of a xylem vessel and a phloem vessel. **2**

(b) Describe the process that transports sugars through a plant. **3**

....

....

....

....

....

....

Question 28 (3 marks)

A pedigree chart of an inherited characteristic is shown. 3

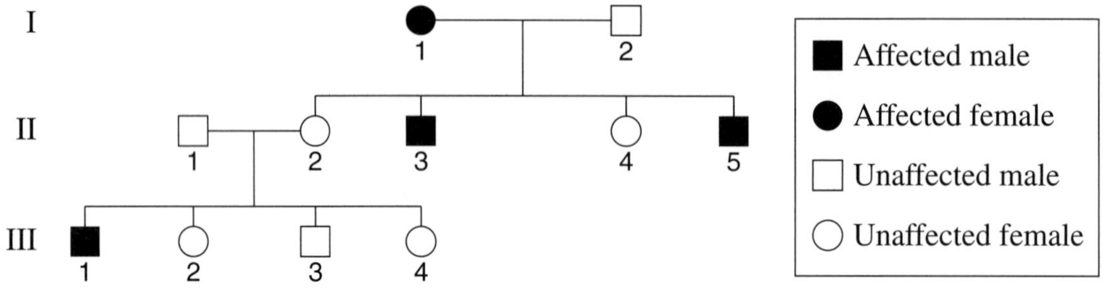

Subsequent genetic analysis showed I-2 does not have the recessive allele.

Explain the inheritance of this characteristic.

....

....

....

....

....

....

....

....

Question 29 (5 marks)

Justify the change in emphasis from treatment to prevention of a named disease. 5

....

....

....

....

....

....

....

....

....

....

Question 30 (6 marks)

(a) What is the role of ONE type of T lymphocyte in the immune response? **2**

....

....

....

....

(b) Explain ONE benefit and ONE limitation of suppressing the immune system in organ transplant patients. **4**

....

....

....

....

....

....

....

....

Question 31 (8 marks)

Assess the importance of the work of Beadle and Tatum to the ability to produce a specific transgenic species. **8**

2017 | HIGHER SCHOOL CERTIFICATE EXAMINATION

Biology

Section II

25 marks
Attempt ONE question from Questions 32–36
Allow about 45 minutes for this section

Answer parts (a)–(e) of one question in the Section II Writing Booklet. Extra writing booklets are available.

Show all relevant working in questions involving calculations.

Question 32 Communication

Question 33 Biotechnology

Question 34 Genetics: The Code Broken?

Question 35 The Human Story

Question 36 Biochemistry (*Not included in this reproduction*)

Question 32 — Communication (25 marks)

Answer parts (a), (b) and (c) of the question on pages 2–4 of the Section II Writing Booklet. Start each part of the question on a new page.

(a) (i) Give an example of a receptor and its stimulus. **2**

(ii) Outline the relationship between the wavelength, frequency and pitch of a sound. **2**

(b) Describe how the lens of the eye changes its shape in order to focus on near and far objects. **3**

(c) (i) Identify the location and function of rhodopsin in the eye. **2**

(ii) Contrast the distribution and function of cone cells and rod cells in the human eye. **4**

Answer parts (d) and (e) of the question on pages 6–8 of the Section II Writing Booklet. Start each part of the question on a new page.

(d) The graph shows the electrochemical changes in the membrane of a neurone when two signals are detected.

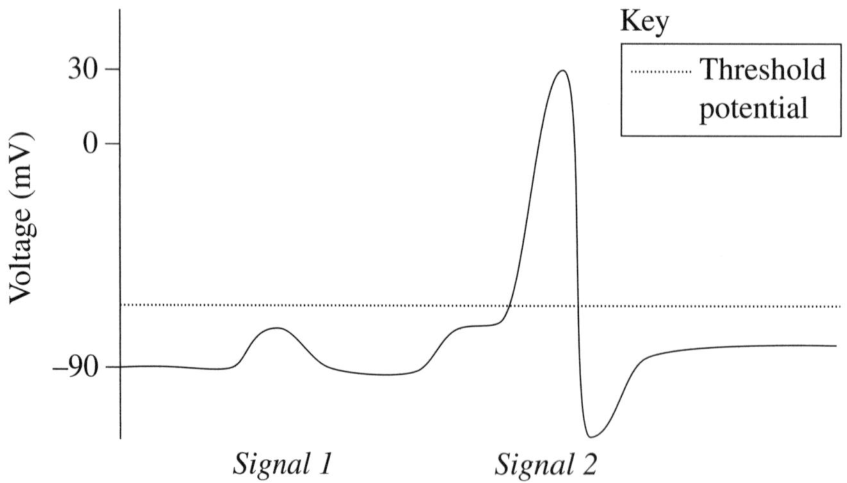

(i) Explain why *Signal 1* does NOT result in the transmission of an impulse. **2**

(ii) Explain the electrochemical changes that would occur in this membrane during *Signal 2*. In your answer, refer to information from the graph. **3**

(e) Assess how our understanding of the path of a soundwave through the ear has led to the development of technologies that assist hearing. **7**

End of Question 32

Question 33 — Biotechnology (25 marks)

Answer parts (a), (b) and (c) of the question on pages 2–4 of the Section II Writing Booklet. Start each part of the question on a new page.

(a) (i) Name TWO biotechnologies used by an early society. **2**

(ii) What are TWO benefits resulting from artificial selection in a specific plant or animal? **2**

(b) Describe the steps involved in the formation of recombinant DNA. **3**

(c) (i) Outline the differences between DNA and RNA. **2**

(ii) Describe the roles of two types of RNA in protein synthesis. **4**

Question 33 continues on the following page

Question 33 (continued)

Answer parts (d) and (e) of the question on pages 6–8 of the Section II Writing Booklet. Start each part of the question on a new page.

(d) The yeast *S. cerevisiae* cannot naturally ferment the sugar xylose. Low value biomass, such as straw and wood fibres, contains up to 20% xylose. *S. cerevisiae* was modified to enable it to produce ethanol from xylose. Information on the two species involved in making the modified *S. cerevisiae* is shown in the table.

Type of organism	*Species*	*Relevant reaction*	*End product*
Bacteria	*Burkholderia cenocepacia*	Utilises xylose in metabolism	Fructose
Yeast	*Saccharomyces cerevisiae*	Utilises fructose in metabolism	Ethanol

(i) Explain why biotechnology was needed to modify *S. cerevisiae*. **2**

Two strains of genetically modified *S. cerevisiae* were produced. The two strains were compared under the same conditions. The results are shown.

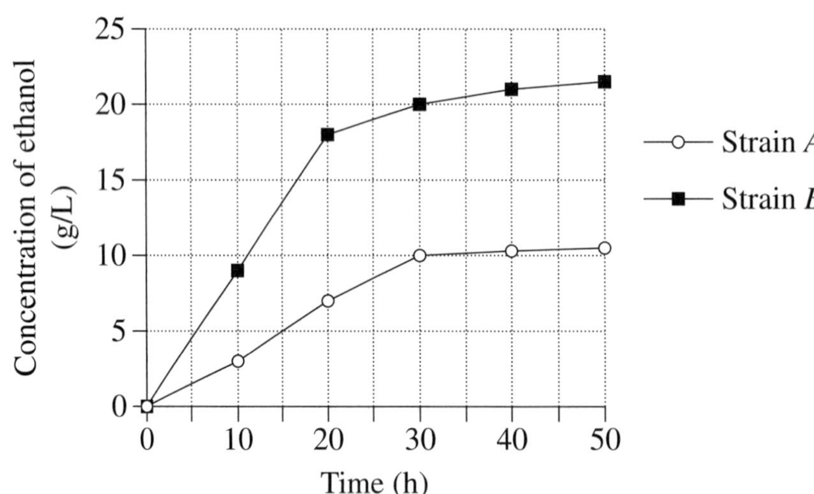

(ii) Justify which of these two strains would be better to use to produce commercial quantities of ethanol using low value biomass. In your answer, refer to information from the graph. **3**

(e) Assess the impact on society of the understanding and application of cell biochemistry. Support your answer with reference to an industrial fermentation process. **7**

End of Question 33

Question 34 — Genetics: The Code Broken? (25 marks)

Answer parts (a), (b) and (c) of the question on pages 2–4 of the Section II Writing Booklet. Start each part of the question on a new page.

(a) The following is a diagram of nucleotides.

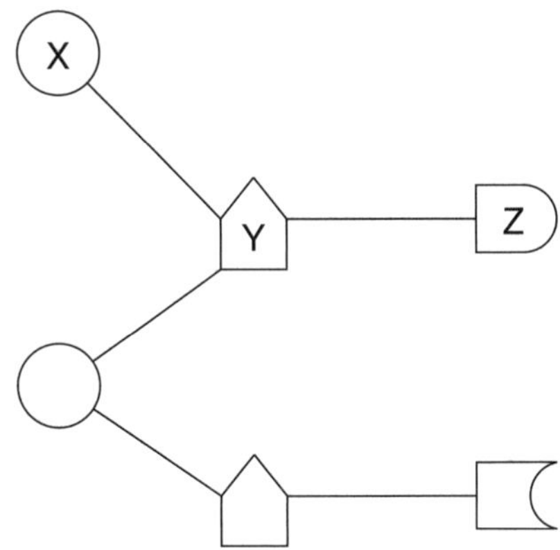

(i) Identify components X and Y. **2**

(ii) Contrast Z in DNA and RNA. **2**

(b) Outline the use of highly variable genes for DNA fingerprinting. **3**

(c) (i) Outline the effect of germ line mutations on species. **2**

(ii) Explain the impact of transposable genetic elements on the genome. **4**

Question 34 continues on the following page

Question 34 (continued)

Answer parts (d) and (e) of the question on pages 6–8 of the Section II Writing Booklet. Start each part of the question on a new page.

(d) The offspring as a result of a dihybrid cross are shown.

RRTT	RRTt	RrTT	RrTt
RRTt	RRtt	RrTt	Rrtt
RrTT	RrTt	rrTT	rrTt
RrTt	Rrtt	rrTt	rrtt

Key
R = red flowers
r = yellow flowers
T = tall plant
t = dwarf plant

(i) Identify the genotype and the phenotype of the parents. **2**

(ii) Explain how the phenotypic ratio would be different depending on whether two genes are carried on the same chromosome or on two different chromosomes. **3**

(e) Analyse the impact of the Human Genome Project on the development of technologies which benefit society. **7**

End of Question 34

Question 35 — The Human Story (25 marks)

Answer parts (a), (b) and (c) of the question on pages 2–4 of the Section II Writing Booklet. Start each part of the question on a new page.

(a) (i) Identify TWO features that classify humans as primates. **2**

(ii) Outline how ONE specific feature is used to distinguish between hominids and other primates. **2**

(b) Describe the process of DNA-DNA hybridisation in determining the evolutionary relationship between one primate and another. **3**

(c) (i) What is the difference between *polymorphism* and *clinal gradation*? **2**

(ii) Explain how polymorphisms have enabled humans to survive in their environment. Use examples in your answer. **4**

Question 35 continues on the following page

Question 35 (continued)

Answer parts (d) and (e) of the question on pages 6–8 of the Section II Writing Booklet. Start each part of the question on a new page.

(d) The graph shows the evolution of the cranial capacity of different hominid species.

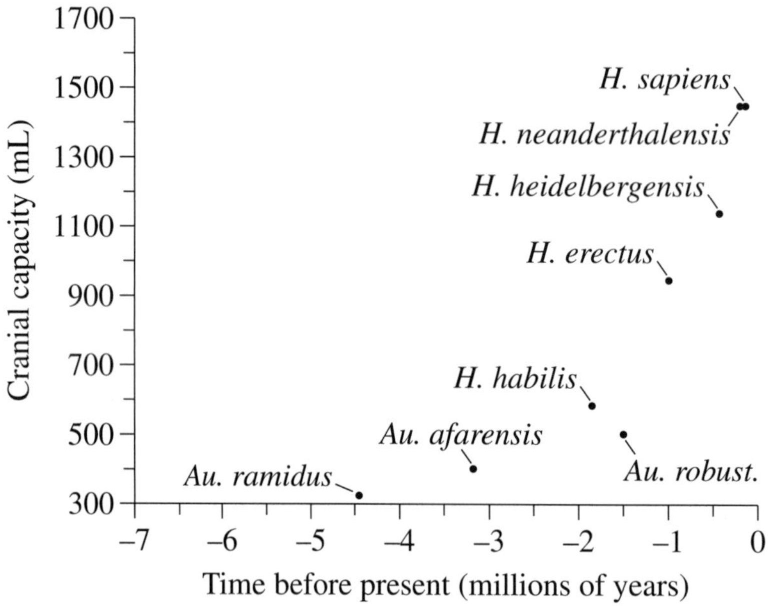

(i) Contrast the change in the cranial capacity of the genus *Australopithecus* with that of the genus *Homo* over time. **2**

(ii) Relate the patterns of migration of genus *Australopithecus* and genus *Homo* to the cranial capacities shown in the graph. **3**

(e) Analyse the impact of modern technologies in the fields of modern medicine and genetic engineering on human evolution. **7**

End of Question 35

End of paper

2017 HSC Examination Paper

Sample Answers

Section I Part A *(Total 20 marks)*

1 C Homeostasis is the process of detecting and responding to changes in the internal environment of organisms.

2 D The nervous system uses sensory nerves to detect changes and responds by activating organs and muscles to respond to environmental changes.

3 D Sir Frank Macfarlane Burnet made a major contribution to understanding how lymphocytes find and destroy foreign antigens.

4 A Lymphocytes include T and B cells and together contribute towards the immune response to fight infection.

5 B Variations of the pentadactyl limb are homologous structures used to compare anatomy as evidence for evolution.

6 C Some white blood cells are called phagocytes because they engulf microorganisms.

7 A Renal dialysis replaces the role of the kidneys to remove toxic urea wastes from the blood.

8 C Dissolving carbon dioxide forms the weak carbonic acid that lowers pH and increases acidity.

9 A The only difference between experiment Y and control W is the presence of the enzyme amylase, ensuring experimental validity.

10 B The diameter of the cell diagram is twice that of the scale that measures 5 microns.

11 C The surface of foreign cells contain antigens to which both T and B cell receptors respond.

12 B The heterozygous pea will produce half its gametes with the T allele and the other half with t while the homozygous short will only produce t which will be dominated by a T allele.

13 D In DNA A matches T and C matches G as bases on complementary strands.

14 C Cells of fungi have cell walls and protozoans are animal-like unicellular organisms without a cell wall.

15 D A colour different from either parent can result if both alleles are fully expressed.

16 B Osmosis is the process by which water moves passively from a weak solution to a strong solution; that is, from where there is a lot of water to where there is less.

17 A Rosalind Franklin's X-ray images of DNA crystals showed the shape of DNA to be a helix.

18 C Figures quoted by reputable sources could be used to check the student's finding that one cell type was 50% larger than the other type. (Repeating the experiment if there was inaccuracy in the methods or equipment would not ensure the accuracy of findings.)

19 A The gametes would carry the haploid number of chromosomes that would then combine together in the zygote of the zebronkey.

20 D The rate of reaction won't change unless the amount of substrate is below the threshold. Given the product is being removed each time, each rate of reaction will be specific for that amount of substrate; therefore if there is less substrate there will be less product.

Section I Part B

21 A: enzyme
B: substrate *(2 marks)*

22 (a)

Type of disease	Name of disease	Cause of disease
Infectious disease	Measles	Airborne virus
Non-infectious disease	Atherosclerosis	Narrowing and hardening of the artery walls caused by smoking, high-fat diets, lack of exercise and stress

(2 marks)

(b) Water is physically screened and filtered. This reduces the risk of infection by removing large and small particles which often carry microorganisms and even algae. The filtration process may be enhanced by the addition of alum to make suspended particles flocculate together for easier removal. Filtration can remove particles and microorganisms smaller than the pore size and greatly reduce numbers of pathogenic organisms such as protozoa (e.g. *Giardia*).

Disinfection is another water treatment that uses chemicals to destroy microorganisms. Oxidising agents such as chlorine, chloroamine and/or ozone are added to kill bacteria. Alternatively irradiation using ultraviolet light is known to kill microorganisms and disinfect the water supply by preventing cell replication through disruption of genetic material (DNA). *(4 marks)*

23

Diagram of vessel	(Thick muscular walls, valves absent)	(Thinner muscular walls, valves present)
Name of vessel	Artery	Vein
Explain how one structural feature of the vessel enables it to carry out its function.	Function: high-pressure blood is carried away from the heart. Structural feature: thick, strong and elastic muscular walls maintain blood pressure within a healthy range so that blood is retained and transported efficiently.	Function: low-pressure blood is returned to the heart. Structural feature: large lumen reduces resistance to low-pressure blood flow. Valves prevent blood flowing backwards.

(5 marks)

24 (a) (i) Genotype: Aa, Bb, Gg *(1 mark)*

(ii)

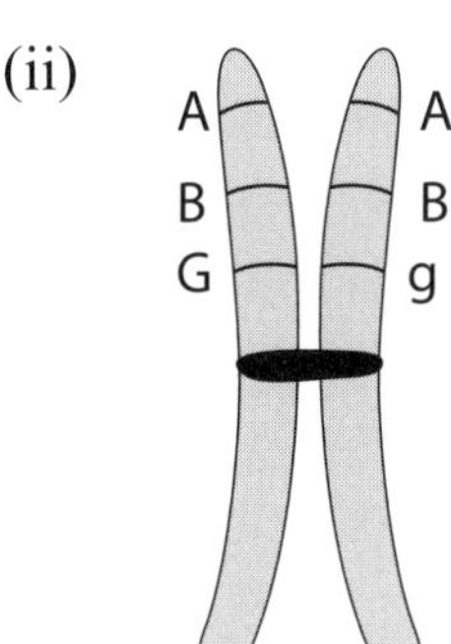

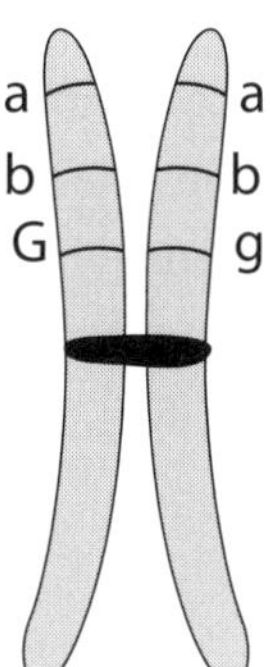

The middle two sequences are the recombined alleles resulting from crossing over. *(1 mark)*

(b) An independent assortment of chromosomes means that homologous pairs of chromosomes randomly segregate to opposite poles of the spindle during meiosis. The result is a large number of different chromosome combinations possible in each gamete. Because independent assortment happens in meiosis in both parents the genotype of the offspring could be highly varied. *(2 marks)*

(c) Reproductive isolation is a key factor in species evolution. While a population continues to interbreed it shares a common gene pool so the genetic differences within groups of the population remain low. If a species becomes geographically and reproductively isolated such as the giant land tortoises in the Galapagos Islands, natural selection may result in changes in gene frequency within different isolated populations. If the isolation continues separate species may evolve. *(3 marks)*

25 Marine fish excrete small quantities of concentrated urine compared to freshwater fish that excrete large quantities of dilute urine. Osmosis occurs in both environments but osmosis tends to make water move into freshwater fish but out of marine fish through their gills and skin. The gills as well as the kidneys play a role in osmoregulation as they secrete salt in marine fish and absorb salt in freshwater fish.

Marine fish, to counter dehydration, excrete concentrated urine. Marine fish kidneys have few, if any, glomeruli so filtration is minimal and may even lack a distal tubule but secretion of salts occurs in the proximal tubule.

Osmoregulation for freshwater fish poses the opposite problem. Water moves into the bodies that contain a relatively higher concentration than freshwater. To keep the balance they excrete large quantities of dilute urine and actively reabsorb salt. Freshwater fish kidneys have many large glomeruli and ions are actively reabsorbed in the tubules, resulting in the production of large quantities of dilute urine. *(4 marks)*

26 (a) The independent variable is substrate concentration. *(1 mark)*

(b) The trend in the graph is that initially an increase in substrate concentration increases the rate of reaction but after a certain concentration the rate of reaction remains unchanged. Enzymes catalyse biochemical reactions by lowering the activation energy because their active site combines with the substrate into an enzyme-substrate complex. The enzyme itself is used repeatedly with additional substrate molecules but each time it must bind with the substrate. Up to a certain increase in concentration of substrate molecules, more and more enzymes can bind and promote the reaction. Beyond that level other limiting factors come into play. It is possible at this point that the enzyme molecules are all combined with substrate molecules so no more are available to increase the rate of reaction. There may be the increase in concentration of products which might inhibit enzyme activity. *(3 marks)*

27 (a)

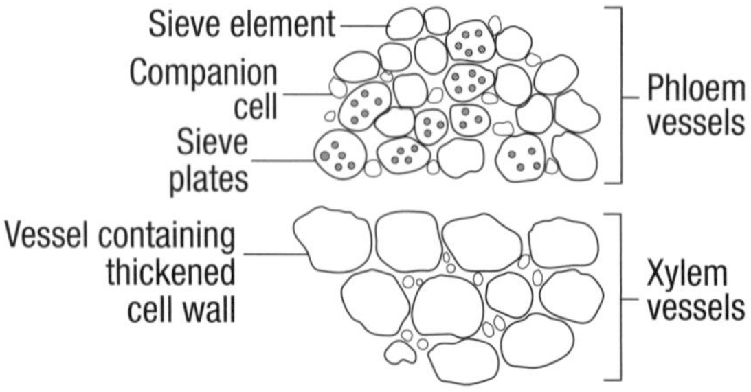

(2 marks)

(b) Sugars travel through the phloem from the leaves, where it is produced, to all parts that need it. They are transported by active transport, i.e. it depends on the expenditure of energy by cells, and the initial loading of the phloem from the leaves is against the concentration gradient.

The hypothesis for sugar movement is the source-to-sink model. Once the sugars are 'loaded' into the phloem near the leaves the osmotic pressure draws water in from surrounding tissues, including the xylem. The increase in pressure pushes the sugar solution to the 'sink' which is the tissues using or storing sugar. *(3 marks)*

28 This pattern of inheritance could be explained as sex-linked, i.e. carried on the X chromosome. For the female grandmother (1 in I) to be affected, she must be homozygous for the condition, i.e. have an allele for the condition on each of her X chromosomes. Supporting evidence is that both (100%) of her sons (3 and 5 in II) are also affected and one daughter (2 in II) is a carrier that is not affected but has one son (1 in III) that is affected. *(3 marks)*

29 There has been an increased emphasis on the prevention of lung cancer. The approach to this disease has benefited from epidemiology that has established a strong link between smoking as a highly correlated risk factor for contracting lung cancer, as well as some other cancers, and increased risks of atherosclerosis and strokes. Changes have occurred in legislation to restrict places where people can smoke and protect workers from second-hand cigarette smoke. Cigarette advertising has been banned, and anti-smoking public awareness and education campaigns have been conducted. Labelling on cigarette packets has also changed.

Justification for the change in emphasis is that there are far better health outcomes from prevention than treatment, both for society and individuals, as prevention is relatively easy, being about behavioural change. The change of emphasis does not imply that patients are no longer treated for lung cancer, surgically or with radio- or chemotherapy, and research is still ongoing to better target chemotherapy drugs, in particular, to cancer cells. *(5 marks)*

30 (a) Killer T cells directly secrete substances to destroy antigens (and also the pathogen with the antigen on its surface), enhance the activity of macrophages engulfing pathogens and inhibit replication of viruses. *(2 marks)*

(b) One benefit of suppressing the immune system in organ transplant patients is that they are less likely to reject the donated organ so that it remains functioning in their body. Suppressing the immune system means that the foreign organs are not recognised as foreign cells and so tissues maintain functioning.

One limitation of suppressing the immune system is that the patient is susceptible to infection as the natural responses to infection through the immune system do not occur. Quite often patients have to avoid contact with likely infectious disease carriers (such as children) or take long-term courses of antibiotics. *(4 marks)*

31 Beadle and Tatum demonstrated how genes worked. They conducted research using bread mould that had had various mutations induced by irradiation with X-rays. For each mutant they were able to identify an enzyme that was inactive by applying various food supplements. Their systematic investigations helped them demonstrate that each gene that became inactive meant that a protein (enzyme) was also inactive. They concluded that one gene made one protein. We have since realised that as many proteins are made of several polypeptide chains combined into the one macromolecular structure there is a link with one gene coding for one polypeptide.

While Beadle and Tatum showed how genes work on a molecular level, there was a lot more research before there was an understanding of the structure and role of DNA and the mechanism by which polypeptides were produced in cells. Franklin, Watkins, Watson and Crick determined the structure of DNA and how the structure supported its replication. The central role of specific enzymes to unzip the DNA and a mechanism for nucleotides to be slotted along both strands to result in two faithful copies needed to be determined. The processes of transcription and translation needed investigation so it was understood how the DNA coded for mRNA and how tRNA carried amino acids into place on the mRNA on a ribosome to accurately make a polypeptide chain.

Creating a specific transgenic species required building on the earlier knowledge to develop methods of identifying specific genes and the enzymes that could 'cut' them out of a DNA sequence, then make multiple copies through the polymerase chain reaction and insert them into the DNA of another species. One strategy was using bacterial plasmids to clone multiple copies of genes. Another problem to be overcome was how to insert the gene into cells of the different species.

Beadle and Tatum made a breakthrough and changed thinking but many more breakthroughs had to be made before genes from one species could be isolated, cut and removed, copied and then re-inserted into a different species. *(8 marks)*

Section II—Options

Question 32—Communication

(a) (i) Rods and cones are photoreceptors in the retina of the eye that respond to the stimuli of light intensity and wavelength. *(2 marks)*

(ii) The longer the wavelength of a sound wave, the lower the frequency and the lower the pitch of a sound. Conversely the shorter the wavelength, the higher the frequency and the higher the pitch of a note. *(2 marks)*

(b) To focus on a distant object the lens is relatively long and narrow, bending the light only slightly so that it focuses on the retina at the back of the eye.

To focus on a near object the light has to bend more so the lens becomes thicker and shorter with a greater curve, in a process called accommodation. *(3 marks)*

(c) (i) Rhodopsin is the photosensitive pigment in the rods of the retina. The rhodopsin is highly sensitive to light in wavelengths between blue and red which stimulate it to break apart, activating a nerve so that the brain detects the light. It works best in low-light levels because in light of high intensity the breakdown is too rapid for its reformation. *(2 marks)*

(ii) The rod cells are more densely concentrated at the edges of the retina and are absent from the fovea and macula at the back of the eyeball. Rods help with night vision, the detection of shape and movement, and discriminate between shades of light and dark.

Cones are highly concentrated in the central fovea and macula. There are three types of cones, each type sensitive to different wavelengths of light. They are specialised for colour vision and visual acuity. *(4 marks)*

(d) (i) Signal 1 does not result in the transmission of an impulse because it did not reach the threshold potential or minimum required. The membrane of the neurone remains impermeable so there is no passage of ions and there is no transmission of a nerve impulse. *(2 marks)*

(ii) In Signal 2 the threshold is achieved (the potential goes higher than the dotted horizontal line in the graph). While resting, the inside of the neurone is more negative than the outside. The action potential is initiated (threshold line reached): the membrane of the neurone becomes permeable, initiating a rapid flow of ions across the membrane, and sodium ions (Na^+), in particular, enter and the membrane becomes depolarised. At the peak the sodium channels close and potassium channels open, with the potassium ions (K^+) leaving the cell. The cell becomes 'hyperpolarised' by the low point on the graph during which the neurone cannot respond to any stimulus and the potassium channels close. The sodium/potassium transporter restores the resting potential. *(3 marks)*

(e) Soundwaves enter through the external ear (ear canal and tympanic membrane) and travel to the middle ear where the vibrations are transmitted through and amplified by the ear ossicles. Sound vibrations are then transferred to the oval window at the junction between the middle and inner ear. From here they travel through fluid in the spiral cochlea to the round window. In the process hairs in the Organ of Corti vibrate and trigger impulses in the auditory nerve.

Hearing aids are relatively simple technologies that use battery power to amplify the sound vibrations that enter through the ear canal. Soundwaves travel normally through the ear to eventually promote an impulse in the auditory nerve. The difficulty is the lack of sensitivity so that some sounds such as background noise are also amplified and some sounds could be painfully loud. They do not improve hearing in the profoundly or totally deaf. This technology makes minimal use of understanding of the path that soundwaves take through the ear as it basically only amplifies soundwaves.

The cochlear implant is designed for any failure of the transmission of the soundwaves through the middle and inner ear or any other failure to stimulate the auditory nerve. In effect it is a replacement ear as it contains a receiver that digitally passes vibrations to a processor that has the ability to discriminate between types of sounds; for example, highlighting important words. Signals are sent to a receiver that directly contacts the auditory nerve to initiate impulses.

An understanding of the passage of soundwaves leads to a better understanding of the causes of deafness, such as the failure of the hair cells in the Organ of Corti. The main aspects of the technologies associated with the cochlear implant are electronics such as microphones and transmitters, and an understanding of how to promote an electrical impulse in the auditory nerve that mimics the impulse that would be generated naturally in a functional ear. *(7 marks)*

Question 33—Biotechnology

(a) (i) Two biotechnologies used by early society were the use of yeast to bring about fermentation to make wine and bread, and the use of bacteria to make cheese and yoghurt. *(2 marks)*

(ii) Artificial selection of varieties of wheat have the benefits of improving the protein content of flour and increasing productivity by resistance to disease and higher yielding grain varieties. This has led to economic and health benefits. *(2 marks)*

(b) Recombinant DNA requires:

1. the DNA from the required gene to be identified and enzymes selected for it to be cut out of its chromosome and joined into a plasmid
2. the cutting and joining to be monitored to ensure the procedure was successful
3. the plasmid containing the inserted DNA to be introduced into a bacterium (transformation). The product of the inserted gene such as human insulin can be extracted from the bacterial colony containing the recombinant DNA. *(3 marks)*

(c) (i) DNA is a double helical molecule made up of chains of nucleotide monomers running in opposite directions. The backbone of the chain is alternating deoxyribose sugars joined to phosphate groups. The two chains are held together by hydrogen bonding between nitrogenous bases that are attached to the sugar molecules. There are four possible bases: adenine, thymine, guanine and cytosine. Adenine from one chain always bonds with thymine on the other chain and cytosine always binds with guanine (A-T and G-C).

RNA is also made of nucleotides but they contain ribose sugars (not deoxyribose) in the backbone along with the phosphate groups. Nitrogenous bases are also attached to the sugar but they include adenine, uracil (not thymine), cytosine and guanine. RNA exists in different forms, e.g. as messenger RNA (mRNA) and transfer RNA (tRNA). *(2 marks)*

(ii) mRNA forms from the template provided when the double-stranded DNA 'unzips' exposing one chain so that the RNA nucleotides slot in to matching bases of the DNA. As mRNA contains uracil instead of the thymine in DNA the complementary bases are uracil pairs with a DNA adenine, adenine pairs with a DNA thymine, guanine pairs with a DNA cytosine and cytosine pairs with a DNA guanine. The mRNA forms from a start sequence in DNA and finishes at a stop sequence. In between the start and stop sequences, the mRNA is modified to carry an accurate record of the coding section of the DNA in sets of triplets of bases called codons. This process is called transcription.

mRNA has an important role as it is able to move out of the nucleus into the cytoplasm and embed on a ribosome.

Translation occurs on the ribosome. tRNA is a short sequence of RNA that binds to a specific amino acid and contains a set of three bases called an anticodon that match the mRNA codon. The role of tRNA is to ensure the correct amino acid is added to the polypeptide chain that forms as a result of transcription and translation. *(4 marks)*

(d) (i) Without biotechnology, i.e. recombinant DNA, the process of obtaining ethanol from sugars in low-value biomass would be a lot more difficult and take many more steps. Yeast (*S.cerevisiae*) with the bacterial gene inserted can carry out both steps from xylose to fructose and then on to ethanol as part of its metabolism. (*2 marks*)

(ii) Strain B is the preferred strain as it produces a higher concentration of ethanol and works more quickly (the initial slope of the graph is higher for strain B) than strain A. (In the first 30 hours strain B produces 20g/L whereas strain A produces 10g/L.) Both strains become less active after 30 hours, so the concentration of the product may act to repress activity. Again strain B is preferred as it appears to tolerate a higher concentration of ethanol than strain A. *(3 marks)*

(e) Industrial fermentation is the intentional use of microorganisms to carry out fermentation reactions to produce useful substances for humans. Traditionally, industrial fermentation was used to produce specific foods and beverages, chemicals such as citric acid, glycerol, lactic acid and ethanol. Understanding the difference between cellular respiration in the presence of oxygen and without oxygen has been the key to achieving a wide variety of products. The use of fermentation by the mould *Aspergillus niger* was developed in 1916. It meant an alternative source of citric acid (rather than depending on citrus extracts) for food flavouring and nutritional benefits of vitamin C retention in processed fruit and vegetables. At the time this technology helped to break the Italian monopoly on citric acid production and meant a more reliable and reasonably priced supply.

Modern industrial fermentation has many twists and turns. It includes using waste in landfill or from abattoirs to extract methane gas and its use in a bioreactor for electricity generation. The methane is the result of bacteria anaerobically digesting carbon sources. The benefits to society are potentially enormous as these type of developments expand as they address problems of greenhouse gases, of which methane is particularly potent, waste disposal, odour and pollution problems, and energy supply. Generation of methane from landfill involves exclusion of air (oxygen) so that maximum fermentation occurs and 'sucking out' the methane without drawing in air. One solution is to pump in carbon dioxide. Percolated rainwater is recycled as it is rich in microbes to promote more fermentation.

Better management of fermentation in sewage treatment plants has resulted in reduced odour and bulk of sewage, production of fertilisers from sludge and, in some gases, the production of biogas.

The pharmaceutical industry is also turning to biotechnology and fermentation to produce therapeutic proteins such as hormones, vitamins, vaccines, antibiotics and enzymes used in the baking and brewing industries, and animal-feed supplements. Many of the products are used for research or diagnosis. Monoclonal antibodies are now being produced in bioreactors as well as in host organisms.

The great variety of products and a wide range of benefits for society and the environment have resulted from the continued use of industrial fermentation. *(7 marks)*

Question 34—Genetics: The Code Broken?

(a) (i) X is a phosphate group. Y is a sugar (e.g. deoxyribose or ribose). *(2 marks)*

(ii) Z is the nitrogenous base of a nucleotide. In DNA they include adenine, thymine, guanine and cytosine whereas in RNA the bases are adenine, uracil, guanine and cytosine. *(2 marks)*

(b) DNA fingerprinting is a technique that involves cutting a sample of DNA into fragments and comparing it with other samples for the purpose of genetically identifying an individual; hence its use in forensics and paternity testing.

Since most of the human DNA is identical for all individuals, it is necessary to use highly variable genes for DNA fingerprinting. These are the non-coding regions of the DNA that do not code for proteins and that are unique to each individual.

The process involves using restriction enzymes to cut the DNA into lengths at specific points, running it through an electrophoresis gel, extracting it onto a membrane and using a radioactive probe (that is, focusing on the variable sections of genes) to match. Excess probe is removed and X-ray sensitive film used to detect the pattern of bands. The result is often expressed as a probability as not all of the DNA is compared. *(3 marks)*

(c) (i) Germ line mutations are changes to the DNA in gametes. These mutations will be passed on to the zygote if fertilisation occurs and so to all cells in the offspring. Germ line mutations have the potential to change allele frequency in a species if enough individuals and their progeny carry the mutation. *(2 marks)*

(ii) Transposable genetic elements are jumping genes or mobile pieces of DNA. They may join pieces of DNA that were initially unrelated. They are well documented in plasmids of bacteria but are now believed to be widespread. They create new nucleotide sequences, cause chromosomal rearrangements and mutations, and can change the way genes are expressed. It is believed that 45% of the human genome is derived from transposable genetic elements. Transposable genetic elements are believed to be related to gene instability and disposition towards cancer. They may play a role in genome evolution and there is still much to be learnt about them. *(4 marks)*

(d) (i) The genotype of each parent is RrTt and their phenotype is red flowers and tall. *(2 marks)*

(ii) If the two genes are on different chromosomes they would assort independently so that the phenotypic ratio would be expected to be: 9 red flowered tall: 3 yellow flowered tall: 3 red flowered dwarf: 1 yellow flowered dwarf.

If the alleles for flower colour and height are on the same chromosome and no crossing over occurred there would be fewer combinations of phenotypes, depending on the actual allele combinations on each homologous chromosome of the parents. If R is linked to T, then the phenotypic ratio to be expected is 3 red tall: 1 yellow dwarf. However, if R is linked to t, then the phenotypic ratio would be 1 red dwarf: 2 red tall: 1 yellow dwarf. *(3 marks)*

(e) The Human Genome Project was completed in 2003 and located the 20 000 (approximate) human genes and sequenced the DNA bases of the total human genome. The Human Genome Project also mapped the genomes of five other species with potential benefits for agriculture and research. Other benefits of the Human Genome Project have derived from the processes involved. This included setting a precedent to consider ethical issues of research, the high level of cooperation—international, public and private—and the production of new technologies, including supercomputers that were needed for data processing.

It has had major ramifications for the benefit of society and individuals, particularly through a better understanding of genetic diseases, health and human evolution. It suggests a change in emphasis from treatment towards genetic testing, early detection, identification and management of risk factors. Much publicity has been given to some women who carry genes for breast cancer and who have had proactive surgery to remove breast tissue and so minimise their risk of contracting the disease.

Forensic science uses DNA to match the DNA profile of crime-scene samples with suspects. DNA is also used to help solve paternity disputes. The Human Genome Project helps better understand the non-coding portions of the human genome that are frequently used in such analyses. Helping exonerate individuals falsely accused of a crime benefits individuals and knowing of the potential use of DNA to solve crime may have a deterrent value for society.

It provides the potential for the development of new gene therapies. For example, Severe Combined Immunodeficiency (SCID) is a genetic disease in which the sufferers previously had to live in a sterile environment because their immune system did not fight infection. Stem cells are harvested from bone marrow and a corrected gene is inserted and injected. Some patients have reportedly been able to produce functional white blood cells and now live a normal life.

One impact of the Human Genome Project is that there is an increased awareness of the benefits and also the limitations of information about the human genome as predictors of the likelihood of contracting a range of diseases such as cancer. The study of gene regulation (epigenetics) has come into greater prominence since the full mapping of the human genome. *(7 marks)*

Question 35—The Human Story

(a) (i) Primates are classified because they have an opposable thumb, binocular vision and nails instead of claws. *(2 marks)*

(ii) Hominids (the great apes) are different from other primates because they have an upright/semi-upright stance (forearms shorter than hind limbs) and lack a tail. *(2 marks)*

(b) DNA-DNA hybridisation is a technique to test how similar DNA is from different organisms. First the DNA is 'unzipped' and the single strands broken into segments using restriction enzymes. Single-stranded DNA from the two species is mixed and allowed to hybridise—depending on the similarity between the base sequence—either loosely or tightly. The degree is measured by the temperature required when the mixture is heated to break the bonds between the bases. The higher the temperature required, the closer the similarity in DNA and presumably the closer the evolutionary ties. These experiments have demonstrated the close relationship between humans and chimpanzees (94.7% similarity). *(3 marks)*

(c) (i) Polymorphism recognises the existence of distinct, genetically determined forms, such as body stature, skin colour, blood types, epicanthic eye folds and sickle-cell haemoglobin (anaemia) in humans. In contrast, a clinal gradation is the gradual change in gene frequency in a population that is often associated with geographic variations or the intermingling or migration of populations with different traits that start interbreeding. Skin colour is an example of a clinal gradation in humans with darker skin colour closely associated with proximity to the Equator and an increase in the frequency of lighter skin further towards the poles. *(2 marks)*

(ii) Darker skin is considered an adaptation that protects from skin cancer and paler skin helps increase the ability to manufacture vitamin D from sunlight. Fair-skinned humans occurred mainly away from the Equator where sun protection was not a problem but the ability to make vitamin D from sunlight was a problem. Darker skin occurred close to the Equator where the greater protection from skin cancer was an asset and the need to make vitamin D was not such a problem.

Forms of haemoglobin vary in humans (the sickle-cell haemoglobin polymorphism) but this is often linked to an incidence of malaria as the HbS form of haemoglobin when heterozygous gives resistance to malaria (the homozygous form often being fatal). The benefit of this polymorphism is evident because of the higher gene frequency of HbS in regions where malaria is prevalent. *(4 marks)*

(d) (i) The cranial capacity of *Australopithecus* increased from just over 300 mL to 500 mL (i.e. by 200 mL) in about 3 million years while the cranial capacity of the genus *Homo* increased from nearly 600 mL to 1450 mL (by 850 mL) in about 1.8 million years. *(2 marks)*

(ii) Fossil evidence suggests that *Australopithecus* did not migrate beyond Africa but that the members of the genus *Homo* migrated into Europe then through Asia to the Americas and Australia. The initial movement out of Africa occurred almost 2 million years ago when *Homo erectus* spread north and eventually as far as Europe, China and Indonesia. The much expanded cranial capacity of *Homo erectus* correlates with an expanded pattern of migration but that in itself is not direct evidence of a causal relationship. The 'Out of Africa' theory of a second migration suggests that modern human migration occurred relatively recently (in the last 100 000 years) with *Homo sapiens* eventually outcompeting other *Homo* species. *(3 marks)*

(e) Human evolution is based on changes in gene frequency in the population leading to changes in phenotypes. For modern technologies to impact on human evolution they need to promote change that is inherited, i.e. transferred through the germ line, and occur while the person has reproductive potential so that the change may impact gene frequencies in offspring. Increasing or decreasing population groups that reproduce have the potential to change allele frequencies and the gene pool and so alter human evolution. Helping sufferers of cystic fibrosis survive longer and to have children could increase the frequency of this condition. Making older people live longer because of heart bypass surgery has no impact on human evolution.

Modern technologies in the field of modern medicine in the prevention and treatment of infectious diseases include antiseptics, antibiotics, antivirals, vaccination and quarantine. A major impact of these technologies is the rapid expansion of the human population. Technologies used for the understanding, prevention and management of non-infectious diseases include genetic research such as the Human Genome Project, the study of gene regulation and expression, and epidemiology correlating factors such as cancer and environmental exposure to mutagens. Lifestyle diseases such as obesity are becoming more prominent. Epigenetic studies are now demonstrating that some changes to gene expression are inherited, with potential for changes in human evolution.

Other modern technologies include improved contraception methods as well as a great expansion in assistance to overcome fertility problems with IVF. Genetic screening of embryos has the potential to reduce the number of people suffering genetic diseases and the frequency of these disease alleles in the population.

Genetic engineering has resulted in the production of human insulin so that diabetics live (and potentially reproduce) with fewer health complications. A new technology called gene drives uses CRISPR-Cas9 which means that scientists can precisely alter, delete and rearrange DNA. Using animal models they have corrected mutations responsible for muscular dystrophy, cystic fibrosis and a form of hepatitis. One area with potential to impact on human health is the use of CRISPR technologies to rid pigs of viruses so that their organs could be used in xenotransplantation. CRISPR technology has the potential to change germ line cells and, in addition, selectively increase the proportion of particular genes in a gene pool. A focus of this research is eradication of the mosquito vectors for diseases such as malaria. There are some ethical concerns about the use of genetic engineering and CRISPR technologies as they can involve changing the genes or gene regulation and may have unforeseen consequences.

The combined impact of these aspects of modern medicine is that the human population has grown and continues to grow at a fast rate, life expectancy has increased and age-related diseases and mental illness are becoming more common. Natural selection as outlined by Darwin and Wallace no longer benefits only those who are physically and reproductively fittest. A significant portion of the effort of modern medicine is directed towards treatments such as artificial body parts and organ transplants that improve the health of people well beyond their reproductive age, so would have little impact on human evolution. *(7 marks)*

Acknowledgements

Illustration Q13, 2003 HSC, *Plant Anatomy*, JD Mauseth, Cummings and Hathaway

Illustration Q33(b), 2003 HSC, PJ Wynne, *Scientific American* (January 2000)

Map Q33(c), 2003 HSC, *The Distribution of the Human Blood Groups and Other Polymorphisms*, 2nd ed. (1976) AE Mourant et al., Oxford University Press

Graphs Q11, 2009 HSC, *Advanced Biology* (1997) M Jones and G Jones, Cambridge University Press

Graphs Q29(c) and 30(c), 2009 HSC, BSE Inquiry, UK Crown

Diagrams Q31(c), 2009 HSC, *Scientific American*, Vol. 299, July 2008

Diagram Q34(a)(i), 2012 HSC, Helen Beare; © Australian Museum, www.australianmuseum.net.au/image/Leg-of-a-modern-human

Text, Q23, 2013 HSC, Adapted extracts from Expedition 360, www.expedition360.com/australia_lessons_science/2001/09/spinifex.html

Diagram, Q33(b), 2013 HSC, Maryland State Department of Education, www.mdk12.org/assessments/high_school/look_like/2007/biology/ftri35.html

Table, Q33(c)(ii), 2013 HSC, Table 5-1, Sweet Pea Phenotypes Observed in the F2 by Bateson and Punnett from *An Introduction to Genetic Analysis*, 7th edition (2000), AJF Griffiths, JH Miller, DT Suzuki et al., New York, WH Freeman

Text, Q33(e), 2013 HSC, 'DNA kits can bring unwanted surprises', Philip Sherwell, *The Telegraph*, London, 7 January 2013

Table, Q34(c), 2013 HSC, World Population Growth History Chart, www.vaughns-1-pagers.com/history/world-population-growth.htm

Diagram, Q34(e), 2013 HSC, Figure 2, Anatomical comparisons of apes, early hominins, *Australopithecus*, *Homo erectus*, and humans, www.nature.com/scitable/knowledge/library/overview-of-hominin-evolution-89010983

Diagram, Q31(c)(ii), 2013 answers, http://en.wikipedia.org/wiki/Hyperopia

Diagram, Q2, 2014 HSC, www.wordpress.as.edu.au/hhine/2013/02/20/diffusion

Chart showing DNA Sequences, 2014 HSC, Q33(d), Homologous genes, www.evolution.berkeley.edu/evolibrary/article/1_0_0/eyes_10

Adapted diagram, Q31(d), 2014 HSC, Human Story, p. 53, *Biology in Context—The Spectrum of Life*, E Kennedy & P Hickman, Oxford University Press, 2002

Adapted diagram, Q32(d), 2015 HSC, A Sound Shadow, www.commons.wikimedia.org/wiki/File:Sound_shadow.svg

Adapted diagram, Q1, 2016 HSC, BBC Bitesize, National 5, Proteins and Enzymes, Revision 4, www.bbc.co.uk/education/guides/zb739j6/revision/4

Adapted diagram, Q7, 2016 HSC, Model of a mammalian nephron, *Inquiry into Biology*, p. 311, fig. 9.4, McGraw-Hill Ryerson, Whitby, Ontario, 2006

Adapted diagram, Q29, 2016 HSC, pea plant flower, www.biology-forums.com/gallery/33_27_06_11_6_39_48.jpeg

Adapted diagram, Q32(a), 2016 HSC, Cross-section of the part of the human body responsible for voice production, www.nathanclarkecommunication.wikispaces.com/file/detail/Anatomy-of-the-larynx.jpg

Adapted diagram, Q32(d), 2016 HSC, Eyeborg, www.printmag.com/imprint/wunderkammer-of-color

Adapted diagram, Q34(a), 2016 HSC, Process used in agriculture, www.learner.org/interactives/dna/images/implications4.gif

Adapted diagram, Q35(a), 2016 HSC, Rock layers, www.keyword-suggestions.com/Zm9zc2lsIGxheWVycw

Adapted diagram, Q36(d), 2016 HSC, Artificial leaf, www.nature.com/news/2011/110929/full/news.2011.564/box/1.html
Diagram adapted by permission from Macmillan Publishers Ltd: *Secrets of Artificial Leaf Revealed*, Nature | doi:10.1038/news.2011.564), © 29 September 2011

Diagram, Q5, 2017 HSC, Four vertebrate forelimbs, https://www.pathwayz.org